THE BIOARCHAEOLOGY OF RITUAL AND RELIGION

THE BIOARCHAEOLOGY OF RITUAL AND RELIGION

Edited by

ALEXANDRA LIVARDA, RICHARD MADGWICK
AND SANTIAGO RIERA MORA

OXBOW | books
Oxford & Philadelphia

Published in the United Kingdom in 2018 by
OXBOW BOOKS
The Old Music Hall, 106–108 Cowley Road, Oxford OX4 1JE

and in the United States by
OXBOW BOOKS
1950 Lawrence Road, Havertown, PA 19083

Hardback Edition: ISBN 978-1-78570-828-2
Digital Edition: ISBN 978-1-78570-829-9 (epub)

A CIP record for this book is available from the British Library and the Library of Congress

Typeset in India by Lapiz Digital Services, Chennai

For a complete list of Oxbow titles, please contact:

UNITED KINGDOM
Oxbow Books
Telephone (01865) 241249, Fax (01865) 794449
Email: oxbow@oxbowbooks.com
www.oxbowbooks.com

UNITED STATES OF AMERICA
Oxbow Books
Telephone (800) 791-9354, Fax (610) 853-9146
Email: queries@casemateacademic.com
www.casemateacademic.com/oxbow

Oxbow Books is part of the Casemate Group

Front cover: Top: Reconstruction of wooden bier, containing a body, flower and other offerings within a hide wrapped with rope from the late Bronze Age funerary cave site Cova des Pas, Minorca, Balearic Islands, Spain. © Alex Solé
Bottom: Individual on a wooden bier buried in the late Bronze Age funerary cave site Cova des Pas, Minorca, Balearic Islands, Spain. © Alex Solé

Contents

List of Tables, Figures and Plates

Tables

Figures

Plates

List of Contributors

Akeret, Örni – Integrative Prehistory and Archaeological Science, Basel University, Spalenring 145, CH-4055 Basel, Switzerland.

Allué, Ethel – Catalan Institute of Human Paleoecology and Social Evolution – University Rovira i Virgili. Campus Sescelades URV (Edifici W3), CP 43007, Tarragona, Spain.

Angle, Micaela – Soprintendenza per i Beni Archeologici del Lazio, Via Pompeo Magno 2, I-00192, Rome, Italy.

Best, Julia – Department of Archaeology, Anthropology and Forensic Science, Faculty of Science and Technology, Bournemouth University, Talbot Campus, Poole, BH12 5BB, UK.
School of History, Archaeology and Religion, Cardiff University, John Percival Building, Colum Drive, Cardiff, CF10 3EU, UK.

Boi, Marzia – Department of Biology-Botany, University of the Balearic Islands, Carretera de Valldemossa Km 7.5, CP07122, Palma, Mallorca, Spain.

Bosi, Giovanna – Laboratorio di Palinologia a Paleobotanica - Dipartimento di Scienze della Vita, Università degli Studi di Modena e Reggio Emilia. Viale Caduti in Guerra 127 41121, Modena, Italy.

Brettell, Rhea – Archaeological Sciences, Faculty of Life Sciences, University of Bradford, Bradford, West Yorkshire, BD7 1DP, UK.

Cabanis, Manon – INRAP. Institut National de Recherches Archéologiques Préventives, 13 bis rue Pierre Boulanger, 63100 Clermont-Ferrand, France. CNRS, Université Clermont Auvergne, GEOLAB, F-63000, Clermont–Ferrand, France.

Calvo Trias, Manuel – ArqueoUIB Research Group, University of the Balearic Islands. Carretera de Valldemossa Km. 7,5, CP 07122, Palma, Spain.

Caracuta, Valentina – Laboratory of Archaeobotany and Palaeoecology (L.A.P.), University of Salento, Via Birago 64, 73100, Lecce, Italy.

Deschler-Erb, Sabine – Integrative Prehistory and Archaeological Science, Basel University, Spalenring 145, CH-4055 Basel, Switzerland.

Fiorentino, Girolamo – Laboratory of Archaeobotany and Palaeoecology (L.A.P.), University of Salento, Via Birago 64, 73100, Lecce, Italy.

Garcia Rosselló, Jaume – ArqueoUIB Research Group, University of the Balearic Islands. Carretera de Valldemossa Km. 7,5, CP 07122, Palma, Spain.

Hayden, Brian – Anthropology Department, University of British Columbia, 213-6303 NW Marine Drive, Vancouver, British Columbia, V6T 1Z1, Canada.

Heron, Carl – Department of Scientific Research, The British Museum, Great Russell Street, London, WC1B 3DG, UK.

Hüster Plogmann, Heide – Integrative Prehistory and Archaeological Science, Basel University, Spalenring 145, CH-4055 Basel, Switzerland.

Iborra Eres, Maria Pilar – C/Conde Olocau, 1–5. 46003, Valencia, Spain.

Jonuks, Tõnno – Department of Folkloristics, Estonian Literary Museum, Vanemuise 42, 51003, Tartu, Estonia.

Koch, Pirmin – University of Cologne/D and Kantonsarchäologie St.Gallen, Rorschacherstrasse 23, CH-9001 St. Gallen, Switzerland.

Livarda, Alexandra – Department of Classics and Archaeology, University of Nottingham, University Park, Nottingham, NG7 2RD, UK.

Madgwick, Richard – School of History, Archaeology and Religion, Cardiff University, John Percival Building, Colum Drive, Cardiff, CF10 3EU, UK.

Martin, William – Chemistry and Forensic Sciences, Faculty of Life Sciences, University of Bradford, Bradford, West Yorkshire, BD7 1DP, UK.

McKenzie, Jo – Archaeological Sciences, Faculty of Life Sciences, University of Bradford, Bradford, West Yorkshire, BD7 1DP, UK.

Miras, Yannick – CNRS, Université Clermont Auvergne, GEOLAB, F-63000 Clermont–Ferrand, France.

Morris, James – School of Forensic and Applied Science, University of Central Lancashire, Preston, PR1 2HE, UK.

Mulville, Jacqui – School of History, Archaeology and Religion, Cardiff University, John Percival Building, Colum Drive, Cardiff, CF10 3EU, UK.

Nadal Lorenzo, Jordi – SERP. Seminary of Prehistoric Studies and Research, Department of History and Archaeology, Section Prehistory and Archaeology. University of Barcelona. C/Montalegre 6. 08001, Barcelona.

Nikolaidou, Daphne – Austrian Archaeological Institut (Österreichisches Archäologisches Institut), Franz Klein-Gasse 1, A-1190, Vienna, Austria.

Picornell-Gelabert, Llorenç – Department of Historical Sciences and Theory of Art, University of the Balearic Islands, Carretera de Valldemossa Km 7.5, CP 07122, Palma, Mallorca, Spain.
Archéozoologie, Archéobotanique: sociétés, pratiques et environnements, UMR 7209 of the CNRS – Muséum National d'Histoire Naturelle – Sorbonne Universités. 55 Rue Buffon, 75005, Paris, France.

Pümpin, Christine – Integrative Prehistory and Archaeological Science, Basel University, Spalenring 145, CH-4055 Basel, Switzerland.

Rannamäe, Eve – Institute of History and Archaeology, University of Tartu, Jakobi 2, 51014 Tartu, Estonia; BioArCh, Department of Archaeology, University of York, Environment Building, Wentworth Way, York, YO10 5DD, United Kingdom.

Riera Mora, Santiago – Seminary of Prehistoric Studies and Research Department of History and Archaeology, Section Prehistory and Archaeology. University of Barcelona. C/Montalegre 6. 08001, Barcelona.

Rolfo, Mario F. – Dipartimento di Storia, Cultura e Società, Università "Tor Vergata", Via Columbia 1, I-00163, Rome, Italy.

Salari, Leonardo – Dipartimento di Scienze della Terra, "Sapienza" Università di Roma, Piazzale Aldo Moro 5, I-00185, Rome, Italy.

Schotsmans, Eline – PACEA, UMR 5199, CNRS-Université de Bordeaux, Bât. B8, Allee Geoffroy St Hilaire, CS 50023, 33615 Pessac Cedex, France.

Servera-Vives, Gabriel – Department of Historical Sciences and Theory of Art, University of the Balearic Islands, Carretera de Valldemossa Km 7.5, CP 07122, Palma, Mallorca, Spain.
Seminary of Prehistoric Studies and Research Department of History and Archaeology, Section Prehistory and Archaeology. University of Barcelona. C/Montalegre 6. 08001, Barcelona.

Silvestri, Letizia – Department of Archaeology, Durham University, South Road, DH1 3LE, Durham, UK.

Skeates, Robin – Department of Archaeology, Durham University, South Road, DH1 3LE, Durham, UK.

Stern, Ben – Archaeological Sciences, Faculty of Life Sciences, University of Bradford, Bradford, West Yorkshire, BD7 1DP, UK.

Veropoulidou, Rena – Museum of Byzantine Culture, Hellenic Ministry of Culture and Sports, Leoforos Stratou 2, 54640, Thessaloniki, Greece.

Wick, Lucia – Integrative Prehistory and Archaeological Science, Basel University, Spalenring 145, CH-4055 Basel, Switzerland.

Acknowledgements

Several people have contributed to the preparation of this volume to which we would like to express our warm thanks. Firstly, to all the authors of the chapters for their excellent collaboration during the often lengthy review process and for adhering to the tight deadlines of the publication process. Also, to all the participants of our session at the 2014 EAA conference for their contribution to stimulating discussions on rituals that formed the springboard of this publication. We are indebted to the 23 reviewers for their insightful comments and the generous work they put into improving this volume. We would also like to thank Kathy Twiss for insightful conversations relating to the volume. We are also grateful to Alex Solé for providing the images for the front cover, to Kirsty Harding and Katie Allen for its design and to Hector A. Orengo for his help with processing the figures of the volume. Finally, many thanks are due to Oxbow Books and particularly to Clare Litt and Julie Gardiner for their guidance and help throughout the publication process.

Preface

The chapters in this volume derive from invited papers and contributions to a session of the same title at the conference of the European Association of Archaeologists in Istanbul in 2014. The broad objectives of the session were threefold. Firstly, to showcase new research on ritual and religion relating directly to perishable material culture, with the bioarchaeological disciplines other than human osteoarchaeology, being at the centre of research. Secondly, to build on recent approaches to ritual research in attempting to transcend the dichotomy between sacred and profane aspects. Rather than explaining signatures that are contrary to the norm, the session centred on rituals as social phenomena, with performative, prescribed and repetitive actions being stressed in the call for papers. Thirdly, to bring together research from a range of bioarchaeological sub-disciplines, that are infrequently presented together, with the aim of crossing boundaries and building links across common research themes and approaches. A total of 33 presentation proposals were submitted from across Europe covering an impressive chronological span, from the Mesolithic to the post-medieval period. The term ritual still engenders a common focus on the sacred and the vast majority of submissions, as expected, focused on contexts with clear ceremonial links, such as mortuary or funerary settings, but with an emphasis on the often neglected bioarchaeological remains. A number of very welcome submissions to this volume provided fine-grained studies on precise practices, revealing patterns of activity that would often be beyond the scope of conventional analyses. The session showcased research from a comprehensive range of bioarchaeological sub-disciplines and included discursive contributions on approaches to ritual and religion. Whilst some of the diversity of the session has been lost in this volume, contributions are still presented from zooarchaeology, archaeomalacology, archaeobotany, anthracology, palynology, organic residue analysis, soil micromorphology and geoarchaeology. Common themes are evident across many of the papers (see also Chapter 1), with analyses often extending beyond the standard approaches employed for a site report in order to reconstruct social practices in detail. The integration of multi-factorial data was also a common thread across many papers. Ultimately, this volume aims to showcase the current state of engagement of bioarchaeology with research on ritual and religion, but also its potential and the need to fully incorporate this line of evidence in order to achieve more holistic readings of past worldviews and lifeways.

Alexandra Livarda, Richard Madgwick
and Santiago Riera-Mora

Chapter 1

Ritual and Religion: Bioarchaeological Perspectives

Alexandra Livarda and Richard Madgwick

Michel Tournier, in his novel *Friday, or, the Other Island* (*Vendredi ou les Limbes du Pacifique*, 1967), rewrites the history of *Robinson Crusoe*. Tournier's Robinson, after a long phase of exhausting loneliness in an island in the Pacific 'losing' himself, constructs an hourglass to control the time. He then enters into a strict regime of activities, such as producing grain and legislating, the latter in front of a lectern while dressed in the best clothes rescued from the shipwreck. All these activities occurred daily until one day he realises that he can get out of his strict schedule of ceremonies by stopping the hourglass, stopping the time, and indulge into dedicating all the time to his inner self and to his reconciliation with the island. The rituals imposed by this Robinson were the backbone, the structure framing his new life on a deserted island, much needed to regain human qualities as he knew them. Whereas food sustained the physical individual, these rituals served to sustain his social being. In the absence of other people and a society to live in, Tournier's Robinson Crusoe at some point felt the need to recreate the norms that he knew, the norms upon which society, as he knew it, was based.

It is how this Robinson Crusoe experienced his reintroduction into what he understood as social life that perhaps best exemplifies how rituals can be viewed: as a performance, a repetitive and formal activity. So far numerous definitions have been attempted to explain the term 'ritual', a divisive and heavily loaded term in archaeological research, bound up with notions of sacredness, symbolism, ceremony and the unexplained. Ritual in archaeology has long been informally defined almost by process of elimination, in opposition to the profane, mundane, optimal or economic. Research has increasingly challenged these dichotomies and there is a common recognition that pursuing a division between sacred and profane is erroneous (see Bremmer 1996 and Bradley 2005 for a brief history).

In strict terminology, ritual pertains to the performance of rites, which are prescribed acts or observances in a religious or other solemn ceremony (ritual, adj./n., *Oxford English Dictionary*). However, the term has frequently been used to mean a habitual or customary activity and is thus as likely to refer to practices in daily routines. It should be noted, however, that many researchers distinguish between custom and ritual by maintaining that the latter is concerned with the supernatural and involves a public aspect (*e.g.* Groot 2008, 99, and references therein). Regardless of the definition, rituals can be seen as projections of what a social unit, no matter how small or big, considers important, its beliefs, moral and behavioural systems within specific cultural and temporal contexts. Even in the case of the 'habitual' and 'customary' definition of rituals, these refer to established actions and procedures that have been accepted as important or simply socially approved to allow for their repetitive function at a personal or social level. Rituals, therefore, may be considered as the materialisation of a form of group values, which result from a series of negotiations that bridge potentially conflicting interests and new realities, and can be experienced and interpreted differently by each participating agent. Consequently, rituals are complex and multifarious in their rationale and manifestation, depending on the specific context and historicity of a given social group, and do not thus sit neatly within a single definition.

A more practical approach to the subject has been teasing out the attributes of ritual as implied in the plurality of its attempted definitions. Bell's (2009) list of six such basic characteristics, described as 'neither exclusive nor definitive' (Bell 2009, 138) is currently still the most comprehensive attempt:

1) formalism;
2) traditionalism that includes the element of repetition either implicitly or explicitly;
3) invariance, which describes a disciplined set of actions imbued with precision and control. Here the elements of repetition, physical control and often spatiality are key;

4) rule-governance;
5) sacral symbolism, which is not related necessarily only to the supernatural but also to other activities and objects that may express certain values, feelings and ideals linked to 'a greater, higher, or more universalized reality' (Bell 2009, 159) of not necessarily religious nature; and
6) performance.

Rituals can, therefore, take many forms, underpinning the ideological structure of societies. It is also generally accepted that rituals are not exclusively associated with religion but, as Verhoeven (2011, 124) aptly stresses, they can be placed at any point within a continuum that has at its two extremes the holy and the quotidian. Bell's (2009) seminal work offers another list of the most prominent types of rituals underpinning most classifications, which include the rites of passage or 'life-cycle' rites; calendrical and commemorative rites; rites of exchange and communion; rites of affliction; feasting, fasting and festivals; and political rites. It is interesting to note that, as Bell acknowledges (2009, 94), these forms of rituals are 'usually associated with clearly defined religious traditions'. Indeed, religion is still the most common framework under which ritual is studied in archaeology, perhaps not surprisingly, as it provides a much easier conceptualisation of rituals and an explicit context of interpretation. This may also partly relate to the universalisation of the dichotomy between the sacred and profane, following the western modern thought, that has often been imposed in archaeological research, resulting in the identification of rituals with religion. Religion also has a somewhat fluid meaning, but can be loosely defined as a particular system of faith and/or worship relating to a collection of beliefs, cultural systems and actions, or world views that are often imbued with narratives, symbols, and sacred histories.

A comprehensive history of the archaeological research on ritual and religion has been provided by Insoll (2004), while several books and papers discuss the various approaches to these subjects and the current developments in their study (*e.g.* Barrowclough and Malone 2007; Bell 2009; Insoll 2011; Kyriakidis 2007; Pauketat 2013; Swenson 2015). The aim of this volume is to specifically explore the positioning of bioarchaeological studies in regards to rituals and religion. In this introductory paper we provide an overview of current trends and highlight selected avenues in which bioarchaeology can contribute to the field. Bioarchaeological research in this volume refers to studies that utilise organic remains or palaeo-environmental data to address archaeological questions. Human osteology has been deliberately excluded from its scope to bring the wider spectrum of the sub-disciplines of environmental archaeology to the forefront. These have traditionally been marginalised in research on worldviews, belief and social systems as understood through rituals and religions. In this context this book is filling a significant gap by demonstrating the great potential of these lines of evidence, which are necessary but not always taken into account in archaeological interpretations.

The bioarchaeology of ritual and religion

Bioarchaeology has to some extent followed the general trend of archaeological research regarding definitions and priorities in work on ritual. The majority of studies has indeed focused on funerary contexts and religious or sacred spaces, areas linked with death, the afterlife and specific deities. The volume of studies and the depth and variety of interpretations offered, however, has generally been limited. This results from the archaeological landscape in which these sub-disciplines developed throughout much of the twentieth century. Bioarchaeological research was rarely integrated, with specialist reports all too often sitting in isolation. Findings were even less frequently embedded in social theory or combined with multiple lines of enquiry. The sub-disciplines of what is usually referred to as environmental archaeology long found themselves entrenched in formulaic modes of analysis designed to address a prescribed suite of questions surrounding diet, subsistence, economy, environment and site formation processes. Recent decades have seen a gradual trajectory of change coupled with new theoretical frameworks and advancement in bioarchaeological methods. Methodological advancements mean that plant and animal husbandry, food processing, mobility, diet, depositional treatment and on-site activities can all be reconstructed to a higher resolution than has previously been possible. Consequently, the potential for the identification of marked patterns of practice, framing (*sensu* Verhoeven 2002) and activities beyond subsistence has never been better and there is now widespread recognition of the under-exploited potential of bioarchaeological research for investigating ritual and religion.

New theoretical approaches have stressed the importance of precise social practices, rather than the identification of atypical archaeological signatures (with only meta-level interpretation), incorporating research on a wider spectrum of past activities and intangible issues (see Handelman 2006; Morris 2011). Attempts of new definitions of bioarchaeology, moving on from 'environmental' archaeologies to 'social' palaeo-ethnobotany, zooarchaeology and so on (*e.g.* Morehart and Morell-Hart 2015; Overton and Hamilakis 2013; Russell 2012), gradually sketch the new paradigms towards which the discipline is moving. The aim of these new directions is to bridge the gap between what has been known as 'specialist' work and current theoretical advances in archaeology, with a direct impact of how ritual and religion is perceived and studied. In this context, an important shift in the examination of bioarchaeological remains in rituals is the extension of the analytical framework to include a consideration of the living organisms and not only their ultimate point of ritual deposition or death.

Morris (2011; 2012; Chapter 9 this volume), for instance, has put forward a biographical approach to the study of animal burials, which shifts the emphasis from the final deposition to the events and the range of transformations that animals undergo during their lifetime (and prior to burial). The point of deposition may be haphazard and of no cultural relevance at all, yet it is almost always that which receives most attention. In other words, it is the full life and post-life history of the animal that needs to be considered through a detailed understanding of taphonomic formation processes to achieve meaningful interpretations of the rituals involved. This is no easy process and a wide range of demographic, taphonomic and contextual variables need to be considered. However, Morris (2011, 172–3) has demonstrated the potential for improved interpretative resolution of Associated Bone Groups (ABGs) by employing this approach on deposits from various sites in the UK and establishing the varied trajectories of life, death and deposition that animals underwent. Sykes (2014), drawing on anthropological and historical evidence, similarly proposes the introduction of more varied interpretations of how living animals could shape daily life in which religion and rituals are an integrated part. She suggests that in moving beyond strictly religious contexts and the compartmentalisation of the sacred and profane that is typical of western societies, the study of animals can shed light on the dynamics of past societies to a much greater extent (*ibid.*, 131). For instance, rituals involving linking humans and animals in key stages of their life cycle, such as birth, as observed in the anthropological record, can help inform on the range of interactions and enhance the interpretative potential of archaeological deposits (*ibid.*). Ethnoarchaeological work also has an important role to play in this process (Broderick 2012). Similarly, a biographical approach in the study of plants has been suggested as an important interpretative framework, taking into account the rich histories of plants to better understand their role in rituals (Morehart and Morell-Hart 2015).

Whilst these new approaches and interpretative considerations are of great value, archaeologists must recognise the inevitability of an imbalance in interpretation towards the final material signature. The richness of information that is preserved in the act of deposition often outweighs the resolution of information on other aspects of the deposit's biography. The opportunity to reconstruct deposition in detail must be maximised, but every effort must be made to redress the interpretative imbalance by drawing on a comprehensive suite of data (pertaining to the deposit, its context and associations) to understand other aspects of an organism's life and post-life trajectory. The current trend away from understanding final deposition towards reconstructing the animal's life, however, misses a major part of the deposit's biography: the period between death and deposition. An increased focus on taphonomy has great potential to address this and has been successfully used to reconstruct social practices at a variety of scales (*e.g.* Madgwick 2008; 2010; 2016; Magnell 2012; 2013; Morris 2011; Chapter 9, this volume). This is particularly important in studies on bone, as bones are robust enough to generally survive in abundance, but also soft and malleable to the degree that they can be altered by a range of processes, thereby taking an imprint of their post-life trajectory (Madgwick and Mulville 2012). This important stage is a focus for Jonuks and Rannamäe (Chapter 12, this volume).

The need for moving beyond the dichotomies of nature *versus* culture and by implication beyond distinctions between, for instance, sacred and profane, and the symbolic and the practical, is stressed by Overton and Hamilakis (2013) in their manifesto for social zooarchaeology. They propose that rather than as an opposing approach to economic and ecological, social zooarchaeology should be seen as a non-anthropocentric framework in which the agency of animals 'both in life and in death' is recognised in co-shaping past practices and decision-making (*ibid.*). As such, animals in ritual (and other) contexts cannot be seen as mere identity markers, representing or symbolising certain ideas, but they need to be understood through their interactions with people that result in a range of sensorial (*ibid.*), and we would add, emotional, responses. The agency of plants through their employment in rituals, actively transforming, creating and reinforcing meaning has been also identified as an important future research direction (Morehart and Morell-Hart 2015). Such an approach has the potential to bring to the forefront new and alternative, theoretically informed (re) interpretations of bioarchaeological assemblages that could open new windows into how we understand the past. In practice the first step in this direction is taking into account and disentangling the formation processes that have led to the creation of an assemblage and its contextual examination that allows consideration of the cultural setting.

These methodological and theoretical changes mean that bioarchaeologists can be at the very centre of new and novel approaches to complex archaeological questions. Yet, it is rare that bioarchaeological work is fully integrated into synthetic studies of rituals. It is often the case that archaeologists turn to bioarchaeological evidence only if other types of material culture are absent, for instance, from burials, considering this by implication as a 'secondary' type of offering. To make matters worse, in many areas the lack of an integrated programme for the controlled collection of bioarchaeological remains means that, often, good contextual information is lacking, or that such types of material are still not regularly recovered. The implication of this is the existence of only a partial record of what is available in certain areas, preventing the emergence of a 'bigger picture'. Examples of this include the rarity of bioarchaeological assemblages from historical periods in Greece, and the relatively sparse studies from the Mediterranean in general. This is largely due to a greater focus on impressive architecture and other material culture remains and the lack of legislation that recognises the need for the collection of all available evidence during archaeological excavation. Silvestri *et al.* (Chapter 10) in this volume highlight this problem in the case of the limited faunal remains recovered and recorded from Middle Bronze Age caves in Central Italy and the meagre information available on assemblages and contexts. Bringing together zooarchaeological and archaeobotanical data recovered in a controlled manner from three new sites Silvestri *et al.* showcase from one hand the value of these lines of evidence towards more holistic approaches to rituals and the use of caves, and on the other hand, the need to interpret bioarchaeological remains considering their overall archaeological context.

Most bioarchaeological work on rituals has intended to reconstruct certain activities and has more rarely focused on in-depth interpretations and the meaning of such activities. In the following sections we discuss various ways in which bioarchaeological evidence can be approached to reconstruct and understand ritualised activity. While doing so, we have deliberately avoided a discussion on *how* to identify ritual practices, as ritual in all its guises is far too multifarious to impose a blueprint for identification. The objective of this introduction is rather to offer some context for the development of bioarchaeological research on ritual and religion and to explore how it can provide insights into past worldviews, beliefs, lived experiences, social relations, transformations, historical changes, and power dynamics. The following sections explore this through case studies, including those presented in this volume. This by no means aims to be exhaustive, but rather to provide a flavour of trends in bioarchaeological research relating to ritual and religion, using case studies and approaches that we consider significant. Studies are loosely clustered under some of the principal themes that underpin these trends: sacred and special contexts, sensory experience and performance, and power, status and social relations. These categories are essentially convenient impositions and much of the research straddles these thematic boundaries.

Sacred and special contexts

The longest tradition of research on the bioarchaeology of ritual and religion focuses on burials or contexts of an overtly sacred or religious nature, such as temples and shrines. At a first level of interpretation, bioarchaeological remains from such contexts have often been explained as the leftovers of meals or offerings, including sacrifices. In order, however, to go one step further and tease out their meaning, it is important to understand the relation of these remains to their broader context (*e.g.* material culture, landscape, depositional environment and so on).

Archaeological work on plant remains across different contexts has evidenced specific associations with ritual settings and, for historical periods, it has been crucial in clarifying the range of plant offerings, providing far more detail than is afforded in written sources. The study of large-scale distributions of plants, for instance, has identified date (*Phoenix dactylifera)* and pine (*Pinus pinea*) as having strong ritual associations during the Roman period (*e.g.* Bakels and Jacomet 2003; Bouby and Marinval 2004; Kislev 1988; Livarda 2011). Analysis of the date and pine distribution in a range of contexts and site types across the northwestern Roman provinces (Livarda 2008; 2013) and of the latter in Roman Britain (Lodwick 2015) has indicated their circulation and trade not only as foods, but also as perishable material culture, integral in certain rituals. Dates, despite their good preservation potential, were found to be relatively rare and largely associated with burials, selected ceremonial activities and mystic cults, including that of Isis (Livarda 2013). The import of this new goddess to the Roman world was thought to have been coupled with the import of a plant native to her country of origin (Egypt), which acquired new significance in these settings. When encountered in burials, dates' value potentially as a symbol of resurrection and the afterlife has further been interpreted as indicative of changing worldviews forming within new religions and socio-political contexts in the provinces (*ibid.*). In the case of pine, its distribution in Roman Britain was examined using detailed taphonomic and contextual criteria, enabling the identification of its ritual use even in some contexts that have no such clear connection (Lodwick 2015). Across the northwestern provinces pine nut appears more often in temples/shrines than in burials, suggesting a different role compared to date (Livarda 2008). In addition, the combustible and aromatic properties of pine, potentially also used as incense, seem to have played an important part in its selection for certain rituals (*e.g.* Bird 2004; see also below). The burning of pinewood (*Pinus* spp.), possibly as torches, in a fashion similar to modern-day candles has also been suggested in the case of ritual ceremonies by the ancient lowland Maya (Morehart *et al.* 2005).

Strong, although again not exclusive, associations with specific ritual settings have been identified for a range of other plants and also animals across time and space. Examples include cacao (*Theobroma cacao*) in pre-Columbian Mesoamerica (*e.g.* Prufer and Hurst 2007), the cockerel (*Gallus gallus*) in Roman Mithraic temples (Lentacker *et al.* 2004), goats (*Capra aegagrus hircus*) at the Roman shrine of Uley, Gloucestershire, UK (Levitan 1993) and fallow deer (*Dama dama*) at temples of Artemis and Diana (Miller *et al.* 2016). There are also many examples of certain animal species having been preferentially selected for deposition in burial contexts, including horses (*Equus caballus*) at Anglo-Saxon Spong Hill, Norfolk, UK (Crabtree 1995) and in Viking Iceland (Leifsson 2012) and horses and dogs (*Canis lupus familiaris*) in Iron Age 'special' deposits at Danebury, Hampshire, UK (Grant 1984). In some cases specific elements are targeted for deposition, such as the mass of cattle (*Bos taurus*) skulls at Bronze Age Irthlingborough, UK (Davis and Payne 1993) and the sheep (*Ovis aries*) humeri in Iron Age Arras burials in East Yorkshire, UK (Stead 1991).

In the Meso-American case cited above, cacao has been traditionally seen as an elite item, used also as a currency, but more thorough research combining archaeological, documentary, iconographic, epigraphic and ethnographic evidence has indicated its use in a range of rites of passage and the initiation of shamans (Prufer and Hurst 2007). In the case of the cockerel in Roman Mithraic temples, an investigation of a feasting deposit at the 3rd century AD temple of Mithras at Tienen, Belgium, identified a large quantity of domestic fowl (*Gallus gallus*), which were thought to be mostly male (Lentacker *et al.* 2004). Similar patterns were identified across other *Mithraea,* providing new insights into this poorly understood and secretive cult (*ibid.*). Mithras has been predominantly depicted killing a bull and in connection with various other animals but not the cockerel. He has, however, been associated with the Invincible Sun, which is in turn interpreted linked to the role of the cockerel as the announcer of the rising sun (*ibid.*). Taking into account data obtained from the age of slaughter of other domestic mammals consumed during the feasting event at Tienen, Lentacker *et al.* (2004) were able to further determine the timing of the banquet as taking place in June or July and suggested the summer solstice, the longest day of the year, as the possible day of celebration, allowing a much more detailed appreciation of the cult's rituals.

An excellent example that can serve to highlight the importance of contextualisation and integration of bioarchaeological evidence is also the case of Mayan dedicatory caches. These are normally found within the construction fill of Pre-Classic and Classic Mayan buildings and their symbolic function is debated (Bozarth and Guderjan 2004). They contain materials such as mineral stones, tools and shells, although organic survival in general is rare (*ibid.*). Bozarth and Guderjan (2004) conducted biosilicate (opal phytoliths, sponge spicules, algal statospores, and diatoms) residue analysis on nine typical Mayan caches at Blue Creek, Belize, each comprising two ceramic vessels placed one on top of the other and inverted in a 'lip-to-lip' manner. The results indicated the presence of maize (*Zea mays*), squash (*Cucurbita*), palm fruits, agave (*Agave*), and heliconia (*Heliconia*), representing food, bags and possibly wrappings, as well as marine sponges in most caches (*ibid.*). Combining these results with other finds, Bozarth and Guderjan (2004) were able to show that each cache contained important elements of the land and the sea, which they suggested to represent the Primordial Sea and the First Mountain of the Mayan mythology. Taking into account the domed lid shape of the caches themselves that could have represented the sky, it was hypothesised that each cache reflected the Mayan Cosmos of Creation, and was potentially used for its re-enactment. The authors also pointed out that all social strata shared the rituals involving the caches, as these were associated to varied social contexts. Therefore, the caches seem to be a kind of a bundle that linked people with the cosmos, each of which had a collection of material culture that could have embodied distinct relations, histories, identities and so on (see Pauketat 2013, 43–58). It is rare that these sort of data are recovered, yet this study demonstrates their potential for shedding light into the way cosmological beliefs and practices can be entwined.

Many of the plants that were used in the caches at Blue Creek were commonly used in Mayan foodways and this is something that has often been observed in ritual contexts. In this volume the work of Caracuta and Fiorentino (Chapter 5) on Roman cemeteries in Apulia, Iborra (Chapter 8) on Iron Age Iberian settlements, and Picornell *et al.* (Chapter 11) on the Son Ferrer prehistoric ceremonial and funerary staggered turriform in Mallorca offer such examples of locally available animal and plant resources incorporated into ritual activities. Best and Mulville (Chapter 13) also discuss the geographical and cultural specificity of avian–human relations in life and how these extended to death ritual. In these studies it is evident that ritual practice is often structured using materials and principles from various spheres of activity and the principal change in these instances is the actual or perceived context. Quantities of produce necessary for the survival and function of a group, the most prized cuts of meat, the most succulent fruits, or perhaps the most esteemed or, in some way special, individuals, can be incorporated into a specifically created context. This may be one way that can legitimise the transformation of common lifeways into ritualised activities that would in turn feed back into a group's social and cosmological system. Therefore, what could be construed as part of the 'everyday' becomes entwined or reproduced in 'ritual' or religious contexts, which further emphasises the problems arising from the universal imposition of a dichotomy between the sacred and profane. This also highlights the need to understand societies as a whole to appreciate how rituals were entangled in specific social meshworks (following Ingold 2011). Swenson (2015, 339) cautions that, for instance:

> changes in the spatial organization of household ritual or burials may have had little to do with shifts in power relations or major alterations in religious worldviews.

He argues that ritualised behaviour can only provide a means to interpret these aspects of society if fully contextualised within the spectrum of activities in a given cultural context (*ibid.*). Contextualised approaches to bioarchaeological assemblages that also integrate these with other lines of evidence (archaeological, historical, *etc.*) are thus essential. The clues to the meaning and the positioning of rituals in social settings may also partly lie in disentangling and deciphering the distinct experiences that structured or dictated ritualised activities, and this is discussed in the following section.

Sensory experiences and performance

A shift in focus to sensory elements has been proposed as a more fruitful research avenue to understand past human experience including that of religious and ritual settings (Hamilakis 2011; 2013). In this regard, the mnemonic properties of sensory stimuli are considered to play a key role in imbuing ritual activities with meaning (*ibid.*). The multi-sensory properties of various plants, animals and their by-products are often the reason for their use in ritual activities. These sensory stimuli can be integral components of the performative and experiential aspects of ritual, which in turn enhance the potency of rituals as vehicles of meaning, whatever their specific role. Bioarchaeological and palaeo-environmental analyses are well suited to contribute to a better understanding of such sensory experiences and performances, although in practice sensory attributes are rarely considered and even more rarely inform interpretations.

Several papers in this volume provide evidence for sensory elements, useful for the reconstruction of a range of past experiences. In combining pollen and charcoal data, Picornell *et al.* (Chapter 11) offer a new dimension to the funerary rituals observed at the Late Iron Age necropolis of Son Ferrer, Mallorca. The authors identified the cultural significance of mastic (*Pistacia lentiscus*), and its selective collection and use in a closed funerary space together with pine, rosemary (*Rosmarinus officinalis*) and other members of the mint family (Lamiaceae). All these have resinous and/or aromatic properties that would have ameliorated and masked the odour of decaying corpses with specific smells and their burning would also have provided light. Similarly, to disguise the smell and temporarily delay the body's decay, a variety of natural and gum resins were employed as offerings and for the treatment of the corpse in the Roman period as identified through chemical analyses (Brettell *et al.*, Chapter 4). Acquiring such resinous substances that often involved their long-distance transport was not only for practical purposes, but also as a mark of status, and for enhancing experience and engendering memories of the ceremonies (*ibid.*). Other sensory effects can be established by identifying the use of flowers, for example through pollen analysis. In the case of Son Ferrer, a variety of wild plants with small and colourful flowers of similar shape were deposited as part of the funerary process (Picornell *et al.*, Chapter 11). In another funerary context on the Balearic Islands, at Bronze Age Cova des Pas on Minorca, detailed micro-contextualisation of pollen indicated the use of different types of plants and flower bouquets at each stage of the corpse's treatment. These practices showed no gender or age distinctions, yet a single adult woman was afforded a different rite, suggesting a particular role in the community (Riera *et al.*, Chapter 3). Several other studies have shown the importance of flowers in burials, such as the case of the dropwort (*Oenathe*) offerings of the Scottish Bronze Age (Clarke 1999; Tipping 1994). In another context, multidisciplinary work at the Roman sanctuary at Kempraten, Switzerland, lead to the identification of a sacred grove, offering a different dimension in the conceptualisation of the experience of the sanctuary landscape (Koch *et al.*, Chapter 6). A quite different sensory effect resulted from animal processing at Iron Age High Pasture Cave, Scotland (McKenzie, Chapter 2). Micromorphological analysis provided detailed insights into the use of space and revealed a set of ritualised activities that possibly involved the killing, defleshing, dismemberment and bloodletting of animals near the cave entrance (*ibid.*). In these instances, the animal blood and flesh would have been evocative components of the ritual performance, central to creating codified emotions and associations in the participants.

The most obvious focus of bioarchaeological research on the senses is on food and taste, yet this has received surprisingly little attention. Food is consumed as part of a wide range of ceremonies, celebrations and other ritualised activities (see *e.g.* section on feasting below). Therefore, the selection of food items, the preparation methods, consumption and deposition manners, are all important factors in unravelling these activities, while the examination of tastes can provide another angle to interpretations. Taste in this context is defined following Livarda (2017) as 'the culturally specific and socially subjective experience of flavour' rather than as simply the chemical sense (gustation). In this framework, flavour refers to the sensation resulting from the combination of different attributes of food (gustatory, olfactory, tactile and thermal) and is influenced by other somatosensual stimuli, such as the visual and auditory (*ibid.* and references therein). An examination, thus, of past tastes in ritual consumption is key to appreciating how this sensory experience was understood and enmeshed in different contexts. It is not only food, however, but also drinks and other intoxicants and hallucinogenic drugs that are often part of ritualised and religious activities, contributing to a range of sensory experiences. Substances that alter the mood or consciousness can engender intense experiences that can be interpreted within cosmological schemes and as gateways to the supernatural (Sherratt 1991, 51–2) as well as impact social structures and relations. A range of psychoactive plants have been identified in a variety of periods and areas as part of rituals or shamanistic practices, including hemp (*Cannabis* sp.), opium poppy (*Papaver somniferum*), blue water lily (*Nymphaea nouchali* var. *caerulea*) and mandrake (*Mandragora officinarum*), by combining archaeobotany with various other lines of evidence (see Merlin 2003 for a comprehensive review). Few studies, however, have explored their sensory contribution within specific ritualised contexts. Hamilakis (1998) was the first to discuss this from an archaeological perspective, suggesting that food and alcohol consumption were significant components of mortuary feasting in Bronze Age Crete, acting as embodied mnemonic devices. He argued that the combination of the acute emotions resulting from a charged context, such as that of burial and death, and those from food and drink would enhance the experience and support the processes of remembering and forgetting in relation to death that would in turn actively contribute to the renegotiation of the power dynamics and social relations within the community. Collard (2012), exploring the consumption of alcohol and opium in Cypriote Bronze Age mortuary rituals, added that these were also important for the individual in suppressing, if temporarily, memories of the deceased, thus, reducing grief and allowing the easier transition to social life.

Animal sacrifice represents another powerful arena in which performance and sensory experience come to the fore. Sacrifice is a relatively common theme in bioarchaeology, particularly relating to classical periods. However, zooarchaeological study has all too often been entrenched in traditional approaches focusing on economy through species, age and element representation. These data have nonetheless often led to insightful interpretations surrounding symbolic selection (*e.g.* Forstenpointer 2003; MacKinnon 2010; 2013). Element (body part) side, for instance, usually seen as principally useful for purposes of quantification, has only rarely been investigated in terms of symbolic selection in prehistoric deposits outside of funerary contexts. Examples include the selection of the right side in certain pig elements in deposits at Neolithic West Kennet Palisade Enclosures, Wiltshire, UK (Edwards and Horne 1997) and of right fore-limb elements of pigs in Iron Age feasting deposits at Llanmaes, Vale of Glamorgan, UK (Madgwick and Mulville 2015a). Studies on sacrifice have also often benefited from integration with historical, epigraphic and iconographic evidence and consequently the reconstruction of social practices, rather than meta-level descriptions of sacrifice, have often been possible (*e.g.* Chenal-Velarde and Studer 2003; Ekroth 2013; Popkin 2013; Trantalidou 2013). However, in spite of more novel pathways to interpretation, sensory and performative elements of sacrifice have invariably been neglected. This represents a profitable future avenue for research.

The ritual use of plants, animals and their products are also linked to experiences of magic, medicine and healing – three closely connected concepts in many past societies – even though this can be more difficult to identify archaeologically. Russell (2012, 392–4) suggested that in the case of animals this is likely due to the fact that researchers do not look for this evidence, while Sykes (2014, 131) also pointed out that animal-based medicines involved a destruction process, for instance pulverisation, that would leave no trace. Nevertheless, although rare there are instances that allow the plausible inference of such practices. Van der Veen and Morales (2014), for instance, discuss the use of spices recovered from the port of Quseir al-Qadim as medicines in the Roman and medieval world. Ciaraldi (2000) interpreted an assemblage of reptiles, amphibians, opium poppy and other plants that were recovered at the bottom of a storage vat at the 1st century Roman Villa Vesuvio, near Pompeii, as potential residues of a medical concoction. Karg *et al.* (2014) found two chewing-gum like objects that were made of a mixture of birch tar and plant oil, and three uncharred cloves of wild garlic (*Allium* sp.) in an amulet box discovered in a female grave at the Late Roman Iron Age site of Vellensby, Denmark. Comparing this with other amulet boxes in graves across Europe, they identified that all derive from wealthy female graves of the same period and contained a variety of materials, which were interpreted as possibly employed for their healing properties. In addition, Miller and Sykes (2016) present possible evidence for the use of fallow deer antlers in medicine in Roman Britain, with some specimens having had the edges of palmate surfaces removed by blades, potentially for use in potions. In the Andes, at the Inca site of Lo Demás, a naturally mummified whole guinea pig (*Cavia porcellus*) was found sacrificed with a slit stomach longitudinally, possibly for divining future events or illnesses in accordance with ethnohistorical evidence (Sandweiss and Wing 1997).

Bone and shell pendants are other examples of animal parts that have often been linked to magic and healing. Such artefacts have been widely found across the world and an examination of their origin (the type of animal from which they derive), properties and biography can shed light into their symbolic value but also their social role. Jonuks and Rannamäe (Chapter 12) in this volume provide an intriguing, critical overview of animal tooth pendants from Estonia, covering a wide timespan from the Mesolithic to the end of the medieval period. They explore the complex relationship between people, animals and worldviews, considering, but also moving beyond interpretations of magic. Several examples of animal-derived pharmacopeia and magic are also discussed by Russell (2012) and Sykes (2014), highlighting once again their complex interactions with people. The control of products with perceived magical or healing properties or the knowledge of their use has an important role in affirming status and power relations, a theme expanded in the next section.

Power, status and social relations

Bioarchaeological remains play an important role in examining aspects of status, power relations, identity, and social change in past societies. Arenas of consumption represent important fora for the negotiation, expression and legitimation of power-relations, both within and between different groups. Food, drink, adornment and various other forms of perishable material culture can also be used to challenge established power structures. Ritualised and religious contexts offer frameworks where these relations can be played out, and their study thus allows insights into such past dynamics. A selection of key areas in this research direction is outlined here.

In certain instances ritual and ceremonial contexts can be treated as 'focal nodes of social networks' (Kyriakidis 2007, 2), an examination of which can shed light on the workings of a society. Examples of papers in this volume that touch upon this subject include the study of the Archaic Sanctuary of Apollo at ancient Zone, Thrace, Greece (Veropoulidou and Nikolaidou, Chapter 7) and of the Gallo-Roman sanctuary of Kempraten, Switzerland (Koch *et al.*, Chapter 6). Animal bone and molluscs recovered and examined from the sanctuary of Apollo at the Greek colony of Zone demonstrated a great variety of terrestrial and marine foods and votive offerings present in the sanctuary (Veropoulidou and Nikolaidou, Chapter 7). The particular selection of these items suggested the interweaving of traditions and beliefs of people from different backgrounds across the Aegean and a mixing of local and colonial ideas and practices, crystallised in ritualised practice at the sanctuary (*ibid.*). In adopting a multidisciplinary approach Koch *et al.* (Chapter 6) succeeded in reconstructing not only a complex range of activities, but also the ritual landscape at the sanctuary at Kempraten, where the cult of Magna Mater appears to have been practised. Much like the Archaic Zone, an amalgamation of practices was identified here, fusing local and regional traditions, and allowing new insights into the plurality of Roman religion (*ibid.*).

One important area for the investigation of power relations and identity through consumption practices are monastic contexts. Here, bioarchaeological research has provided a critical counterpoint to historical narratives, stressing the gap between actual practice and projected image. According to St Benedict's rule for instance, monastic diet had to be regulated and meat was not to be eaten, except in special occasions and circumstances, such as by the ill. Bioarchaeological work, however, has provided a more nuanced view of food consumption practice, identifying variations and transformations in relation to monastic dietary regimes and shedding light on the dynamic nature of these institutions and their changing role within society. For example, archaeobotanical data from the Late Antique monasteries of Kom el-Nana, Epiphanius and Phoebammon in Egypt are in stark contrast to the bland and frugal diet indicated by texts (Harlow and Smith 2001). A range of garden crops were identified in these monasteries that would render the everyday diet more variable and nutritious, which in turn has ramifications for how the dynamics of these institutions are understood. Ervynck's (2004) work, which compared zooarchaeological evidence of consumption patterns between monks, peasants and the nobility across medieval and post-medieval Belgium, is also noteworthy. Patterns were interpreted according to differences in religious ideology, land ownership and land access rights. In this case, the archaeological data showed that meat was generally less common in monastic diet, but when present it differed to that consumed by the nobility, reflecting the specific position of each group within society. While the elite had access to wild game and pork, the monks would instead consume beef, mutton and fish in an attempt to maintain a distinction to the former group (*ibid.*).

Food rules and the adherence to or ignorance of taboos has long been an important topic in studies of power relations and social change in anthropology (*e.g.* Douglas 1966; Tambiah 1969) and has become increasingly valued in bioarchaeological research. A classic example that has been widely researched is the avoidance of fish by Neolithic farmers in Britain, as demonstrated by zooarchaeological and stable isotope research (Richards and Schulting 2006; Thomas 2003). However, as methods advance, more nuanced interpretations are achieved, with sporadic marine food consumption evidenced in times of hardship through incremental isotope analysis (Montgomery *et al.* 2013). Other key studies include research on horse consumption in medieval England that has demonstrated that horses were consumed in spite of a religious taboo (Poole 2013; Serjeantson 2000) and the avoidance of pork as a cultural marker in the southern Levant (*e.g.* Hesse 1990; Sapir-Hen *et al.* 2015).

Feasting is one of the most common ritualised practices that provides a focal point for intra- and inter-community social interaction. There remains no consensus on what constitutes a feast (see Dietler and Hayden 2001; Jones 2007), nor how to identify feasting archaeologically (see Dietler and Hayden 2001; Kansa and Campbell 2004; McCormick 2009; Miracle 2002; Twiss 2008). In spite of this, numerous researchers have discussed the wide-ranging social role of feasting, particularly in terms of the negotiation, legitimation and consolidation of power relations, the organisation of production and distribution and the mobilisation of labour. As Hastorf (2008) states, feasts condense sociality and contain a political edge. Feasting research has a much longer history in anthropology (Hayden and Villeneuve 2011), but recent years have seen a proliferation of archaeological studies on the topic, far too many to recount here. Feasting research in archaeology has generally been hindered by the frequent inability to identify discrete events and to access certain performative patterns of practice that remain beyond archaeological recovery. The use of a broad suite of scientific techniques and greater contextualisation and integration with ethnographic and historical research has been central to countering this problem and advancing feasting studies in archaeology.

A major focus of this development has been striving to reconstruct production, preparation, consumption, performance and deposition in greater detail. For example, the application of isotope analysis has demonstrated the role feasting played in supporting long distance inter-community networks in Late Neolithic Britain, with feasts drawing people and animals from afar (Viner *et al.* 2010). The application of specific theoretical frameworks combined with network analysis has also proved fruitful in characterising 'communities of consumption' sustained by feasting in the pre-Hispanic American southwest (Mills 2016). New statistical approaches to bone taphonomy have provided improved resolution on the frequency and scale of feasting events at an intra- and inter-site level in Early Iron Age Britain (Madgwick 2016; Madgwick and Mulville 2015b). Zooarchaeological analysis has revealed socially circumscribed practices in the age, species and parts of animals consumed in a variety of contexts (Hamilakis and Harris 2011; Kansa and Campbell 2004; Madgwick and Mulville 2015a; McCormick 2002; Whitley and Madgwick 2018). Similarly, lipid residue analysis has established prescribed locations within the site of Durrington Walls for the consumption of certain products in Late Neolithic Britain (Craig *et al.* 2015). In addition, detailed contextualisation and integration of evidence has enhanced the detail with which feasts can be reconstructed and has convincingly separated domestic and feasting deposits at Neolithic Çatalhöyük in Anatolia (Twiss 2012). Combining evidence from historical sources and epigraphy has also provided a more nuanced understanding of the role of feasting in expressing and reinforcing new identities in Archaic Greece (*e.g.* Brisart 2015; Huber and Méniel 2015). Ethnographic research is also of great value and Hayden (Chapter 14) draws on this to explore feasting and power from a novel perspective – the phenomenon of secret societies. Multidisciplinary approaches to feasting have indeed shown particularly strong potential in illuminating socio-political processes. Examples include the examination of feasting in Viking Age Iceland that was linked to changes in the chiefly economy and the production of barley (*Hordeum vulgare*) as a political choice (Zori *et al.* 2013), and the identification of feasting in Neolithic Makriyalos, Greece, which was interpreted as a regional gathering where social hierarchies were consolidated (Pappa *et al.* 2004).

These studies have demonstrated that novel and multi-factorial approaches are of great value in providing higher resolution information on how feasts played out in practice. Moving beyond single dataset analyses can thus provide a richness of evidence that extends well beyond traditional interpretations. Explorations of wider social and cultural issues that are reified through or embedded in feasting are increasingly common. The role of feasting in maintaining power structures has been argued in wide-ranging contexts from Neolithic Mesopotamia (Emberling 2016) to Archaic Greece (Blok *et al.*, 2018). There are many more recent examples in which feasting is explored primarily from a socio-cultural perspective, with economy and subsistence being supplementary themes. Feasting studies have matured considerably in recent years and the gulf between research in anthropology and archaeology has narrowed markedly. There is still a long way to go, but this trajectory is sure to continue as greater inter-disciplinarity and methodological advancement means feasting practices can be reconstructed with ever greater resolution.

Concluding remarks

This chapter set the scene of the volume by providing context for the development of bioarchaeological research on ritual and religion. It offered an overview of selected key themes, contexts and approaches that have dominated studies. It also highlighted constraints that have hindered the progress of research, and identified specific weaknesses and profitable future directions. Bioarchaeological research on ritual and religion arguably lags behind many other sub-disciplines of archaeology. Great progress has been made nevertheless and bioarchaeological research on these themes is no longer in its infancy. It is perhaps best described as in its adolescence, yet to reach full maturity, but progressing rapidly, exploring new themes and employing and combining new approaches.

This volume showcases a range of new research from the traditional to the novel, from the macroscopic to the molecular. Some studies explore well-trodden paths in ritual studies and others go well beyond the beaten track. Several other volumes address these themes from a bioarchaeological perspective. Ryan and Crabtree (1995) showcased papers on the symbolic role of animals in a book that was ahead of its time. Miracle and Milner (2002) presented a collection of papers on the social context of food consumption and more recently Hastorf (2016) published a thought-provoking monograph on the social archaeology of food. The volume by Jones-O'Day *et al.* (2004), though just zooarchaeological, represented a step change in the breadth of studies thematically, chronologically and geographically that had ritual and religion as a focus. Campana *et al.* (2010) show further advancement on the topics of colonialism, complexity and transformation. More recently zooarchaeological volumes have had a more defined focus including Pluskowski (2012) on ritual animal killing and Ekroth and Wallensten (2013) on animals in an ancient Greek context. In addition, Sykes (2014) and Russell (2012) have both produced excellent monographs on social aspects of faunal studies, including sections on ritual. The papers by Palmer and van der Veen (2002) and Morehart and Morell-Hart (2015) focusing on plants in archaeology are very good examples of reviews of social approaches to these data, including ritual aspects.

These volumes and papers have been instrumental in advancing the discipline. However, they have all been confined largely to single sub-disciplines, generally zooarchaeology, or to specific themes, such as the social context of food. This volume, therefore, fills an important gap by bringing together and showcasing a cross-section of new research from across the sub-disciplines of bioarchaeology, some of which rarely focus on these themes. The scope is intentionally broad and by drawing together this spectrum of research in one volume, it is hoped that it might act as a springboard for more interdisciplinary studies in the future. Ultimately, this volume endeavours to highlight the significance of bioarchaeological data in interpretations of ritual contexts, and to contribute to the paradigm shift towards more holistic interpretations of past lifeways, embracing a comprehensive range of past experiences, worldviews and social approaches.

References

Bakels, C. and Jacomet, S. (2003) Access to luxury foods in Central Europe during the Roman period: the archaeobotanical evidence. *World Archaeology* 34(3), 542–57.

Barrowclough, D. A. and Malone, C. (eds.) (2007) *Cult in Context: Reconsidering Ritual in Archaeology*. Oxford, Oxbow.

Bell, C. (2009) *Ritual. Perspectives and Dimensions*. Oxford–New York, Oxford University Press.

Bird, J. (2004) Incense in Mithraic ritual: the evidence of the finds. In M. Martens and G. De Boe (eds.) *Roman Mithraism, the Evidence of the Small Finds*, 191–9. Brussels, Museum He Toreke.

Blok, J., Van den Eijnde, F., and Strootman, R. (eds.) (2018) *Feasting and Polis Institutions*. Leiden, E. J. Brill.

Bouby, L. and Marinval, P. (2004) Fruits and seeds from Roman cremations in Limagne (Massif Central) and the spatial variability of plant offerings in France. *Journal of Archaeological Science* 31, 77–86.

Bozarth, S. R. and Guderjan, T. H. (2004) Biosilicate analysis of residue in Maya dedicatory cache vessels from Blue Creek, Belize. *Journal of Archaeological Science* 31, 205–15.

Bradley, R. (2005) *Ritual and Domestic Life in Prehistoric Europe*. London, Routledge.

Bremmer, J. N. (1996) Religion, 'Ritual' and the Opposition 'Sacred vs. Profane'. In F. Graf (ed.) *Ansichten Griechischer Rituale*, 9–32. Stuttgart, B. G. Teubner.

Brisart, T. (2015) Le Banquet orientalisé mise en forme de la commensalité et pratiques identitaires en Grèce proto-archaïque. In A. Esposito (ed.) *Autour de Banquet: Modèles de Consommation et Usages Sociaux*, 55–69. Dijon, EDU.

Broderick, L. (2012) Ritualisation (or the four fully articulated ungulates of the apocalypse). In A. Pluskowski (ed.) *Animal Ritual Killing and Burial: European Perspectives*, 22–32. Oxford, Oxbow.

Campana, D., Crabtree, P., de France, S., Lev-Tov, J. and Choyke, A. (eds.) (2010) *Anthropological Approaches to Zooarchaeology: Complexity, Colonialism, and Animal Transformations*. Oxford, Oxbow.

Chenal-Velarde, I. and Studer, J. (2003) Archaeozoology in a ritual context: The case of the sacrificial alter in Geometric Eretria. In E. Kotjabopoulou, Y. Hamilakis, P. Halstead, C. Gamble and P. Elefanti (eds.) *Zooarchaeology in Greece: Recent Advances*, 215–20. London, British School at Athens.

Ciaraldi, M. (2000) Drug preparation in evidence? An unusual plant and bone assemblage from the Pompeian countryside, Italy. *Vegetation History and Archaeobotany* 9, 91–8.

Clarke, C. (1999) Palynological Investigations of a Bronze Age Cist Burial from Whitsome, Scottish Borders, Scotland. *Journal of Archaeological Science* 26, 553–60.

Collard, D. (2012) Drinking with the dead: psychoactive consumption in Cypriote Bronze Age mortuary ritual. In D. Collard, J. Morris and E. Perego (eds.) *Food and Drink in Archaeology* 3, 23–32. Blackawton, Prospect Books.

Crabtree, P. (1995) The symbolic role of animals in Anglo-Saxon England: evidence from burials and cremations. In K. Ryan and P. Crabtree (eds.) *The Symbolic Role of Animals in Archaeology*, 20–6. Philadelphia PA, MASCA.

Craig, O., Shillito, L-S., Albarella, U., Viner-Daniels, S., Chan, B., Cleal, R., Ixer, R., Jay, M., Marshall, P., Simmons, E., Wright, E. and Parker Pearson, M. (2015) Feeding Stonehenge: cuisine and consumption at the Late Neolithic site of Durrington Walls. *Antiquity* 89, 1096–109.

Davis, S. and Payne, S. (1993) A barrow full of cattle skulls. *Antiquity* 67, 12–22.

Dietler, M. and Hayden, M. (2001) Digesting the feast – good to eat, good to drink, good to think. In M. Dietler and B. Hayden (eds.) *Feasts: Archaeological and Ethnographic Perspectives on Food, Politics, and Power*, 1–22. Washington DC, Smithsonian Institution.

Douglas, M. (1966) *Purity and Danger: An Analysis of Concepts of Pollution*. New York, Praeger.

Edwards, A. and Horne, M. (1997) Animal bone. In A. Whittle (ed.) *Sacred Mound, Holy Rings*, 117–28. Oxford, Oxbow Books.

Ekroth, G. (2013) What we would like the bones to tell us: a sacrificial wish list. In G. Ekroth and J. Wallensten (eds.) *Bones, Behaviour and Belief. The Zooarchaeological Evidence as a Source for Ritual Practice in Ancient Greece and Beyond*, 15–30. Stockholm, Acta Instituti Atheniensis Regni Sueciae, Series in 4/55.

Ekroth, G. and Wallensten, J. (eds.) (2013) *Bones, Behaviour and Belief. The Zooarchaeological Evidence as a Source for Ritual Practice in Ancient Greece and Beyond*. Stockholm, Acta Instituti Atheniensis Regni Sueciae, Series 4/55.

Emberling, G. (2016) Structures of authority: feasting and political practice in the earliest Mesopotamian states. In G. Emberling (ed.) *Social Theory in Archaeology and Ancient History*, 35–59. Cambridge, Cambridge University Press.

Ervynck, A. 2004. Orant, pugnant, laborant. The diet of the three orders in the feudal society of medieval north-western Europe. In S. Jones O'Day, W. Van Neer and A. Ervynck (eds.) *Behaviour behind Bones. The Zooarchaeology of Ritual, Religion, Status and Identity*, 215–23. Oxford, Oxbow Books.

Forsternpointer, G. (2003) Promethean legacy: Investigations into the ritual procedure of the 'Olympian' sacrifice. In E. Kotjabopoulou, Y. Hamilakis, P. Halstead, C. Gamble and P. Elefanti (eds.) *Zooarchaeology in Greece: Recent Advances*, 203–14. London, British School at Athens.

Grant, A. (1984) Survival or sacrifice? A critical appraisal of animal burials in Britain in the Iron Age. In C. Grigson and J. Clutton-Brock (eds.) *Animals and Archaeology 4: Husbandry in Europe*, 221–8. BAR International Series 227. Oxford, Archaeopress.

Groot, M. (2008) *Animals in Ritual and Economy in a Roman Frontier Community: Excavations in Tiel-Passewaaij. Amsterdam Archaeological Studies 12*. Amsterdam, Amsterdam University Press.

Hamilakis, Y. (1998) Eating the dead: mortuary feasting and the politics of memory in the Aegean Bronze Age societies. In K. Branigan (ed.) *Cemetery and Society in the Aegean Bronze Age Societies*, 115–32. Sheffield, Sheffield Academic Press.

Hamilakis, Y. (2011) Archaeologies of the senses. In T. Insoll (ed.) *Oxford Handbook of The Archaeology of Ritual and Religion*, 208–25. Oxford, Oxford University Press.

Hamilakis, Y. (2013) *Archaeology and the Senses. Human Experience, Memory and Affect*. Cambridge, Cambridge University Press.

Hamilakis, Y. and Harris, K. (2011). The social zooarchaeology of feasting: the evidence from the 'ritual' deposit at Nopigeia-Drapanias. In *Proccedings of the 10th Cretological Congress* 1, 199–218. Khania, Filologikos Syllogos Chrystosomos.

Handelman, D. (2006) Conceptual alternatives to 'ritual'. In J. Kreinath, J. Snoek and M. Stausberg (eds.) *Theorizing Rituals: Issues, Topics, Approaches, Concepts*, 37–49. Leiden, Brill.

Harlow, M. and Smith, W. (2001) Between fasting and feasting: the literary and archaeobotanical evidence for monastic diet in Late Antique Egypt. *Antiquity* 75, 758–68.

Hastorf, C. A. (2008) Food and feasting, social and political aspects. In D. Pearsall (ed.) *Encyclopedia of Archaeology*, 1386–95. San Diego CA–Oxford, Academic Press.

Hastorf, C. A. (2016) *The Social Archaeology of Food. Thinking about Eating from Prehistory to the Present*. Cambridge, Cambridge University Press.

Hayden, B. and Villeneuve, S. (2011) A century of feasting studies. *Annual Review of Anthropology* 40, 433–49.

Hesse, B. (1990) Pig lovers and pig haters: patterns of Palestinian pork production. *Journal of Ethnobiology* 10, 195–225.

Huber, S. and Méniel, P. (2015) Pratiques sacrificielles et commensales à Érétrie au VIII siècle avant notre ère. In A. Esposito (ed.) *Autour de Banquet: Modèles de Consommation et Usages Sociaux*, 71–84. Dijon, EDU.

Ingold, T. (2011) *Being Alive: Essays on Movement, Knowledge and Description*. London, New York, Routledge.

Insoll, T. (2004) *Archaeology, Ritual, Religion*. London, New York, Routledge.

Insoll, T. (ed.) (2011) *Oxford Handbook of the Archaeology of Ritual and Religion*. Oxford, Oxford University Press.

Jones, M. (2007) *Feast: Why Humans Share Food*. Oxford, Oxford University Press.

Jones O'Day, S., Van Neer, W. and Ervynck, A. (eds.) (2004) *Behaviour behind Bones. The Zooarchaeology of Ritual, Religion, Status and Identity*. Oxford, Oxbow Books.

Kansa, S. and Campbell, S. (2004) Feasting With the Dead? – A ritual bone deposit at Domuztepe, south eastern Turkey (c. 5550BC). In S. Jones-O'Day (ed.) *Behaviour behind Bones: The Zooarchaeology of Ritual, Religion, Status and Identity*, 2–13. Oxford, Oxbow Books.

Karg, S., Hansen, U. L., Walldén, A. M., Glastrup, J., Pedersen, H. Æ. and Nielsen, F. O. S. (2014) Vegetal grave goods in a female burial on Bornholm (Denmark) from the Late Roman Iron Age period interpreted in a comparative European perspective. *Danish Journal of Archaeology* 3(1), 52–60.

Kislev, M. E. (1988) *Pinus pinea* in agriculture, culture and cult. In H. J. Küster (ed.), *Der Prähistorische Mensch und seine Umwelt. Festschrift für Udelgard Körber-Grohne zum 65. Geburtstag*, 73–9. Stuttgart, Theiss.

Kyriakidis, E. (ed.) (2007) *The Archaeology of Ritual*. Los Angeles CA, Cotsen Institute of Archaeology, University of California.

Leifsson, R. 2012. Evolving traditions: horse slaughter as part of Viking burial customs in Iceland. In A. Pluskowski (ed.) *Animal Ritual Killing and Burial*, 184–94. Oxford, Oxbow Books.

Lentacker, A., Ervynck, A. and Van Neer, W. 2004. Gastronomy or religion? The animal remains from the mithraeum at Tienen (Belgium). In S. Jones O'Day, W. Van Neer and A. Ervynck (eds.) *Behaviour behind Bones. The Zooarchaeology of Ritual, Religion, Status and Identity*, 77–94. Oxford, Oxbow Books.

Levitan, B. (1993) Vertebrate remains. In A. Woodward and P. Leach (eds.) *The Uley Shrines: Excavation of a Ritual Complex on West Hill, Uley, Gloucestershire: 1977–9*, 257–63. London, British Museum Press.

Livarda, A. (2008) Introduction and Dispersal of Exotic Food Plants into Europe during the Roman and Medieval Periods. Unpublished PhD thesis, University of Leicester.

Livarda, A. 2011. Spicing up life in northwestern Europe: exotic food plant imports in the Roman and medieval world. *Vegetation History and Archaeobotany* 20, 143–64.

Livarda, A. (2013) Date, rituals and socio-cultural identity in the northwestern Roman provinces. *Oxford Journal of Archaeology* 32(1), 101–17.

Livarda, A. (2017) Tastes in the Roman provinces: an archaeobotanical approach to socio-cultural change. In K. C. Rudolph (ed.) *Taste and the Ancient Senses*, 179–96. London: Routledge.

Lodwick, L. (2015) Identifying ritual deposition of plant remains: a case study of stone pine cones in Roman Britain. In T. Brindle, M. Allen, E. Durham and A. Smith (eds.) *TRAC 2014. Proceedings of the Twenty Fourth Annual Theoretical Roman Archaeology Conference*, 54–69. Oxford, Oxbow Books.

MacKinnon, M. (2010) When left is right: the symbolism behind side choice among ancient animal sacrifices. In D. Campana, P. Crabtree, S. de France, J. Lev-Tov and A. Choyke (eds.) *Anthropological Approaches to Zooarchaeology: Complexity, Colonialism, and Animal Transformations*, 252–60. Oxford, Oxbow Books.

MacKinnon, M. (2013) 'Side' matters: Animal offerings at ancient Nemea. In G. Ekroth and J. Wallensten (eds.) *Bones, Behaviour and Belief. The Zooarchaeological Evidence as a Source for Ritual Practice in Ancient Greece and Beyond*, 129–48. Stockholm, Acta Instituti Atheniensis Regni Sueciae, Series in 4/55.

Madgwick, R. (2008) Patterns in the modification of animal and human bones in Iron Age Wessex: Revisiting the excarnation debate. In O. P. Davis, N. M. Sharples and K. E. Waddington (eds.) *Changing Perspectives on the First Millennium BC*, 99–118. Oxford, Oxbow Books.

Madgwick, R. (2010) Bone modification and the conceptual relationship between humans and animals in Iron Age Wessex. In J. Morris and M. Maltby (eds.) *Integrating Social and Environmental Archaeologies: Reconsidering Deposition*. BAR International Series 2077, 66–82. Oxford, Archaeopress.

Madgwick, R. (2016) New light on feasting and deposition: Exploring accumulation history through taphonomic analysis at later prehistoric middens in Britain. *Archaeological and Anthropological Sciences* 8, 329–41.

Madgwick, R. and Mulville, J. (2012) Investigating variation in the prevalence of weathering in faunal assemblages in the United Kingdom: A multivariate statistical approach. *International Journal of Osteoarchaeology* 22, 509–22.

Madgwick, R. and Mulville, J. (2015a) Feasting on fore-limbs: Conspicuous consumption and identity in later prehistoric Britain. *Antiquity* 89, 629–44.

Madgwick, R. and Mulville, J. (2015b) Reconstructing depositional histories through bone taphonomy: Extending the potential of faunal data. *Journal of Archaeological Science* 53, 255–63.

Magnell, O. (2012) Sacred cows or old beasts? A taphonomic approach to studying ritual killing with an example from Iron Age Uppåkra, Sweden. In A. Pluskowski (ed.) *Animal Ritual Killing and Burial*, 195–208. Oxford, Oxbow Books.

Magnell, O. (2013) The taphonomy of ritual bone depositions. In G. Ekroth and J. Wallensten (eds.) *Bones, Behaviour and Belief. The Zooarchaeological Evidence as a Source for Ritual Practice in Ancient Greece and Beyond*, 115–28. Stockholm, Acta Instituti Atheniensis Regni Sueciae, Series in 4/55.

McCormick, F. (2002) The distribution of meat in a hierarchical society: The Irish evidence. In P. Miracle and N. Milner (eds.) *Consuming Passions and Patterns of Consumption*, 25–31. Cambridge, McDonald Institute for Archaeological Research.

McCormick, F. (2009) Ritual feasting in Iron Age Ireland. G. Cooney, K. Becker, J. Coles, M. Ryan and S. Sievers (eds.) *Relics of Old Decency: Archaeological Studies in Later Prehistory*, 405–12. Dublin, Wordwell.

Merlin, M. D. (2003) Archaeological evidence for the tradition of psychoactive plant use in the Old World. *Economic Botany* 57(3), 295–323.

Miller, H., Carden, R., Lamb, A., Madgwick, R., Osborne, D., Symmons, R. and Sykes, N. 2016. Dead or alive? Investigating long-distance transport of live fallow deer and their body-parts in Antiquity. *Environmental Archaeology* 21, 246–59.

Miller, H. and Sykes, N. (2016) Zootherapy in archaeology: the case of the fallow deer *(Dama dama dama)*. *Jounal of Ethnobiology* 36, 257–76.

Mills, B. (2016) Communities of consumption: Cuisines as constellated networks of situated practice. In A. Roddick and A. Stahl (eds.) *Knowledge in Motion, Constellations of Learning Across Time and Place*, 248–70. Tucson AZ, University of Arizona Press.

Miracle, P. (2002) Mesolithic meals form Mesolithic middens. In P. Miracle and N. Milner (eds.) *Consuming Passions and Patterns of Consumption*, 65–88. Cambridge, McDonald Institute for Archaeological Research.

Miracle, P. and Milner, N. (eds.) (2002) *Consuming Passions and Patterns of Consumption*. Cambridge, McDonald Institute for Archaeological Research.

Montgomery, J., Beaumont, J., Jay, M., Keefe, K., Gledhill, A., Cook, G., Dockrill, S. and Melton, D. (2013) Strategic and sporadic marine consumption at the onset of the Neolithic: increasing temporal resolution in the isotope evidence. *Antiquity* 87, 1060–72.

Morehart, C. T., Lentz, D. L. and Prufer, K. M. (2005) Wood of the gods: the ritual use of pine (*Pinus* spp.) by the ancient lowland Maya. *Latin American Antiquity* 16(3), 255–74.

Morehart, C. T. and Morell-Hart, S. (2015) Beyond the ecofact: toward a social paleoethnobotany in Mesoamerica. *Journal of Archaeological Method and Theory* 22(2), 483–511.

Morris, J. (2011) *Investigating Animal Burials: Ritual, Mundane and Beyond*. BAR British Series 535. Oxford, Archaeopress.

Morris, J. (2012) Animal 'ritual' killing; from remains to meanings. In A. Pluskowski (ed.) *Animal Ritual Killing and Burial: European Perspectives*, 8–21. Oxford, Oxbow Books.

Mulville, J., Madgwick, R., Powell, A. and Parker Pearson, M. (2012) Flesh on the bones: animal bodies in Atlantic roundhouses. In A. Pluskowski (ed.) *Animal Ritual Killing and Burial*, 205–19. Oxford, Oxbow Books.

Orton, D. (2012) Taphonomy and interpretation: an analytical framework for social zooarchaeology. *International Journal of Osteoarchaeology* 22, 320–37.

Overton, N. J. and Hamilakis, Y. (2013) A manifesto for a social zooarchaeology. Swans and other beings in the Mesolithic. *Archaeological Dialogues* 20(2), 111–36.

Palmer, C. and van der Veen, M. (2002) Archaeobotany and the social context of food. *Acta Palaeobotanica* 42(2), 195–202.

Pappa, M., Halstead, P. Kotsakis, K. and Urem-Kotsou, D. (2004) Evidence for large-scale feasting at Late Neolithic Makriyalos, N. Greece. In P. Halstead and J. Barrett (eds.) *Food, Cuisine and Society in Prehistoric Greece*, 16–44. Oxford, Oxbow Books.

Pauketat, T. R. (2013) *An Archaeology of the Cosmos. Rethinking Agency and Religion in Ancient America*. London–New York, Routledge.

Pluskowski, A. (ed.) (2012) *Animal Ritual Killing and Burial: European Perspectives*, 22–32. Oxford, Oxbow Books.

Poole, K. (2013) Horses for courses? Religious change and dietary shifts in Anglo-Saxon England. *Oxford Journal of Archaeology* 32, 319–33.

Popkin, P. R. W. (2013) Hittite animal sacrifice. Integrating zooarchaeology and textual analysis. In G. Ekroth and J. Wallensten (eds.) *Bones, Behaviour and Belief. The Zooarchaeological Evidence as a Source for Ritual Practice in Ancient Greece and Beyond*, 101–14. Stockholm, Acta Instituti Atheniensis Regni Sueciae, Series in 4/55.

Prufer, K. M. and Hurst, W. J. (2007) Chocolate in the underworld space of death: cacao seeds from an early Classic mortuary cave. *Ethnohistory* 54(2), 273–301.

Randall, C. (2010) More ritual rubbish? Exploring the taphonomic history, context formation processes and 'specialness' of deposits including human and animal bones in Iron Age pits. In J. Morris and M. Maltby (eds.) *Integrating Social and*

Environmental Archaeologies: Reconsidering Deposition. BAR International Series 2077, 83–102. Oxford, Archaeopress.

Richards, M. and Schulting, R. (2006) Touch not the fish: The Mesolithic-Neolithic change of diet and its significance. *Antiquity* 80, 444–58.

'ritual, adj. and n.' *Oxford English Dictionary Online*. September 2015. Oxford University Press. http://www.oed.com/view/Entry/166369?redirectedFrom=ritual (accessed 20 October 2017).

Russell, N. (2012) *Social Zooarchaeology: Humans and Animals in Prehistory*. Cambridge, Cambridge University Press.

Ryan, K. and Crabtree, P. (eds.) (1995) *The Symbolic Role of Animals in Archaeology*. Philadelphia, MASCA.

Sandweiss, D. H. and Wing, E. S. (1997) Ritual rodents: the guinea pigs of Chincha, Peru. *Journal of Field Archaeology* 24(1), 47–58.

Sapir-Hen, L., Meiri, M. and Finkelstein, I. (2015) Iron Age pigs: New evidence on their origins and role in forming identity boundaries. *Radiocarbon* 57, 307–15.

Serjeantson, D. (2000) Good to eat and good to think with: classifying animals from complex sites. In P. Rowley-Conwy (ed.) *Animal Bones, Human Societies*, 179–89. Oxford, Oxbow.

Sherratt, A. (1991) Sacred and profane substances: the ritual use of narcotics in later Neolithic Europe. In P. Garwood, D. Jennings, R. Skeates and J. Toms (eds.) *Sacred and Profane. Proceedings of a Conference on Archaeology, Ritual and Religion, Oxford 1989*, 50–64. Oxford, Oxford University Committee for Archaeology Monograph 32.

Stead, I. (1991) *Iron Age cemeteries in East Yorkshire*. London, English Heritage.

Swenson, E. (2015) The Archaeology of Ritual. *The Annual Review of Anthropology* 44, 329–45.

Sykes, N. (2014) *Beastly Questions. Animal Answers to Archaeological Issues*. London, Bloomsbury.

Tambiah, S. (1969) Animals are good to think and good to prohibit. *Ethnology* 8, 423–59.

Thomas, J. (2003) Thoughts on the 'repacked' Neolithic Revolution. *Antiquity* 77, 67–74.

Tipping, R. (1994) Ritual Floral Tribute in the Scottish Bronze Age. Palynological evidence. *Journal of Archaeological Science* 21, 133–9.

Trantalidou, K. (2013) Dans l'ombre du rite: vestiges d'animaux et practiques sacrificielles en Grèce antique. In G. Ekroth and J. Wallensten (eds.) *Bones, Behaviour and Belief. The Zooarchaeological Evidence as a Source for Ritual Practice in Ancient Greece and Beyond*, 61–86. Stockholm, Acta Instituti Atheniensis Regni Sueciae, Series in 4/55.

Twiss, K. C. (2008) Transformations in an early agricultural society: feasting in the southern Levantine pre-pottery Neolithic. *Journal of Anthropological Archaeology* 27, 418–42.

Twiss, K. C. (2012) The complexities of home cooking: public feasts and private meals inside the Çatalhöyük house. *Journal for Ancient Studies* 2, 53–73.

Van der Veen, M. and Morales, J. (2015) The Roman and Islamic spice trade: new archaeological evidence. *Journal of Ethnopharmacology* 167, 54–63.

Verhoeven, M. (2011) The many dimensions of ritual. In T. Insoll (ed.) *Oxford Handbook of the Archaeology of Ritual and Religion*, 115–32. Oxford, Oxford University Press.

Verhoeven, M. (2002) Ritual and ideology in the pre-pottery Neolithic B of the Levant and south-east Anatolia. *Cambridge Archaeological Journal* 12(2), 233–58.

Viner, S., Evans, J., Albarella, U. and Parker Pearson, M. (2010) Cattle mobility in prehistoric Britain: Sr isotope analysis of cattle teeth from Durrington Walls (Wiltshire, UK). *Journal of Archaeological Science* 37, 2812–20.

Whitley, J. and Madgwick, R. (2018) Consuming the wild: more thoughts on the andreion. In J. Blok, F. van den Eijnde and R. Strootman (eds.) *Feasting and Polis Institutions*. Leiden, E.J. Brill.

Zori, D., Byock, J., Erlendsson, E., Martin, S., Wake, T. and Edwards, K. (2013) Feasting in Viking Age Iceland: sustaining a chiefly political economy in a marginal environment. *Antiquity* 87, 150–65.

Chapter 2

Sacred to the Soil: Micromorphology, Geoarchaeology, and the Bioarchaeology of Ritual and Religion, with Reference to the Iron Age Site of High Pasture Cave, Scotland

Jo McKenzie

Introduction

Geoarchaeological techniques are an increasingly important contributor to the exploration of ritual behaviour, often via analyses that complement bioarchaeological investigations (see *e.g.* Koch *et al.* this volume). A powerful tool in the investigation of 'special events' on site is soil micromorphology, with its potential to provide rich, uniquely targeted datasets, inclusive of a range of bioarchaeological entities such as fuels, bone, and botanical remains, crucially within their taphonomic context (Matthews 2010, 98) and potentially resolved to the physical space of a single human agent. The following sections discuss how micro-archaeological techniques can facilitate the exploration and interpretation of bioarchaeological remains as holistic information groups, able to contextualise stratigraphic relationships, address questions of taphonomy, equifinality and event sequences, and sometimes inform directly upon ritual behaviours by defining specific events, such as heating or disturbance, or by illustrating micro-stratigraphies within the construction process. As the engagement of environmental archaeology with questions relating to ritual and religious behaviour increases (Morris and Maltby 2010), geoarchaeological techniques have an ever-bigger role to play in facilitating the finely nuanced interpretations of bio-assemblages which help us understand the social agency behind their creation.

Geoarchaeological techniques are also crucial to our understanding of the concept of archaeological sediments themselves as actively created entities, as able to 'encode social memory and biographies, evoke an understanding of human agency, resonate with social and ritual practice and play an active role in social networks and figured worlds' as artefacts (McAnany and Hodder 2009, 2). The physical and chemical makeup of soil and sediment structures from mounds to middens can illuminate the social and ritual context informing and structuring their placement, creation and maintenance, helping us explore sediment accumulations and earthen structures as 'both monumental architecture, and the location of ritual and ceremonial activity' (Johansen 2004). The immovability of sediment structures creates a new nexus of interpretative potential: objects which maintain a physical relationship with the landscape of their creation and curation in a way that, mostly, individual artefacts and ecofacts do not. Structured archaeological sediments have unique potential to illuminate the exploration of ritual and religion in the past, also representative of 'not so much as the record of something, but as the material conditions structuring past existences and the reproduction of the conditions that brought them into being' (Jusseret 2010, 681). Micromorphological techniques in particular can play an important role in such contextualisation, not only at the site scale (as seen in several of the examples in the following sections) but also as part of studies considered at the landscape scale, such as that of the Bronze Age midden sites of southern England (*e.g.* Waddington and Sharples 2011) and recent Orcadian investigations into midden curation and redeposition, and its role in symbolic remodelling of both site and landscape at the sites of Ness of Brodgar (*e.g.* Shillito *et al.* 2015) and Mine Howe (Card *et al.* forthcoming).

Geoarchaeological signatures for ritual activity

Fire

Perhaps archaeology's most visible signifier for ritual activity, this is an area in which geoarchaeology has greatly contributed to understanding. Micromorphological studies in particular have refined and expanded identification and also the potential for interpretation of fuels and their residues. An example of recent experimental work is Braadbaart *et al.*'s (2012) detailed comparison of the microscopic characteristics of fuel types from the three key resource bases (contemporary vegetation, fossil fuels and animal by-products), expanding the potential for investigation of the social dimension of fuel resource choice and investigating differences in thermal, physical and chemical properties. The supporting context provided by site-focused studies can also facilitate advances in material identification. Villagran *et al.*'s (2013) study of occupation layers at sealing sites in Antarctica produced a geoarchaeological dataset enabling identification of the presence of burned fats, probably seal blubber, and investigating the relationship of the use of this fuel resource both with other fuel types and hearth structures. Although not directly addressing ritual behaviour, such studies – advancing techniques of identification and the contextualisation of sometimes exotic or unusual resource types – make important contributions to our understanding of ritualised acts of burning. Microscopic investigations into the physical traces of combustion sequences have also shed light on biographies of fire loci and the human agency behind them, such as Miller and Siever's (2012) investigation into the micromorphology of occupation deposits at the South African rock shelter at Sibudu. Here, experimental grass burning demonstrated that fired bedding material, a common feature of the rockshelter deposits, represented intentional and structured activity, with experiments showing that ignition of these materials and the construction of such ashed lenses required considerable effort. By contrast, Mallol *et al.* (2013) also used experimental reconstruction to demonstrate how precarious such signatures may be. Their investigation into Palaeolithic combustion structures indicated that key actions, such as relighting, and the inclusion of specific residues, such as charred fats (as explored by Villagran *et al.* 2013), may often leave little trace in the archaeological record, with disturbance features, including trampling and sweeping, leaving a far more distinguishable signature.

Multi-disciplinary studies featuring similar experimental and reconstruction techniques have facilitated direct investigations into ritual practice, such as Fiorentino and D'Oronzo's (2010) study into the use of burning as part of cult practice at the Sanctuary of Apollo in Hierapolis, Turkey. Here, experimental hearth constructions were used to investigate activity within the *escharon* – a traditional location for the residues of material burnt during religious activities, which is sometimes used as the locus for particular rituals – identified next to the temple structure. Experiments designed to replicate and characterise the specific micro-archaeological character of the heat-affected surfaces and suite of fuel residues revealed by excavation concluded that the *escharon* had been the location for a series of ground-level combustions. These fire-spots were cleaned of their ash residues and the same location used for a repetition of the burning process.

Multi-disciplinary, including microscopic, scientific investigations into combustion features have also identified materials providing enigmatic glimpses of the practicalities of ritualised activity. Photos-Jones *et al.*'s (2007) investigation into the chemical and physical character of 'cramp' – a glassy fuel ash slag formed from dung and seaweed and a common component of Orcadian cremation burials – indicated that cramp production may have been a key component of the mechanics of the funeral pyre, with a bed of sandy seaweed laid underneath and around the elevated pyre, possibly as a medium for trapping small fragments of cremated bone. These would adhere to the partly molten mix during the latter stages of burning and could then be retrieved for interment by collection of the cramp along with the larger bone fragments (Photos-Jones *et al.* 2007).

Structures

Both the construction and alteration of ritual structures through acts of cordoning, enclosing, removing, and reopening leave geoarchaeological and/or microstratigraphic signatures which can inform on ritual behaviours. Sherwood and Kidder's (2011) microstratigraphic study into the soils and sediments forming Mississippian pre-Colombian mounds illustrated the significant social and ritual meaning potentially embedded within not only the visible elements of the mound (such as the deliberate colour choices made for creation of surface veneers) but also the very fundamentals of construction technique – highlighting, for example, the use of a thin layer of light-coloured silt to cover a dark, marshy substrate prior to mound construction proper as an act potentially both practical and symbolic. Karkanas *et al.*'s (2012) micromorphological investigation of infill sequences within Mycenaean chamber tombs demonstrated the value of geoarchaeology in identifying and characterising episodes of tomb reopening and exposure, as well as the identification of ephemeral structural evidence for ritual practice, such as the finely cemented floors formed from calcareous sediment found in association with not only the original tomb construction, but also episodes of reopening. At Tell Qarassa, Syria, geoarchaeology was used alongside bioarchaeological and osteological analyses to investigate the stratigraphic and social complexity of the re-use of domestic structures within funerary rituals, indicating that interments only took place after comprehensive abandonment and partial collapse of house structures (Santana *et al.* 2015). Finally, a compelling example of the value of multi-disciplinary geoarchaeological investigation to understanding ritual activity within structures is seen in Van Keuren and Roos' (2013) investigation into the sediment infill of a ceremonial Pueblo '*kiva*' structure at Fourmile Ruin, Arizona. Here, micromorphological and geochemical characterisation of the stratigraphy of the *kiva* deposits drew a vivid picture of a ritualised closure sequence featuring structured acts of deposition, burning, exposure and redeposition. The finely detailed observations made possible through micromorphology are especially informative, with the identification of significant inputs of non-local, deliberately transported sediments indicating a well organised closing ritual, and extensive lenses made up of wall plaster fragments facing upwards, sitting above a thin layer of char materials, indicating a planned and controlled series of burning and dismantling events focused on the structure.

Spaces and places

The study of occupation surfaces is an area of geoarchaeological analysis in which multi-disciplinary studies have produced detailed and highly resolved evidence for specific activities, such as Milek's (2012) analysis of the spatial organisation of activity areas within a Viking age house at Reykjavik, Iceland. Such studies may inform on ritual behaviours both directly and indirectly. At the Neolithic tell site of Makri, Greece, microstratigraphic study of floor construction and renovation explored the relationship between formalised floor construction activities and significant social events. In this study micromorphology characterised 'formal' floor surfaces featuring elements suggestive of ritualised practice, such as the use of specific material combinations indicative of specialised construction techniques. Similar sequences of floor regeneration were seen within different dwellings on the site, suggesting that floor re-making may have formed a part of community events (Karkanas and Efstratiou 2009). This was explored in more detail at the Bronze Age site of Mitrou, Greece, where the significance of such patterns of building regeneration to changing social organisation was studied using micromorphology. Changes in the pattern of floor maintenance observed during the Late Bronze Age contributed to analysis of wider changes in settlement structure at this time, which in turn may reflect wider social change related to the emergence of a political elite (Karkanas and Van de Moortel 2014). Such studies are closely linked to, and informed by, ethnoarchaeological research, where information from geoarchaeology can play a central role in interpretation, as for instance in Boivin's study of floor maintenance practices in Rajasthan (2010), or Rondelli *et al.*'s (2014) geochemical investigations into the use of space within traditional Gujarat houses.

Geoarchaeological analyses have also explored identification of public space in archaeological contexts. This has focused on construction elements, such as Shahack-Gross *et al.*'s (2005) investigation of deposits relating to a Phoenician monumental structure at the site of Tel Dor, Israel, which related differing mineralogical signatures for floor constructions with public versus private spaces, and on sediment analyses, such as Shillito and Ryan's (2013) use of comparative micromorphological and bioarchaeological (phytolith) datasets to identify open spaces at the prehistoric settlement of Çatalhöyük, Turkey. These and other geoarchaeological analyses have stimulated critical assessment of traditional differences in archaeological perceptions of private versus public space. While the latter is typically examined in terms of performance loci, the former is usually characterised as domestic, a separation that is unlikely to facilitate nuanced understanding of contexts of ritual activity (Fleisher 2014).

Taphonomic processes

A central consideration in any bioarchaeological study that infers ritual or social practice from the physical remains of biological features is the potential effect of diagenesis. Post-depositional processes may compound difficulties in characterising often ephemeral traces of ritualised behaviour in archaeological contexts, within which the issue of equifinality is also a key consideration. Microscopic techniques play an ongoing role in addressing such issues, for example in Huisman *et al.*'s (2012) study into the chemistry of combustion products and their taphonomic effects, particularly the potential for alkaline ash deposits to contribute to charcoal disaggregation in the burial environment. Similar contributions have also been a by-product of site-focused studies, such as Miller and Sievers' (2012) burning experiments discussed above. Noting that experimental ash residues were dominated by articulated phytoliths, they suggest that calcareous ash, previously assumed to have been an original component of the sedge and grass residues but lost through diagenesis, was in fact unlikely to have formed a major part of this particular type of ash deposit in the first place. Matthews' (2010) summary of advances in the study of taphonomic processes and understanding of depositional pathways for archaeological plant remains, and the ongoing challenges this poses, calls for integration of geoarchaeological and archaeobotanical evidence as 'best practice' for analysis. It is self-evident that the potential of both disciplines is further strengthened when also undertaken in the context of an effective research and excavation framework.

Micromorphology, geoarchaeology, and bioarchaeology in context: High Pasture Cave

The use of geoarchaeological, particularly micromorphological, analysis in conjunction with evidence from bioarchaeological datasets are examined here within a holistic research framework to investigate the social and ritual context of events at the site of High Pasture Cave, located on the Isle of Skye, Scotland.

The site

The subterranean focus of the High Pasture site is a cave passage above the active streamway of a *c.* 320m limestone cave system winding through southern Skye. The natural hollow forming the cave entrance lies in the imposing shadow of the Cuillin Mountains, on the lower northern slopes of Beinn an Dubhaich (236m) at *c.* 58m above Ordnance Datum (NG 5943 1971) (Plate 1). The topography of High Pasture is suggestive of the liminal zone represented by the cave itself, with Beinn an Dubhaich representing a tertiary granite intrusion splitting the western end of an outcrop of the Cambrian Durness Limestone (Ryder 1974). At this geological junction, fertile, well-drained arable and pasture ground upon brown earths derived from limestone drift gives way abruptly to rough peat and heathland. Here, a long-ago diversion in the cave stream created a 12m-long, dry, level cave extending into a darkness from which still emanates the hiss and boom of water from the tumbling streamway below. Disturbance by modern-day cavers prompted excavation of the site between 2003 and 2010.

The archaeology

Ephemeral signs of Neolithic activity at High Pasture precede more substantial evidence for some Bronze Age use of the site (Phase 1), with ard marks and a range of cut features associated with Beaker ceramics (Birch and Wildgoose 2013). However, activity at the site overwhelmingly dates to the Early–Middle Scottish Iron Age, *c.* 800 cal BC–cal AD 150 (Hamilton 2014). This activity focuses on two linked but clearly delimited areas which appear to have formed perceptually different zones: the cave itself, and a 'precinct' area established within and around the natural karstic hollow forming the cave entrance point. The stratigraphy of use of both cave and precinct is complex, with large accumulations of midden within the precinct necessitating continual remodelling of the site throughout its Iron Age use in order to retain access between cave and precinct and also, it appears, to structure the terms of such access (see below).

The earliest Iron Age evidence for structured use of the cave is a paved walkway constructed through the natural hollow. Low revetting walls and terraces were created, flanking the descent of this pathway across the boulder fill of the hollow and into the cave (Phase 2: Plate 2a–c). Within the narrow cave passage, deposits rich in artefacts and bioarchaeological material began to accumulate (Plate 2d). Outside the cave, hearths were constructed on the terraces beside the walkway, and the hollow became a focus for intensive activity, with repeated burning events creating large accumulations of ash and carbonised residues (Plate 3a–b). These were left *in situ*, and new hearths and pits were constructed upon them. Ash and midden began to fill the hollow and the original walkway was backfilled. Further Early Iron Age activity in Phase 3 saw the construction of a low retaining wall and a stairwell in order to maintain access to the cave (Plate 3c), and paved walkways re-established along both cave and precinct. Outside of the precinct, a series of burnt spreads date from this point through the Middle Iron Age to the end of Phase 5, some of which are associated with ephemeral structures. Ongoing midden accumulation appears to have then driven a cycle of repeated reconfiguration of precinct and stairwell. Burning activity focused on a point adjacent to the walkway where several extensive hearths were superimposed, with evidence for smaller hearths throughout the precinct. Complex sequences of midden and fuel residue-rich deposits associated with these events formed a deep, artefact- and biofact-rich stratigraphy which blanketed the precinct and necessitated further revetting and extension of the stairwell to keep the cave access open (a process maintained through Middle Iron Age Phases 3 into 6; Plate 3d). An example of the layout of the precinct, and its relationship to stairwell and cave during these phases is seen at Fig. 2.1. Within the cave, a parallel sequence of deposits rich in artefacts and bioarchaeological material built up, their distribution suggesting a focus on the point at which the cave passage narrows towards its rear. In Phase 4, an enclosing wall was built around part of the now transformed hollow, defining and shielding the cave entrance.

Gradual subsidence of midden, now even burying the enclosure wall, appears to have prompted deposition of a thick clayey levelling layer across the precinct in Phase 5. Upon this, further hearths were constructed and the now deep vertical stairwell was covered by a corbelled roof. Within the cave, a partially paved floor was established. Finally, with only a suggestion of the original hollow remaining and a rough boulder wall now defining the precinct, a last phase of later middle Iron Age activity (Phase 6) features the most intensive and unusual episodes of in-cave activity: large deposits of pig and cattle displaying overwhelming evidence for ritualised treatment (Drew and Horton 2014, and below). After this, following *c.* 800 years of activity, the stairwell was backfilled and the entrance to the cave closed. A final episode may suggest that the cave location remained embedded in social memory long after active use ceased. A 'closing' inhumation of an adult female along with a 9-month perinate, a 3–6-month foetus, and associated animals was placed into the backfilled stairwell *c.* cal AD 5–135: *c.* 95–270 years after the end of activity at the site.

Structures and artefacts: a ritual arena?

Structural features do not indicate that the cave or precinct were, at any point, a domestic setting. A key focus is the

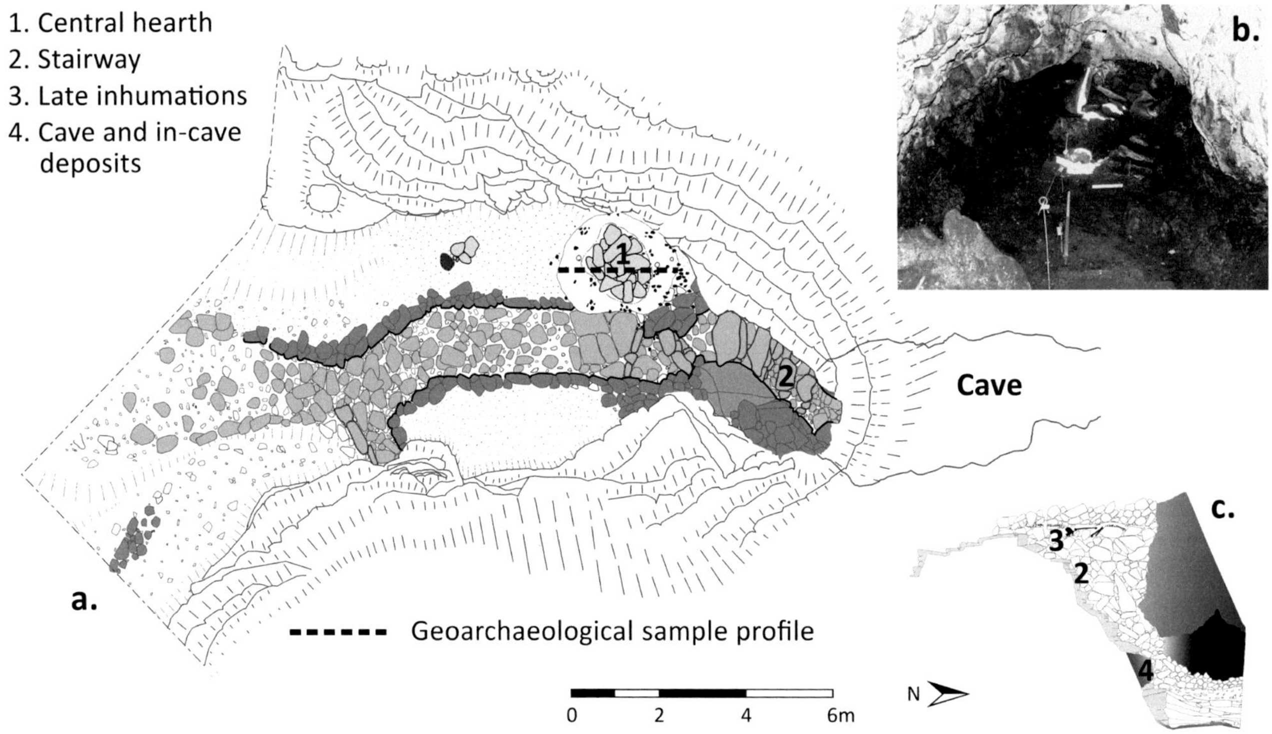

Figure 2.1 a) Plan view of precinct in Phase 3: walkway, stairwell and central hearth; b) View of stairwell base from cave interior; c) Schematic section of the stairwell as continuing from (a), depicting the link between cave and precinct, the stairwell backfill stratigraphy and the position of the late stairwell inhumation.

choreographing of access to the cave. This is seen both in episodes of the stairwell restructuring (for example, the increased darkness created by the corbelled roof) but also in features, such as a large slab-built hearth constructed in a position which partially obstructs the primary walkway near the cave entrance (seen partially extant in Plate 2b). This may have represented an intentional demarcation point preventing easy access to, or viewing of, the cave entrance (Birch, pers. comm.). Outside the precinct, ephemeral stone-built structures indicate surface activity for which evidence has not survived, but the precinct and cave itself appear devoid of structural features indicative of habitation.

Artefact assemblages show that High Pasture Cave was a focus for the deposition of significant objects, with strong indications that this was within a ritual context. The *c.* 2500-artefact assemblage is the largest from any recorded cave excavation in the UK (Birch; Dowd, pers. comm.) and includes antler, bone and stone objects and tools, decorative objects in steatite, cannel coal and bronze; iron tools, knives (one deposited in its organic sheath) and pins. Ephemeral evidence for some craft activity involving these materials ranges from stone manufacturing waste to evidence for metalworking. Evidence for structured deposition of unusual and special objects is ubiquitous. Multiple caches of stone tools are seen within the debris of individual precinct hearths, and groupings of antler and steatite objects within the cave – such as a cache of steatite whorls representing a range of stages of object manufacture, and a group of seven antler pegs interpreted as textile working tools. Quernstones show highly ritualised contexts of deposition. Almost all are highly fragmented, indicating deliberate breakage of these resilient objects. Grinding surfaces show defacement, and few fragments refit, indicating either that querns were smashed elsewhere and fragments brought to High Pasture, or that they were broken on site and the fragments separated (Cruickshanks and Hunter 2015a). Many querns are re-used in foundation or closing deposits and within hearths, steps and walls. The largest assemblage of ring-headed pins from a single Scottish site includes an example broken into two pieces and placed in a leather receptacle before deposition in the cave, with residues of organic material traced on the refitting pin fragments. Three spear-heads are unique within Iron Age contexts on Skye. One shows incised decoration, and another was hafted when deposited, the remaining mineralised wood secured with a copper alloy pin. Its detached spear-tip refits at an angle across a large corrosion blister, indicating that it was attached but bent when deposited – suggestive of deliberate damage. The spearhead's weight in iron (328g) would have made it a valuable object (Cruickshanks and Hunter 2015b). A discrete accumulation of charcoal debris included tabular and turned bowl fragments and the remains of a charred wooden bridge from a musical instrument. Identified as a lyre, this is the earliest physical evidence for such an instrument in Western Europe (Lawson, in Birch and Wildgoose 2013, 83). This range of significant object types is reminiscent of the observation made by Manem (2012) in his study of the Bronze Age caves of the Duffaits culture in western France. He draws attention to the fact that

> all the craft and economic activities that one cultural group could practice are represented in these caves – weaving, metallurgy, amber trading, bone industry, pottery, hunting, livestock framing, harvesting and agriculture – a concentration of activities which seems incompatible with a household context. (Manem 2012, 145)

The artefact assemblages from High Pasture Cave strongly indicate that this was a special place, their volume and richness indicating that this site was a focus for significant communal events, certainly a place for deposition of significant objects, and ultimately for undertaking ritual and special activities. The peculiarity of the cave environment gives a unique insight into the potential complexity of such acts. Four of the stone tools recovered from the precinct show calcite deposits overlying worked areas, indicating that they had lain in the cave after use long enough for calcite deposits to form before being removed to the precinct (Cruickshanks and Hunter 2015a).

Bioarchaeological assemblages

The ritual character of the High Pasture site is also strongly indicated by the rich and diverse zooarchaeological and botanical assemblages recovered. These provide a powerful tool for understanding events within both the cave and the precinct, illustrating a focus on the deposition of foodstuffs and the use of fire technology.

The High Pasture Cave faunal assemblage alerted archaeologists not only to the site's presence but also to its potential significance. Over 54,000 bones were recovered from excavation, two-thirds from the cave, and almost all the remainder from within the enclosed area of the precinct. The most striking of these deposits are dated to the final phase of activity within the cave (Phase 6) and differ markedly not only to the rest of the site assemblage, but also to material seen throughout the rest of Iron Age Scotland. The latest in-cave midden layer within Phase 6 features a large (3500+ fragments) bone deposit, the identifiable proportion of which consists of almost 74% pig. This is an unprecedented concentration in a Hebridean assemblage, of which an average pig representation is roughly 3–4%, and the largest concentrations 15% at most (Drew 2014, 1). Excellent bone preservation allowed identification of near complete skeletons, indicating probable killing and butchery on site. Age analysis identified a grouping of *c.* 7–8 months in age, with further individuals killed at approximately a year later, at around 20 months. Both groups indicate concentrations of juvenile pigs killed within a discrete winter period (*ibid.*, 39). Butchery patterns show an unusual pattern of treatment, with great attention paid to the separation of carcasses into left and right along the sagittal line, particularly noticeable in the butchery of the skulls. However, despite such intensive treatment, no attempt was made at breaking the long bones to access the marrow. This is highly unusual in an Iron Age context, where bone fat was routinely exploited for a variety of purposes, and may suggest a glut of meat which, alongside the age data, suggests a single feasting deposit. Roughly contemporary with this are three individual deposits of near-complete cattle skeletons, two placed in compact bundles along the cave passage, and the third recovered from a ledge above the streamway further into the cave complex. These also show sagittal line splitting of a skull, but the carcasses show comprehensive breakage and evidence for marrow extraction (*ibid.*).

Earlier phases see a greater emphasis on activity within the cave precinct, and here, the social and/or ritual context of activity is harder to interpret. Deposits of more fragmentary faunal material displaying more typical patterns of Iron Age species composition show huge spikes in concentration within and around the main hearth loci, particularly within Phase 3 (Plate 4e) – however, this more typical domestic pattern is belied by the association of these deposits with the intensity of burning events and the pattern of object deposition within the precinct, as well as with large deposits of cleaned burned barley grain which echo similar, likely ritual deposits seen within the cave (see below).

Over 50,000 charred plant macrofossils were recovered from the High Pasture excavation, an assemblage dominated by hulled barley grains with a smaller hazelnut shell fraction and a limited suite of weed seeds. Its distribution parallels that of the animal bone, with most material recovered from the cave (29,000+ entities) and an increase in material from later phases, particularly in the cave and in the later Phases 5 and 6 (Plate 4d) (Drew and Horton 2014). Assemblage characteristics strongly suggest a directly ritual element to the majority of deposits, particularly within the cave. Seven of the clearly defined Phase 4–5 and 6 in-cave midden layers produced assemblages of over 1100 charred barley grains each, with four showing 1500+ grains. These very large assemblages are notable for their lack of associated contaminants, with very few weeds seeds or chaff fragments present for the number of grains recovered. This observation, which holds true for both the in-cave and precinct assemblages, indicates that the grain deposited was burnt at a late point of processing where almost all the contaminants had been removed. It is hard to see how very large assemblages of this kind could represent domestic activity, and it seems unlikely that accidental burning of cleaned grains can explain such large, repeated depositions (Horton, in Birch and Wildgoose 2007). These deposits strongly suggest the votive deposition of grain within the cave, alongside (especially in Phase 6) the similarly ritualised context of animal bone deposition.

Within the precinct, grain recovery is strongly biased towards a small number of key contexts: mainly hearths and their associated residues, but also contexts associated with paving (Drew and Horton 2014). Almost 1/3 of the grain recovered from the precinct is associated with the Phase 4 hearth feature F15.20 (mainly as part of the fuel residue deposit context C15.38 – Plate 4e). This and several other precinct contexts associated with peaks in cleaned burned grain also show intensive bone inputs (Plate 4e). The latest grain rich deposit is placed within the last phase of the stairwell, extending down the steps into the cave and along its length. Again, it seems that domestic activity is an insufficient explanation for these patterns of deposition when considered alongside precinct-wide patterns of artefact, bone and grain accumulation.

Georchaeological analysis

Investigation of the potential for geoarchaeological, especially micromorphological analysis, to contextualise these more ambiguous suites of bioarchaeological data in the precinct is the focus of the following study. Using the stratigraphic block system (see below) as a tool for linking related event sequences, a series of geoarchaeological samples taken through the sequence of fuel residue deposits at the central hearth location are discussed in context with key features of the faunal and botanical assemblages.

Methods

Sampling

Sixteen Kubiëna samples were taken in a vertical sequence through a west facing section bisecting the central hearth sequence (Fig. 2.1a and Plate 4). Alongside each Kubiëna sample, bulk samples were taken for a range of analyses including pH and total phosphate.

Micromorphology

Thin section slides were prepared at the University of Stirling, UK. Water was removed from all samples by acetone exchange and confirmed by specific gravity measurement. Samples were impregnated using polyester 'crystic resin 17449' and catalyst Q17447, thinned with acetone and a standard composition of 180ml resin/1.8ml catalyst/25ml acetone. Samples were impregnated under vacuum to ensure complete outgassing, cured for 3–4 weeks, then heated for 4 days at 40°C before slicing, precision lapping to 30 microns and cover-slipping. Slides were analysed using an Olympus BX-50 petrological microscope, following the procedures of Bullock *et al.* (1985) and Stoops (2003).

Bulk soils were air dried prior to analysis of soil pH, using the method of Bascomb (1974), and total phosphate, following the methodology of Smith and Bain (1982), with colorimetric determination then undertaken via ammonium molybdate/ascorbic acid against standards ranging from 0–10 micrograms P. From this total phosphorus (mg/100g) was calculated for each sample and then converted to total phosphate (P2O5/P2 = 9141.96/61.96 = × 2.29).

Stratigraphic relationships: relating geoarchaeological and bioarchaeological datasets

Post-excavation analysis has subdivided the large and complex phases of activity at High Pasture Cave into a series of 'stratigraphic blocks' (SBs), representing discrete sequences of events, such as the construction and use of a sequence of hearths, stages of the stairwell construction, or precinct-wide events, including the closing of the primary walkway (Birch, pers. comm.; Plate 4e). The stratigraphic blocks enable post-excavation linkage of groups of related contexts excavated at different times due to the complex topography of the site, and the consideration of sub-groups of artefact and bioarchaeological assemblages relating to specific event sequences. These subdivisions also allow detailed comparison of the relationship between the geoarchaeological dataset (which here discusses a single vertical sequence of deposits) and features of interest in the bioarchaeological assemblage (which may relate directly to sections of this geoarchaeological sequence, but not represent the same numbered context). Plate 4 and Table 2.1 present these data links. Note that the High Pasture Cave context descriptors are prefaced 'F', denoting a construction feature (such as a hearth), or 'C' denoting a deposit context. Peaks in bone and/or grain assemblages recovered from hearth and associated deposit sequences are identified by stratigraphic block, enabling a link with the geoarchaeological data. The 21 sampled contexts making up the geoarchaeological profile subdivide into 13 stratigraphic blocks representing all 7 phases. This study focuses on the nine blocks spanning Phases 3–5 in this location, which are associated with some of the most significant bioarchaeological assemblages in the precinct area. Key observations are discussed below.

Micro-stratigraphy and the intensity of activity at High Pasture Cave

The majority of the 21 sampled contexts of the geoarchaeological profile represent fuel residue and midden deposits associated with the use of adjacent hearths. Exceptions to this are the walkway backfill deposit C2.34 at the profile base and the Phase 6/7 dump deposits at the top (Plate 4d–e). Within these contexts, 62 'micro-lenses' were identified, ranging from broad divisions within the dumps to events only a few millimetres thick (Plate 4c–d, Table 2.1). Many of the latter are defined by their fuel residue characteristics, and are interpreted as specific events associated with hearth use within the precinct. The orientation of both fuel residue and other inclusions within these lenses is often horizontal, successive lenses giving a finely laminated appearance suggestive of the trampling and compression of active surface, or near-surface, deposits (Plates 4a and 5a).

Plate 4d and Table 2.1 illustrate this complexity of subdivision within the sequence, with Plate 4e showing how this correlates, via stratigraphic block, to key hearth structures and associated contexts with significant bone and charred plant assemblages. Several contexts show numerous subdivisions, particularly C2.15b, C2.16a, C2.16b, C2.26a; and to a lesser extent, C2.05b, C2.05d and C2.06c – all of which (with the exception of a fairly sterile silt and clay relating to levelling layer C2.06c) are associated with contexts showing elevated levels of bone, charred plant material, or both.

These spikes in bioarchaeological material seem to correlate well with apparent intensification of activity in the precinct. However, micromorphology indicates that not all of this relates directly to combustion. C2.16b represents one of the most finely stratified contexts in the sequence, yet its subdivisions are defined not by evidence for burning events, but by successive thin, fairly sterile depositions featuring degraded plant materials and, most noticeably, varying degrees of iron enrichment. These indicators for some potential hiatus in the sequence sit within stratigraphic block 7e, featuring C15.56, one of the most bone rich contexts after the grand accumulation at hearth F15.20. The material from this context showed little evidence of heating, unlike most of the particularly bone rich contexts (Drew 2014, 57–60). Did events at the cave entrance include significant activities related to carcass processing, consumption, and/or discard with only an indirect relationship to combustion activity and hearth use, despite the ubiquity of hearth structures and fuel residue deposits? It is clear that the intensive, extremely finely laminated depositional history of the High Pasture precinct deposits demands caution in interpretation – where episodes of apparent hiatus in activity of one kind may not mean cessation of another, within complex cycles of highly structured creation and discard within a restricted and precisely managed arena.

Microscopic analysis also illustrates what may be missing from the sequence, with evidence for truncation seen clearly near the top of the sequence in context C2.05 at slide 4 (Plate 4d). Here, it would appear that C2.05a accumulates upon a horizontal truncation cutting across the sharply sloping accumulations of C2.05b, C2.05c, and C2.05d, with the two C2.05a deposits repeating this pattern of sharply sloping accumulation. With this deposit adjacent to paving relating to the stairwell entrance (Birch and Wildgoose 2013), it is not surprising that there may have been removal and reorganisation of material as well as accumulation. Such indicators have clear implications for the interpretation of the patterning of ritual activity at the cave entrance, replete with evidence for the structural reorganisation and management of accumulated materials. It is notable that bone-rich contexts C2.27 and C2.32 (Plate 4e) are both in stratigraphic blocks whose thin-sectioned contexts show few separations into individual micro-lenses in thin section (C2.18, 2.22, 2.23; C2.26b: Plate 4e, Table 2.1). This could indicate that material may have been lost from these phases of activity, perhaps through cleaning of the active area around a hearth, or as a result of reorganisation within the precinct as a whole.

Fuels and fires

Fuel residues are a major element of the thin section dataset, with over 2/3 of identified micro-lenses rich in carbonised materials and several contexts showing a multiplicity of intense, discrete burning episodes. Fuel residues were categorised to type where possible into wood charcoal, carbonised material with a fibrous, 'peaty' structure, and black, carbonaceous residue showing no internal structure conforming to the 'char' categorisation described by Braadbaart *et al* 2012. These were then divided by comparative analysis from very high to low or absent concentrations (Table 2.1). While wood charcoal predominates (Plate 5a), it seems that visitors to High Pasture also fed their fires with peat, with regular inputs of stringy, fibrous carbonised material (Plate 5b), and patches of heated silt, which may represent ashed peaty inputs. These are more frequent in Phases 3–4. However, biases of both preservation and identification are possible. Amorphous 'char' is recorded throughout the sample sequence, indicating that combustion and/or taphonomic processes may have rendered an unknown proportion of both fuel types unclassifiable. With wood charcoal seen throughout, and with most deposits showing high fuel residues having at least some peaty input, there is no clear 'switch' from one fuel type to another. Rare vitrified materials are also present, as is wood ash, though there is no strong correlation between micro-lenses showing wood ash preservation, and contexts with higher pH (Table 2.1).

Identification of fuel types within and between micro-lenses of the more 'intensive' contexts shows interesting patterning, with several abrupt changes in the relative frequency of charcoal to peat-like inclusions: for example, micro-lenses d-h within context C2.26a, and a-c within C2.16a (Table 2.1). It would seem that these very distinct, yet sometimes millimetre thin sequences may represent very differently composed fires – yet sharp, clear boundaries indicate immediate burial and minimal disturbance, and therefore a rapid sequence of events. This may suggest that ritual burnings at High Pasture were highly specific events that required the complete cessation of a single fire before another was begun, with either incidental or deliberate variations in the ratio of charcoal to peat between fires preserved in the fuel residue sequence.

*Table 2.1 Table showing presence (rare *, common **, frequent ***) or absence of key micromorphological indicators against pH, total P and spikes in associated grain/bone content, resolved to micro-lens and by phase, content and stratigraphic block.*

Thin section	Phase	Stratigraphic block & related contexts	Context	Micro-lens	Spike in bone (B) &/ or grain (G)	Wood char	Fibrous carbonised	Char	Fuels low/absent	Burnt bone	Unburnt bone	Bone low/absent	pH	P_2O_5 (mg/100g soil)
2–1	7	11	C2.03a	a					*			*	6.8	541
		11		b					*			*		
2–2	6	10	C2.03b	a		**		*		*	**		6.9	537
		10		b					*			*		
	5	9c: hearth F2.07, grain-rich	C2.05a	a	G				*			*	6.7	637
2–3				b	G	**		*				*		
				c	G				*			*		
2–4				d	G				*			*		
				e	G	**					*			
		9b: hearth F2.13; grain-rich; related context 15.25 grain rich	C2.05b		G	**		*		*	*		6.4	303
			C2.05c		G	**		*			*		6.4	367
		9a: hearth F2.05; grain rich	C2.05d	a	G	**				*	*		6.6	873
2–5				b	G			*				*		
				c	G	*	*					*		
		9a: clay levelling deposits	C2.06a	a					*			*	6.6	428
				b		**		*		*	**			
			C2.06b			***	*	*			**			
			C2.06c	a		*		*				*	6.4	531
				b		**		*				*		
2–6				c		*						*		
2–7	4	8c: hearth F15.20; bone-rich; related context F15.38 grain-rich	C2.15a		B, G			**				*	6.4	722
			C2.15b	a	B, G	*	*	*		*	*		6.6	1597
				b	B, G				*			*		
				c	B, G	***	*	**		*	**			
				d	B, G	***		*				*		
				e	B, G				*			*		
				f	B, G	**	*	*		**				
				g	B, G				*	**	**			
			C2.15c	a	B, G	**					*		6.6	479
2–8			C2.15b	h	B, G	**	*	*		***	*		6.6	1597
			C2.15c	b	B, G	**	**			**	*		6.6	479
		8a: hearth F2.27 bone-rich	C2.16a	a	B	*	*	*				*	6.7	1374
				b	B	*	***					*		
				c	B	*		**				*		
				d	B	*		*				*		
2–9				e	B	***	**	*				*		
				f	B	**	*					*		

(Continued)

*Table 2.1 Table showing presence (rare *, common **, frequent ***) or absence of key micromorphological indicators against pH, total P and spikes in associated grain/bone content, resolved to micro-lens and by phase, content and stratigraphic block. (Continued)*

Thin section	Phase	Stratigraphic block & related contexts	Context	Micro-lens	Spike in bone (B) &/ or grain (G)	Wood char	Fibrous carbonised	Char	Fuels low/absent	Burnt bone	Unburnt bone	Bone low/absent	pH	P_2O_5 (mg/100g soil)
2–10	3	7e: hearths F2.15/ F15.28; related context C15.56 very bone-rich	C2.16b	a	B				*			*	6.6	287
				b	B				*			*		
				c	B				*			*		
				d	B				*			*		
				c	B				*			*		
				e	B				*			*		
		7c: hearth F2.10 grain-rich; related context C2.27 bone-rich	C2.18		B				*			*	6.4	1355
2–11			C2.22		B, G	*	*						6.8	484
			C2.23		B, G	**	**					*		
		7b: hearth F15.30 related context C15.64 very bone-rich	C2.26a	a	B				*			*	6.8	1050
2–12				b	B	*	*					*		
				c	B		*					*		
				d	B		*							
				e	B	*	*					*		
				f	B		***					*		
2–13				g	B	*		*				*		
				h	B		*					*		
				i	B	**	**					*		
				j	B	*	*					*		
		7a: hearth F15.35 grain-rich; related context C2.32 bone-rich	C2.26b	a	B, G	**	**						7	3057
2–14				b	B, G	**	**							
	2	6	C2.34	a		**	*					*	7.2	419
2–15	1a	2		b		*								
2–16				c					*			*		
			C2.35						*			*	6.9	671

Bone and grain, burning and cleaning: butchery and feasting

Despite the recovery of large amounts of bone from throughout the precinct area, the majority of micro-lenses show relatively low or even absent bone (Table 2.1). While unburnt bone predominates slightly, most bone-rich horizons show a mixture of both burnt and unburnt fragments, the former usually smaller, the latter appearing to show markedly good preservation (Plate 5c–e). This is interesting, as despite the protective alkaline effect of both the limestone bedrock and the quantities of ash within the sequence, excavators described smaller fragments of unburnt bone as soft and difficult to retrieve (Birch, pers. comm.). In addition, the features of iron movement seen throughout the sample profile indicate that, not surprisingly for this exposed northern location, illuviation processes should result in poorer survival of smaller bone outside the cave. The fairly limited instances of alkaline wood ash seen within the sample sequence also do not correlate with particularly bone-rich micro-lenses. The potential reason for this unusually good preservation, and its likely relationship to other ritual additions to the deposit sequence, are discussed in the following section.

There is a variable correlation between the amount of bone recovered from key contexts (Plate 4e) and the levels of bone seen in related thin sections: while the numerous micro-lenses of context C2.15 (relating to the main bone 'peak' focused on hearth F15.20) show the highest concentrations of both burnt and unburnt bone, the distribution of this is uneven, with not all micro-lenses showing high bone content, and below context C2.16a, none of the bone-rich contexts from Phase 3 are represented: significant bone presence in thin section ceases abruptly at context C2.16a.

It is interesting to consider that this may relate in part to the point made above with regard to fuel residue patterning: if not all rituals focused on bone processing and deposition involved combustion and hearth use, then perhaps not all rituals focused on the hearths involved the burning of bone. In the absence of microscopic evidence for charred grain the following remains speculative, but it is interesting to consider that some of the fuel-residue rich yet low in bone micro-lenses seen particularly in the Phase 3 deposits may represent activities within the hearth sequences unrelated to bone processing, such as the final, possibly ritualised stages of processing and burning of the large caches of barley grain from these deposits described by Drew and Horton (2014, 14). Detailed interpretations of activity involving the grain are impossible to decipher: despite 'elevated' levels of material coming from contexts associated with several key hearths (F2.07, F2.13, F2.05 and especially the very large grain concentrations associated with hearth F15.20) (Plate 4e), individual contexts show relatively low grain counts, with even 'high' grain concentrations representing less than 2.5% of the whole precinct assemblage (*ibid.*). However, if comprehensive cleaning of the hearths after individual combustion episodes, suggested above as a potential interpretation of the dramatic fuel type variations seen in some adjacent deposits, did take place, then it is reasonable to suppose that fires lit in order to burn grain would indeed be low in bone, even if high bone counts are seen within the stratigraphic 'block' representing activity around the hearth as a whole. Other primary purposes for individual burnings are also possible, such as the creation of dramatic effects involving smoke or aroma within the precinct and/or cave. A single fragment of resin retrieved from within the cave has been identified as Pinaceae sp., possibly spruce, indicating a potential aromatic purpose (Heron 2017), and the High Pasture Cave charcoal assemblage includes several species unusual for the region, such as holly, cherry and pine – the latter of which provides a smoky burn with a distinct aroma (Cressey 2014). However, individual burning events do not explain the wholesale decrease in microscopic bone presence towards the base of the deposit. For this, and a likely interpretation for the level of bone preservation in the sequence as a whole, we turn to the final key feature of the High Pasture Cave micromorphological dataset.

Phosphatic features, bone degradation, and the specifics of ritual activity

Although soil pedofeatures relating to illuviation as well as low-level movement and disturbance (such as organic, silty and dusty clay infills and coatings) are seen throughout the sample sequence, by far the most significant group of pedofeatures are those relating to phosphate movement and accumulation. Two types of these can be observed. Amorphous phosphate appears in thin section as a fine, featureless yellow material which infills soil void space and is isotropic in crossed polarised light. Significant accumulation of this material is unusual in wet, usually acidic Scottish soils, where illuviation processes normally see phosphate that has passed into the soil from degrading organic materials (such as plants or, more usually, bone) cycled rapidly. The pH of the High Pasture sample profile is neutral to slightly acidic (Table 2.1) and yet these features are frequent throughout the sequence, present in most contexts and significant in 29 out of 62 micro-lenses. All but one of the contexts high in bone are also high in phosphatic features, and the majority of phosphate rich sub-lenses are also high in fuel residues. Amorphous phosphatic infills are distinctive, often forming part of the laminated appearance of the micro-lenses (Plate 5f). Others are more amorphous, and often reminiscent of partially decayed organic materials, especially bone.

The second phosphatic pedofeature seen in the micromorphological sample sequence is one more directly linked to bone decomposition. Calcium-iron-phosphate (Ca-Fe-P) pedofeatures are distinctive features that are formed during the release of Ca and P from bone hydroxyapatite during decomposition, appearing as small radial crystallisations (Plate 5g). These have been identified in numerous geoarchaeological investigations featuring bone-rich contexts and phosphate-rich environments (*e.g.* Simpson *et al.* 2000). Ca-Fe-P pedofeatures are observed in seven micro-lenses, almost all clustered in the lower half of the profile between contexts C2.16b and C2.26a. A lone Ca-Fe-P rich lens is seen at the top of the profile in C2.05a-b. None of these contexts show good bone preservation.

Total P analysis (Table 2.1) indicates significant elevation of soil phosphate throughout the profile, and identifies the cave precinct deposits as representative of a highly phosphate-saturated environment. It is suggested that although many of the smallest fragments of bone once present in individual micro-lenses have degraded (as evidenced by the presence of Ca-Fe-P pedofeatures), the majority of especially the larger fragments seen in thin section have been protected from physical dissolution by this phosphate saturation, in a similar process to that examined by Correa *et al.* (2013) in their study of phosphate-rich shell mounds. Here, P enhancement was attributed to inputs of organic materials and ash, creating an apatite-rich soil matrix that retarded bone dissolution, ensuring preservation of larger, denser bone fragments. This may explain the increased proportion of smaller fragments of burnt to unburnt bone seen throughout the bone-richer micro-lenses, and perhaps also the superficially well-preserved but soft and difficult to excavate precinct bone, which is also noted in the faunal analysis as less well preserved than the in-cave bone (Drew 2014, 3). In thin section, we can see that the internal structure of these 'soft' bone fragments seem perfectly preserved, but that chemical alteration may in fact be fairly advanced (Plate 5h). This is a significant assemblage preserved in unusual conditions, providing a fascinating insight into processes of bone decomposition.

Where does this enhancement come from? The evidence provided by the bone assemblages and, particularly, the unusual butchery and depositional treatments of its larger 'special' deposits may provide an answer. Although ash may be responsible for some P-input, it is suggested that a key purpose of the cave precinct was as a locus for rituals forming a precursor to the dramatic in-cave depositions noted in the faunal assemblage: ritualised activity involving slaughter, defleshing, the breaking of bones, and the spilling of blood. A key observation supporting this is the poor correlation between sediment lenses showing the highest phosphate levels and those with the most frequent bone and fuel residue inclusions. Even allowing for some down-profile movement of phosphate (Table 2.1), there is an indication that non-bone sources of phosphate, also not directly correlated with burning events, may be responsible for the high P input. Concentrations of blood, flesh and marrow are an obvious source, with similar phosphatic features and enhancements noted at sites where fleshy wastes represent a significant input into sediment contexts, such as fish processing middens (Simpson and Barrett 1996). It is easy to imagine the potential for symbolic or ritual significance in a nominally similar sequence of events at High Pasture Cave: perhaps the ceremonial killing and blood-letting of animal(s) within a focal part of the precinct, or the collection of blood during this process which could have then been spilled onto the ground at the mouth of the cave or around the hearth. Clear evidence for the importance of butchery to these events colours the picture further: dismemberment as performance within this small arena would have resulted in the bloody addition of significant amounts of flesh to the ground, even if much of this flesh and perhaps even associated viscera were then at least partly consumed or removed.

Conclusions

This chapter has illustrated the diverse and wide-ranging contribution of geoarchaeology to our understanding of ritual activity in the archaeological record: from construction to destruction, and from modification of materials to that of space and place. By fine-tuning understandings of the context of deposition of such sediments and the materials within them, geoarchaeological analyses make a unique contribution to our capacity for visualising ritual activity. At High Pasture Cave, holistic assessment of the rich geoarchaeological and bioarchaeological dataset provided by the precinct deposits facilitated the reconstruction of potential scenes of ritual activity within this key area of the site.

We can hypothesise that those entering the enclosed cave precinct would have stepped into a space set apart from the world outside. The geography of the precinct area indicates that activities which took place there were precisely choreographed, with a focus on the large central hearth, and clear references to the prime significance of the cave beyond. The dark bulk of the cave with its dramatic roofed stairwell may have only just been visible, the cave entrance screened by the stone bulk of a hearth seemingly placed to create both a physical and symbolic barrier.

Essential to the precinct ritual were activities involving large amounts of grain and meat, with a focus on the large central hearth. Micromorphology indicates that the fires of this hearth were not only carefully constructed, perhaps using specific fuel types, but were likely brought to an equally deliberate close before another burning event could take place. In what may have represented a relatively confined space, such precise management of the ritual process may have been both expedient and represented a series of finely nuanced symbolic set-pieces, potentially involving the control of various senses and the display and ultimate donation of the riches of individuals or groups in the form of tools, treasures and food.

However, not all of these activities appear to have centred on the hearth. The large input of phosphate into the precinct sediments could indicate that a core part of the precinct ritual may have focused on blood and viscera. These geoarchaeological findings chime with the faunal material recovered from inside the cave to create a picture of bloodletting and butchery at and around the cave entrance, and to illustrate the likely close association between events in the precinct and ritual activity within the cave itself.

Micromorphological and bioarchaeological evidence for ritual activity within the cave precinct suggests a complex, multi-purpose arena where a range of activities and events would take place, marking the importance of this special, part-hidden place to the social and cultural landscape of the participants. This integrated analysis has highlighted the ability of geoarchaeology to shed light upon 'invisible' aspects of the archaeological record, helping to create a vivid picture of the ways in which the Iron Age visitors to High Pasture Cave may have expressed the cosmological significance of this very special place within their community.

Acknowledgements

The author would like to thank Historic Environment Scotland for providing support for the presentation of this research at EAA Istanbul 2014. All photographs are taken from the High Pasture Cave site archive, with permission of Mr. Steven Birch. All images prepared by Jo McKenzie and Mary Peteranna using data taken from the site archive, with permission.

References

Bascomb, C. L. (1974) Physical and chemical analyses of <2mm samples. In B. W. Avery and C. L. Bascomb (eds.) *Soil Survey laboratory methods*, 14–41. Harpenden, Soil Survey of England and Wales.

Berna, F., Behar, A., Shahack-Gross, R., Berg, J., Boaretto, E., Gilboa, A., Sharon, I., Shalev, S., Shilstein, S., Yahalom-Mack, N., Zorn, J. and Weiner, S. (2007) Sediments exposed to high temperatures: reconstructing pyrotechnological processes in Late Bronze and Iron Age Strata at Tel Dor, Israel. *Journal of Archaeological Science* 34, 358–73.

Birch, S. and Wildgoose, M. (2007) Uamh an Ard Achadh (High Pasture Cave) & Environs Project, Strath, Isle of Skye: Data Structure Report HPC003. Unpublished report, West Coast Archaeological Services.

Birch, S. and Wildgoose, M. (2013) Uamh an Ard Achadh (High Pasture Cave) & Environs Project, Strath, Isle of Skye: Data Structure Report HPC007. Unpublished report, West Coast Archaeological Services.

Boivin, N. (2010) Life rhythms and floor sequences: excavating time in rural Rajasthan and Neolithic Çatalhöyük. *World Archaeology* 31, 367–88.

Braadbaart, F., Poole, I., Huisman, H. D. and van Os, B. (2012) Fuel, Fire and Heat: an experimental approach to highlight the potential of studying ash and char remains from archaeological contexts. *Journal of Archaeological Science* 39, 836–47.

Bullock, P., Fedoroff, N., Jongerius, A., Stoops, G., Tursina, T. and Babel, U. (1985) *Handbook for Soil Thin Section Description*. Wolverhampton, Waine Research.

Correa, G. R., Schaefer, C. E. and Gilkes, R. J. (2013) Phosphate location and reaction in an archaeoanthrosol on shell-mound in the Lakes Region, Rio de Janeiro State, Brazil. *Quaternary International* 315, 16–23.

Cressey, M. (2014) High Pasture Cave: charcoal analysis. Unpublished report for the High Pasture Cave Project, West Coast Archaeological Services. Edinburgh, CFA Archaeology.

Cruickshanks, G. and Hunter, F. (2015a) *Coarse Stone Artefacts from High Pasture Cave: Interim Report*. Edinburgh, National Museums Scotland.

Cruickshanks, G. and Hunter, F. (2015b) *The Iron Artefacts from High Pasture Cave: Interim Report*. Edinburgh, National Museums Scotland.

Drew, C. A. (2014) High Pasture Cave: The Faunal Assemblage. Unpublished interim report for the High Pasture Cave Project, West Coast Archaeological Services. Cromarty, University of Durham.

Drew, C. A. and Horton, E. (2014) High Pasture Cave: palaeoenvironmental analysis of cave and surface deposits. Unpublished interim report for the High Pasture Cave Project, West Coast Archaeological Services. Cromarty, University of Durham.

Fiorentino, G. and D'Oronzo, C. (2010) An archaeobotanical and experimental approach to identifying successive fire events in hearth structures in the sanctuary of Apollo in Hierapolis (Turkey). *Palethnologie* 2, 59–68.

Fleisher, J. (2014) The complexity of public space at the Swahili town of Songo Mnara, Tanzania. *Journal of Anthropological Archaeology* 3, 1–22.

Fleisher, J. and Sulas, F. (2015) Deciphering public spaces in urban contexts: geophysical survey, multi-element soil analysis, and artifact distributions at the 15th–16th century AD Swahili settlement of Songo Mnara, Tanzania. *Journal of Archaeological Science* 55, 55–70.

Hamilton, D. (2014) Preliminary modelling results for the radiocarbon dates from High Pasture Cave, Isle of Skye, Scotland, U.K. Unpublished interim report, High Pasture Cave Project.

Heron, C. (2017) Analysis of the Resin (Find 6605) from High Pasture Cave. Unpublished report. Department of Scientific Research, BM. London, British Museum.

Huisman, D. J., Braadbaart, F., van Wijk, I. M. and van Os, B. J. H. (2012) Ashes to ashes, charcoal to dust: micromorphological evidence for ash-induced disintegration of charcoal in Early Neolithic (LBK) soil features in Elsloo (The Netherlands). *Journal of Archaeological Science* 39, 994–1004.

Johansen, P. G. (2004) Landscape, monumental architecture, and ritual: a reconsideration of the South Indian ashmounds. *Journal of Anthropological Archaeology* 23, 309–30.

Jusseret, S. (2010) Socialising Geoarchaeology: Insights from Bourdieu's Theory of Practice Applied to Neolithic and Bronze Age Crete. *Geoarchaeology* 25, 675–708.

Karkanas, P. and van de Moortel, A. (2014) Micromorphological analysis of sediments at the Bronze Age site of Mitrou, central Greece: patterns of floor construction and maintenance. *Journal of Archaeological Science* 43, 198–213.

Karkanas, P., Dabney, M. K., Smith, R. A. K. and Wright, J. C. (2012) The geoarchaeology of Mycenaean chamber tombs. *Journal of Archaeological Science* 39, 2722–32.

Karkanas, P. and Efstratiou, N. (2009) Floor sequences in Neolithic Makri, Greece: micromorphology reveals cycles of renovation. *Antiquity* 83, 955–67.

Madgwick, R., Mulville, J. and Evans, J. (2012) Investigating diagenesis and the suitability of porcine enamel for strontium (87Sr/86Sr) isotope analysis. *Journal of Analytical Atomic Spectrometry* 27, 733–42.

Mallol, C., Hernández, C. M., Cabanes, D., Machado, J., Sistiaga, A., Pérez, L., Galván, B. (2013) Human actions performed on simple combustion structures: An experimental approach to the study of Middle Palaeolithic fire. *Quaternary International* 315, 3–15.

Manem, S. (2012) The Bronze Age Use of Caves in France: Reinterpreting their Functions and the Spatial Logic of their Deposits through the *Chaîne Opératoire* Concept. In K. A. Bergsvik and R. Skeates (eds.) *Caves in Context: The*

Cultural Significance of Caves and Rockshelters in Europe, 138–65. Oxford, Oxbow Books.

Matthews, W. (2010) Geoarchaeology of plant remains and microarchaeological residues in early urban environments in the Ancient Near East. *Quaternary International* 214, 98–113.

McAnany, P. A. and Hodder, I. (2009) Thinking about stratigraphic sequence in social terms. *Archaeological Dialogues* 16, 1–22.

Milek, K. B. (2012) Floor formation processes and the interpretation of site activity areas: An ethnoarchaeological study of turf buildings at Thverá, north-east Iceland. *Journal of Anthropological Archaeology* 31, 119–37.

Miller, C. E. and Sievers, C. (2012) An experimental micromorphological investigation of bedding construction in the Middle Stone Age of Sibudu, South Africa. *Journal of Archaeological Science* 39, 3039–51.

Morris, J. and Maltby, M. (2010) *Social and Environmental Archaeologies: Reconsidering Deposition*. BAR International Series No. 2077. Oxford, Archaeopress.

Photos-Jones, E., Ballin Smith, B., Hall, A. J. and Jones, R. E. (2007) On the intent to make cramp: an interpretation of vitreous seaweed cremation 'waste' from prehistoric burial sites in Orkney, Scotland. *Oxford Journal of Archaeology* 26, 1–23.

Rondelli, B., Lancelotti, C., Madella, M., Pecci, A., Balbo, A., Perez, J. R., Inserra, F., Gadekar, C., Ontiveros, M. A. C. and Ajithprasad, P. (2014) Anthropic activity markers and spatial variability: an ethnoarchaeological experiment in a domestic unit of Northern Gujarat (India). *Journal of Archaeological Science* 41, 482–92.

Ryder, P. F. (1974) The Caves of the Beinn an Dubhaich Area, Isle of Skye. *Transactions of the British Cave Research Association* 1, 101–25.

Santana, J., Velacso, J., Balbo, A., Iriarte, E., Zapata, L., Teira, L., Nicolle, C., Braemer, F. and Ibáñez, J. J. (2015) Interpreting a ritual funerary area at the Early Neolithic site of Tell Qarassa North (South Syria, late 9th millennium BC). *Journal of Anthropological Archaeology* 37, 112–27.

Shahack-Gross, R., Albert, R-M., Gilboa, A., Nagar-Hilman, O., Sharon, I. and Weiner, S. (2005) Geoarchaeology in an urban context: the uses of space in a Phoenician monumental building at Tel Dor (Israel). *Journal of Archaeological Science* 32, 1417–31.

Sherwood, S. C. and Kidder, T. R. (2011) The Da Vincis of dirt: Geoarchaeological perspectives on Native American mound building in the Mississippi basin. *Journal of Anthropological Archaeology* 30, 69–87.

Shillito, L.-M., McKenzie, J., Card, N., Pike, S. and Chan, B. (2015) Subsistence, technology and resource use in Neolithic Orkney at the Ness of Brodgar. *Antiquity Project Gallery* 348.

Shillito, L-M. and Ryan, P. (2013) Surfaces and streets: phytoliths, micromorphology and changing use of space at Neolithic Çatalhöyük (Turkey). *Antiquity* 87, 684–700.

Simpson, I. A. and Barrett, J. H. (1996) Interpretation of midden formation processes at Robert's Haven, Caithness, Scotland using thin section micromorphology. *Journal of Archaeological Science* 23, 543–56.

Simpson, I. A. Perdikaris, S., Cook, G., Campbell, J. L. and Teesdale, W. J. (2000) Cultural sediment analyses and transitions in early fishing activity at Langenesvaeret, Vesteralen, Northern Norway. *Geoarchaeology* 15, 743–63.

Smith, B. F. and Bain, D. C. (1982) A sodium hydroxide fusion method for the determination of total phosphate in soils. *Communications in Soil Science and Plant Analysis* 13(3), 185–90.

Stoops, G. (2003) *Guidelines for the Analysis and Description of Soil and Regolith Thin Sections*. Madison, Soil Science Society of America.

Van Keuren, S. and Roos, C. I. (2013) Geoarchaeological evidence for ritual closure of a kiva at Fourmile Ruin, Arizona. *Journal of Archaeological Science* 40, 615–25.

Villagran, X. S., Schaefer, C. E. G. R. and Ligouis, B. (2013) Living in the cold: geoarchaeology of sealing sites from Byers Peninsula (Livingston Island, Antarctica). *Quaternary International* 315, 184–99.

Waddington, K. and Sharples, N. (2011) *The excavations at Whitchurch 2006–2009: An Interim Report*. Cardiff Studies in Archaeology, Specialist Report 31. Cardiff, University of Cardiff.

Chapter 3

Pollen Signatures of a Ritual Process in the Collective Burial Cave of Cova des Pas (Late Bronze Age, Minorca, Balearic Islands, Spain)

Santiago Riera Mora, Gabriel Servera-Vives, Llorenç Picornell-Gelabert, Manon Cabanis, Marzia Boi and Yannick Miras

Introduction

Pollen analyses started to be more commonly applied to the study of funerary sites from the 1960s onwards (Girard 1968; Roux 1967). The first such studies dealt mainly with collective prehistoric burials with the aim to identify pollen assemblages contained in sediments covering the corpses, with both palaeo-environmental and cultural objectives (Girard 2006). Gradually, however, more emphasis was being placed on the cultural dimension of pollen assemblages when interpreting the use of plants in funerary practices (*e.g.* Bertolani *et al.* 1983; Bui Thi Mai and Girard 2003; 2010). This interest in the cultural dimension of plants lead to the application of pollen studies to individual burials in order to better understand which plants were used as offerings and for corpse treatments. Consequently, a strategy based on higher spatial sampling resolution of sites and corpses became more frequent in such approaches (*e.g.* Clarke 1999; Tipping 1994; Whittington 1993).

During the last couple of decades, the interest in corpse treatment lead to the use of pollen analyses more commonly for non-buried corpses from closed contexts, such as sarcophagi (*e.g.* Girard and Maley 1999). Such studies were carried out on mummified and embalmed corpses (*e.g.* Ciuffarella 1998; Girad and Maley 1987), where approaches and sampling methods from forensic palynology were largely employed (*e.g.* Fornaciari *et al.* 2008; Giuffra *et al.* 2011a; 2011b; Marchesini and Marvelli 2009). The role of plants in funerary rituals has often been studied using multi-proxy approaches, including both plant macroremains (*e.g.* Corbineau 2014; Hadjouis *et al.* 2011; Vermeeren and van Haaster 2002) and microremains, like pollen, phytoliths (Cabanes and Albert 2011; Kvavadze *et al.* 2015) and more recently, non-pollen palynomorphs (NPP), mainly fungal spores. The latter although very rarely applied to the study of burials, they have furnished valuable information on the presence of clothes and vegetal materials, such as at Bronze Age graves from Scotland (Clarke 1999) and Georgia (Kvavadze *et al.* 2015).

The present study of corpses buried in Cova des Pas combines approaches and techniques from archaeopalynology, the study of pollen assemblages contained in sediments from archaeological contexts, objects and stratigraphies, and forensic palynology, which refers to the study of pollen signatures at scene crimes in legal contexts (Wiltshire 2006a). Taking into account that both subdisciplines share objectives, approaches, materials and methods (Mercuri *et al.* 2009) they can be applied to the study of funerary archaeological sites (Wiltshire 2006a; 2006b). Employing this combined approach, this study endeavours to reconstruct the ritualisation of the dead, considering that corpses constitute the main material evidence of burial practices. In addition, it aims to trace the practices involving plants in Cova des Pas in order to achieve a better understanding of the social relationships between plants and this Bronze Age community.

Geographical setting

The cave of Cova des Pas (39° 57' 50" N, 4° 00' 30" E, 80m.a.s.l.) is located at the southern part of the island of Minorca, the northernmost island of the Balearic archipelago (Plate 6a). The cavity opens at a cliff of the Trebalúger ravine (Plate 6b) (Fullola *et al.* 2007).

Minorca has Mediterranean climate with four-month-long summer droughts, mean annual temperature of 17.7°C and mean annual precipitation of 555mm at the Mercadal station (Franquesa *et al.* 2011; Pons and Gómez-Pujol 2003). *Quercus ilex* (evergreen oak) forests grow throughout the centre of the island expanding in deep ravines and valleys, while machia communities, mainly formed by *Olea europaea* var. *sylvestris* (wild olive tree), *Pistacia lentiscus* (mastic tree), *Juniperus phoenicea* (Phoenicean juniper) and *Pinus halepensis* (Aleppo pine) are dominant in coastal areas (Pons and Gómez 2003).

Southern Minorca is characterised by a set of parallel deep ravines, cutting through the southern limestone platform. Numerous burial caves and hypogea have been reported on the cliff-walls of these ravines, such as Cova des Càrritx, Cova des Mussol and Cova de S'Alblegall (Plate 6a) (Arnau *et al.* 2003; Lull *et al.* 1999), while settlements are mainly located on the upper platform. The location of cemeteries indicates that these canyons played a key role in the symbolic world of Minorcan prehistoric groups.

Archaeological and geoarchaeological contexts

Cova des Pas is a hard-to-access small cavity of about 6.5m × 4.5m, which opens on a cliff-wall, 15m above the thalweg of the ravine (Plate 6b), below the karstified plains of Son Mercer de Baix (Bergadà *et al.* 2015). The cave was used as a collective burial for 66 corpses (Plate 6c) (Fullola *et al.* 2007; Guerrero *et al.* 2006). Of these, 23 individuals were males, 24 were females and 19 remained indeterminate in terms of gender (Armentano *et al.* 2010; Armentano *et al.* 2012). Radiocarbon dates show that the main period of the burial activity was between 900 and 800 BC (van Strydonck *et al.* 2010), indicating that the cave was probably in use by a small community for a short time during the Late Bronze Age. Palaeo-demographic data and the absence of traumatic perimortem lesions suggest attritional deaths for the studied individuals (Armentano *et al.* 2012). The anatomical connection of bones indicates that corpses remained in primary position and any disturbance or removal of previous buried bodies was undertaken with the introduction of new individuals (Armentano *et al.* 2012).

One of the main characteristics of the cave's archaeological record is the good preservation of organic remains, which include human soft tissues (Prats-Muñoz *et al.* 2013), hair, animal skins, plant fibre ropes and wood (Cho and Chinchilla 2008; Guerrero *et al.* 2006; Solé *et al.* 2016). The stable dry environment inside the cave, the diagenetic sedimentary processes, the anaerobic conditions and the adipocere formation due to the superimposition of bodies have been suggested as the main causes for the preservation of these organic tissues and materials (Bergadà *et al.* 2015; Cabanes and Albert 2011; Prats-Muñoz *et al.* 2013).

The sediment infilling the cave is scarce, reaching a maximum depth of 30cm. According to the geoarchaeological study three sedimentological layers can be distinguished as follows (Bergadà *et al.* 2015): a) The upper sedimentological Level I (15–25cm thick), which is a dark brown silty and sandy layer, has clear evidence of biological activity, such as microfaunal bones and excremental phosphatic masses; b) Level II (8–15cm thick), which corresponds to the burial layer, is characterised by brownish silty clays, contains evidence of vegetal remains, such as charcoal fragments, humified plant debris, phosphatised plant residues and silica phytoliths. The formation of this layer was the result of human activities, the decomposition of corpses and the decarbonisation of the bedrock; and c) The lower Level III is intermittently present in the cave and can only be found as infilling of rock fissures and near the walls. Composed by reddish clays, it corresponds to endokarstic sediments. This sedimentary level had almost completely disappeared due to natural or anthropogenic causes by the time when the corpses were deposited inside the cavity.

Both geoarchaeoelogical and anthropological data suggest that the cave remained closed during its use as burial site (Level II) and after its abandonment until the recession of the cliff wall (Bergadà *et al.* 2015; Prats-Muñoz *et al.* 2013). Skeletons were not disturbed and no evidence of action of carnivores or rodents has been reported on human bones (Armentano *et al.* 2012). A recession of the cliff wall occurred subsequently during the formation of Level I but the preservation of the corpses and the scarcity of external inputs in sediments indicate that the environmental conditions were maintained undisturbed inside the cave. The existence originally of a smaller entrance instead of the current one, as suggested by Bergadà *et al.* (2015), may have also contributed to these conditions.

The high degree of preservation and anatomic articulation of skeletal remains in spite of the successive burials during the cave's use indicated that funerary practices of corpse preparation were carried out outside the cave, before these were introduced and finally deposited inside the cavity (Armentano *et al.* 2012). This also suggests that previous corpses were not removed when new ones were introduced and deposited. The corpses were probably prepared after the *rigor mortis* (24–36 hours after death), with their upper and lower extremities being strongly flexed to adopt a foetal position and tied up using plant fibre ropes as shown in Plate 6d (Armentano *et al.* 2012; Guerrero *et al.* 2006). At least some of the individuals were also wrapped up with an animal skin shroud and tied up with plant fibre ropes, resulting in a mortuary package (Fullola *et al.* 2007; Solé *et al.* 2016). The presence of charcoal, branches, leaves and seeds (Picornell-Gelabert 2012) indicates the use of plants in the corpse preparation and in specific ritual practices (Plate 6e). When the corpses were prepared, wooden biers and ropes were probably employed to descend these to the cave's entrance (Solé *et al.* 2016). Once inside the cave, the corpses were deposited on the bedrock and successive burials were piled up during the time span of the cave's use, creating on average three layers of individuals (Armentano *et al.* 2012). This provides further support to the hypothesis that the cave was only an inhumation space and that no voluntary manipulations of corpses were carried out inside the cavity (*ibid.*).

In this archaeological deposit the favourable taphonomic and sedimentological conditions and the exceptional preservation of organic remains, created unique conditions that rendered the site ideal for a detailed pollen study.

Objectives and approach

The overall aim of the pollen study was to determine the role of plants in funerary practices of this Minorcan Late Bronze Age community. The specific objectives were to:

a) determine which plants were used in corpse preparation and as funerary offerings;
b) establish in which parts of the corpses these plants were placed and for what purpose;
c) trace the sequence of actions of this funerary ritual;
d) determine common and different funerary patterns between individuals in relation to gender, age and social condition; and
e) discuss the selection of plants in relation to their sensorial features and cultural meanings.

It must be taken into account that the interpretation of pollen assemblages in a collective grave is a complex task due to the successive overlap of corpses and plant materials. In this sense, the accumulation of burials into the cave and the fluid mediums of corpse decomposition could cause a certain degree of post-depositional pollen mobility. However, it must be stressed that skeletons remained undisturbed and only few removed skeletal remains have been observed (Armentano *et al.* 2012).

In order to overcome these taphonomical issues we designed a research strategy, which was adapted both to the conditions of this collective grave and the specific objectives of the study. Our approach was based on the integration of principles and techniques used in archaeopalynology and forensic palynology (Mercuri *et al.* 2009; Wiltshire 2006a; 2006b). This approach is based on the following principles:

a) microsampling in precise and well-defined archaeological and anthropological microcontexts;
b) high-resolution mapping of pollen assemblages from the cave and the corpses using the forensic palynology principle of *multiple evidential samples* strategy (high variety of locations, objects and microcontexts) (Wiltshire 2006a); and
c) the study of control samples from the profile and a moss cushion located at the exterior of the cave entrance that furnishes data for the characterisation of the natural pollen background, allowing the detection of human induced pollen enrichments in burials (Wiltshire 2006a). These samples allow the detection of both the natural pollen supply and the taphonomical processes related with the decomposition of pilled corpses and post-depositional material mobility.

Considering that pollen can be an airborne particle, the interpretation of the pollen assemblage in terms of presence of plants inside the cave was also discussed using the evidence provided by the presence of plant macroremains, such as seeds and branches (Picornell-Gelabert 2012). This approach renders possible the detection of pollen enrichments or anomalies in specific microcontexts that can reveal pollen supplies indicative of deliberate actions and plant uses by humans.

In the framework of forensic palynology each pollen signature contains in itself a story of the context (Wiltshire 2006b) as well as links between spaces and human actions both in spatial and temporal dimensions (Wiltshire 2006a; 2006b). As a result, the application of this approach on Cova des Pas allowed us to track the different plant uses in a temporal succession of actions that configure the whole funerary ritual sequence.

Materials and methods

In total, 40 palynological samples, corresponding to 12 individuals, were analysed. The corpses studied had been placed in the main burial space of the cave (Plate 6c). Detailed archaeological and anthropological information for each sample and the main features of the studied individuals are outlined in Table 3.1.

Pollen samples were obtained under microscope from the sticking sediment adhered to human bones and animal skin shrouds in order to locate pollen assemblages in highly precise microcontexts. A detailed study of the different parts of the corpses was carried out for three individuals (22, 33 and 41), in order to locate with precision the placement of different plants in the corpse. Samples from both the internal and external faces of the skin shrouds from individuals 4, 33 and 37 were also analysed (following Wiltshire 2006c).

Additionally, pollen assemblages of samples at successive depths in a sedimentological profile were studied as comparative material to the pollen signatures obtained from corpse microcontexts. A profile was selected from near the southern wall, at an area of the cave with fewer human remains and less disturbance (Plate 6c). Similarly, the cushion moss at the exterior of the cave was fully analysed in order to characterise the natural pollen rain.

The moss and sediment samples were treated following standard palynological extraction procedures (Faegri and Iversen 1989; Heim 1970), which include successive digestions with KOH, HCl and HF. All samples were mounted in glycerine. The weight of the analysed samples varied between 1–3g of dry sediment. *Lycopodium clavatum*

Table 3.1 Main features of the studied individuals, including gender, age and radiocarbon dating (after Armentano et al. *2012 and van Strydonck* et al. *2010), and brief description of the microcontexts of samples.*

Sample number	*Code*	*Individual*	*Microcontext*	*Gender*	*Age (yr old)*	*C14 (yr BP)*	*C14 calibrated 2 σ (yr cal BC)*
4		1	Thoracic cavity	F	Adult (25–35)	2575±30	810–740
5		1	Between legs			2585±40	830–740
53		1	Adhered to skull			2730±30	930–810
59	3781	3	Adhered to skull			2610+30	820–770
60	B 125	3	Adhered to ulna				
64	B 13	3	Under the skull, on the bedrock	F	Adult (30–40)		
9	8972/8973 (B 69)	4	Under thorax	M?	Young (15–16)	2775±25	1000–840
43	8972	4	Rope adhered to the external face of shroud			2695±30	900–810
44	8972	4	Internal face of shroud				
46	8972	4	Between rope and the external face of shroud.				
1	A 1	9	Adhered to calcaneous	F?	Child 2 (11–13)		
2	A 599 (B 147)	9	Adhered to left tibia				
3	B 135	9	Between ribs				
61	1816 (B 148)	13	Adhered to right femur	M	Adult/Mature (43–55)		
62	5053	13	Adhered to right clavicle				
66	7947	22	Adhered to skull	M	Adult/Mature (40–50)		
67	7948	22	Thoracic cavity				
68	7744	22	Between legs				
105	B144 (7954)	22	Left side of coccyx				
106	B 75 (7983)	22	Left femur				
107	B 107	22	Thoracic cavity				
108	B 98	22	Between patellas				
8	8001	33	Between rope and external face of shroud	F	Adult (24–29)	2705±30	910–800
10	7638 (B 46)	33	Between mandible and plant branches			2675±30	895–795
55	8001	33	Between shroud and hands				
69	7946	33	Adhered to skull				
70	7945	33	Thoracic cavity				
71	7944	33	Between legs, close to coccyx				
101	B 90	33	Under ribs/under corpse				
102	B40	33	Under corpse				
14		37	External face of shroud	M	Child 1 (2)		
95		37	Under corpse				
96	B 39	37	Under corpse				
35	7037	41	Adhered to skull	F	Adult/Mature (35–45)		
38	6965	41	Adhered to vertebral column				

(Continued)

Table 3.1 Main features of the studied individuals, including gender, age and radiocarbon dating (after Armentano et al. *2012 and van Strydonck* et al. *2010), and brief description of the microcontexts of samples. (Continued)*

Sample number	*Code*	*Individual*	*Microcontext*	*Gender*	*Age (yr old)*	*C14 (yr BP)*	*C14 calibrated 2 σ (yr cal BC)*
103	B112 (6954)	41	Adhered to left femur				
104	B44	41	Under corpse				
97	B 49	42	Under corpse	F?	Child 2/ Young (11–13)		
98	6888	44	Under vertebral column/Under corpse	F?	Adult/Mature (45–60)		
99	B 151	44	Under the skull				
56	7530 (B72)	47	Adhered to left tibia	M	Adult/Mature (40–45)		
74	7938	Profile	Level I				
75	7939	Profile	Level I				
76	7949	Profile	Level II				
77	7941	Profile	Level II (contact with bedrock)				

nb dates have been rounded out to 10 years as per international convention

spore tablets were added to calculate pollen concentration per gram of dry sediment (Stockmarr 1971) and a minimum count of 350 pollen grains per sample was reached. Pollen clumps were counted as a potential indicator of the presence of flower heads and they were included in pollen sums as individual grains (Clarke 1999; Corbineau 2014).

Pollen identifications were conducted with the aid of atlases and morphological keys, such as Beug (2004), Boi and Llorens (2007), Moore *et al.* (1991) and Reille (1992–8), as well as the pollen reference collection available at SERP (University of Barcelona) and the Botany Department of the University of Balearic Islands (Spain). The Cerealia (cereal) pollen type was determined using morphometric criteria established by Andersen (1979).

Ordination by principal component analyses (PCA) was conducted using XLSTAT (Addingsoft 1995–2015) to display the relationship between pollen taxa (variables) and pollen spectra (objects) with the Euclidean method (ter Braak 1987). In addition, a similarity cluster was performed based on a Pearson correlation coefficient. The barycentres of the four groups obtained in the cluster were plotted on PCA. The variables used included a large number (43) of major pollen taxa and groups while rare pollen types (with values <1% of the total pollen sum) were excluded from the statistical analyses.

Results

Overall, the sediments of the cave were rich in pollen grains and fern spores, with an average pollen concentration of 332,000 palynomorphs/g and a maximum value of 1,124,000 palynomorphs/g recorded in sample S61. The good preservation of pollen grains was indicated by their high concentration, the low degree of damaged pollen and the high diversity of pollen types (a total of 192 pollen types were identified in all samples).

Control pollen samples: the sedimentological profile (11-L) and the moss cushion

Samples S76 and S77 from the sedimentological Level II of the profile, belonging to the burial layer (Bergadà *et al.* 2015), had the highest pollen concentration of all the profile samples with a mean value of 260,000 palynomorphs/g (Fig. 3.1). These pollen assemblages were characterised by high percentages and concentrations of the entomophilous (insect pollinated) taxon Brassicaceae (mustards), with values up to 54.5% and 119,000 pollen grains/g. In S76 anemophilous (wind-pollinated) pollen taxa, such as *Plantago* (plantain, combining *Plantago lanceolata* type and *P. major/media*) and Poaceae (grasses), also show noticeable percentages of *c.* 6% and 13% respectively, while Cerealia type frequencies are also noticeable (5.2%). *Plantago* reaches its highest percentage and concentration (20.5% and 53,060 pollen grains/g) at the bottom of level II in sample S77, which is in contact with the bedrock.

The upper level I (samples S74 and S75) is characterised by an increasing presence of anemophilous woody taxa, such as *Pistacia, Olea* and *Pinus* (up to *c.* 21%, 12% and 4.5% respectively), while the concentrations and percentages of Brassicaceae and *Plantago* decline: Brassicaceae decrease to 37% and 27,000 grains/g (S75), and *Plantago* to 3.7% and 3,740 grains/g (S74). Poaceae concentrations decrease to 96,000 grains/g while Cerealia type disappears from this level.

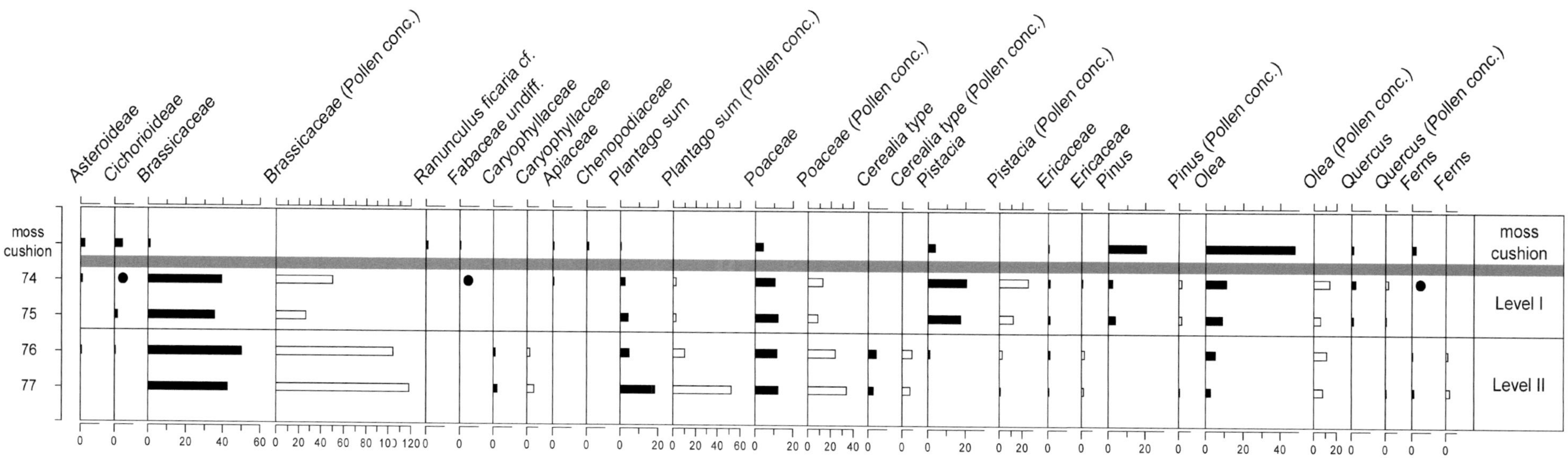

Figure 3.1 Selected pollen taxa from the sedimentological profile 11L and of the moss cushion located near the cave's entrance. Selected percentage and pollen concentration curves are shown. Black bars = percentage values; White bars = pollen concentration values (pollen grains/g).

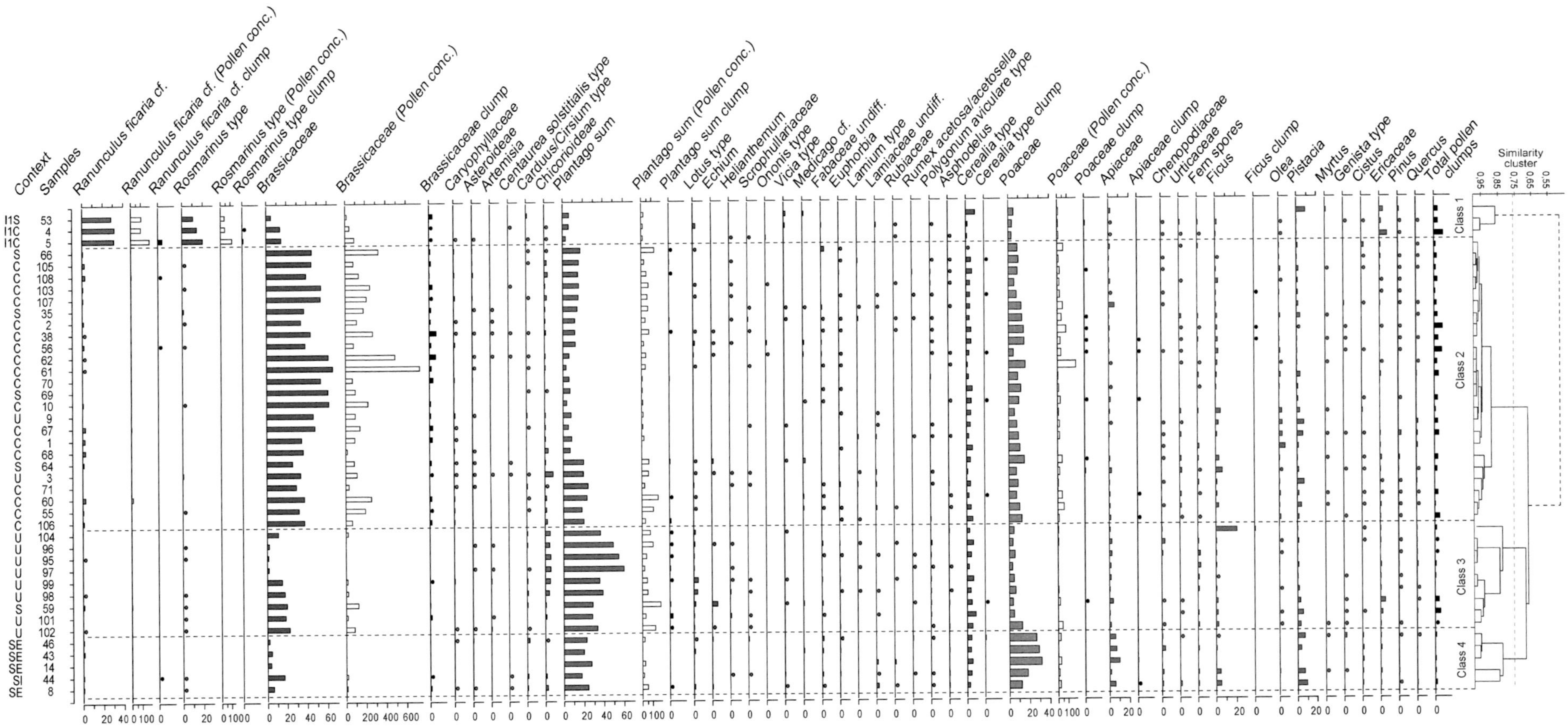

Figure 3.2 Selected pollen curves of 40 studied samples ordered following the cluster analysis. Grey bars = percentage values; white bars = pollen concentration values (pollen grains/g); black bars = percentages of pollen clumps. Samples have been organised in 4 classes following the results of the cluster analysis. Context sample codes are: S = skull; C = corpse; U = under corpses; SI = shroud inner face; SE = shroud external face; I1 = individual 1; for sample numbers see Table 3.1.

The pollen changes observed between the sedimentological levels II and I confirm the former geoarchaeological conclusions (Bergadà *et al.* 2015). In this sense, the pollen assemblage of level II was dominated by entomophilous taxa, the pollen grains of which have very restricted dispersal under natural conditions. For this reason, the pollen amounts of these plants in level II must have been the result of their intentional introduction inside the cavity as part of the burial package (Bergadà *et al.* 2015; Cabanes and Albert 2011). The higher contribution of anemophilous taxa in the upper level I could be explained by the presence of an opening at the cave entrance during the formation of this level according to sedimentological and micromorphological studies (Bergadà *et al.* 2015). Pollen grains of anemophilous plants are widely airborne dispersed and their presence inside the cave must have occurred under natural conditions, which corroborates the cave opening during the formation of level I.

The cushion moss, located near the cave's entrance on the ravine thalweg close to the wall-cliff, recorded the current pollen rain, which was characterised by high values of *Olea* (48%) and *Pinus* (21%), moderate presence of *Pistacia* (4%) and Poaceae (5%), and low amounts of Brassicaceae (1.5%) and *Plantago* (<1%) (Fig. 3.1). Therefore, the current pollen assemblage observed in the moss cushion clearly differed from that recorded on the surface sediments inside the cave (S74). Although it has been suggested that the entrance of the cave is currently larger than that in the past (Bergadà *et al.* 2015), no significant contribution of modern pollen rain has been recorded on the top of the cave's sediment.

Pollen assemblages from corpses

The results of the pollen analyses, including percentages, pollen clumps and pollen concentration of selected taxa, are shown in a diagram where samples have been ordered following the cluster analysis (Fig. 3.2). Cluster analysis resulted in four classes of pollen assemblages, which were represented by different symbols on the PCA scatter plot (Fig. 3.3). The results of this classification agree almost completely with those obtained by the PCA (Fig. 3.3). The two first axes of the PCA have eigenvalues of 5.685 (axis 1) and 5.233 (axis 2) and account for the 25.4% of the total variance (13.3% of the variation for axis 1 and 12.1%, for axis 2). The first axis separates pollen samples that have high presence of Brassicaceae, *Olea* and Caryophyllaceae (pink family) on the negative side (Class 2) from pollen samples with higher diversity of taxa, such as Poaceae (grasses), Chenopodiaceae (goosefoot family), Apiaceae (parsley family), *Asphodelus* (asphodels) *and Carduus/Cirsium* (thistles), on the positive side (Class 4). The second axis divides pollen samples dominated by *Rosmarinus* type (rosemary) and *Ranunculus ficaria* cf. (fig buttercup) on the negative side (Class 1) from pollen samples with high amounts of *Plantago,* Cichorioideae and fern spores on the positive side of the plot (Class 3).

Samples from Classes 1 and 3 are mainly distributed along axis 2. On the negative side, Class 1 includes all samples from individual 1 whereas Class 3 that clusters on the positive side includes samples located under the corpses of most individuals. Samples from Classes 2 and 4 are mainly ordinated along axis 1 (Fig. 3.3). On the negative side of axis 1 Class 2 includes all samples located in corpses. Inside this Class, two sub-Classes can be distinguished in the PCA distribution: (i) sub-Class 2a, mainly composed of samples located in different parts of the corpse (thorax, pelvic cavity and legs) that are in fact in contact because of its foetal position, and (ii) sub-Class 2b, including samples from skulls (Fig. 3.3). On the positive side of axis 2 Class 4 includes samples adhered to skin shrouds.

Accordingly, PCA analysis indicates that the location of the samples on the corpse and the particularity of individual 1 are the explanatory variables of the distribution of the pollen samples. This also explains the intermediate position on the scatter plot of some samples (*e.g.* S71 and S55) as these were placed at the interface of two different microcontexts (Fig. 3.3).

Discussion

Flower and fruit offerings

Brassicaceae is the main taxon present in 11 out of the 12 studied individuals, associated with corpses, with values reaching 64%, whatever the age or the gender of the individual and its location in the cave (Fig. 3.2). High pollen percentages, concentrations and regular occurrences of pollen clumps of Brassicaceae indicate that this taxon was the main component of the floral tributes to most individuals, the exception being individual 1 (see below). The presence of these plants in corpses is corroborated by the identification of Brassicaceae small branches through wood analysis (Picornell-Gelabert 2012). The tradition of depositing floral tributes on corpses during the Neolithic and Bronze Age and later has been demonstrated by numerous other pollen studies that revealed high concentrations of pollen and occurrence of pollen clumps, indicating the offerings of eye-catching flowers on corpses and heads (*e.g.* Clarke 1999; Dickson 1978; Tipping 1994; Whitttington 1993).

High-resolution analyses of individuals 33 (Fig. 3.4), 22 (Fig. 3.5) and 41 show that the maximum percentages of pollen grains and clumps of Brassicaceae are located at the thorax area. In this part of the corpse, pollen reaches 44% in individual 41 (S38) (Fig. 3.2) and around 54% in individuals 33 (S70) (Fig. 3.4) and 22 (S107) (Fig. 3.5). Samples from the lower limbs also reveal high Brassicaceae percentages: between 37% and 45% in individual 22 (S105 and S106) (Fig. 3.5) and 54% in individual 41 (S103) (Fig. 3.2). Due to the forced foetal position of the corpse, the flexed limbs remained in contact with the thorax and this position explains the high percentages of Brassicaceae recorded also in samples from the legs. These data suggest that during the preparation of the corpse, Brassicaceae flower offerings were probably deposited between the thorax and the flexed legs.

The percentages of Brassicaceae pollen are lower in the skulls. Values of 45% (S66) and 38% (S35) were found, for example, at the heads of individuals 22 and 41 respectively (Figs. 3.2 and 3.5), suggesting that this plant offering was mainly concentrated at the thorax. These distinctive pollen signatures between the skull and the thorax are highlighted by the discrimination of two subclasses within Class 2 of the PCA (Fig. 3.3). Nevertheless, slight differences exist in the case of the thorax and head of individual 33, where high percentages of Brassicaceae were also found on the skull with values of *c.* 61% in samples S10 and S69 (Fig. 3.4). In this individual Brassicaceae plants were deposited between the mandible and the hands as indicated by the presence of branches of this taxon in this part of the corpse (Picornell-Gelabert 2012) (Plate 6e).

Taking into account that corpses were wrapped with animal skins, microcontextual samples were also obtained to identify whether Brassicaceae plants were placed inside the shroud. With this aim, three shroud fragments were studied in detail, analysing samples from the internal and

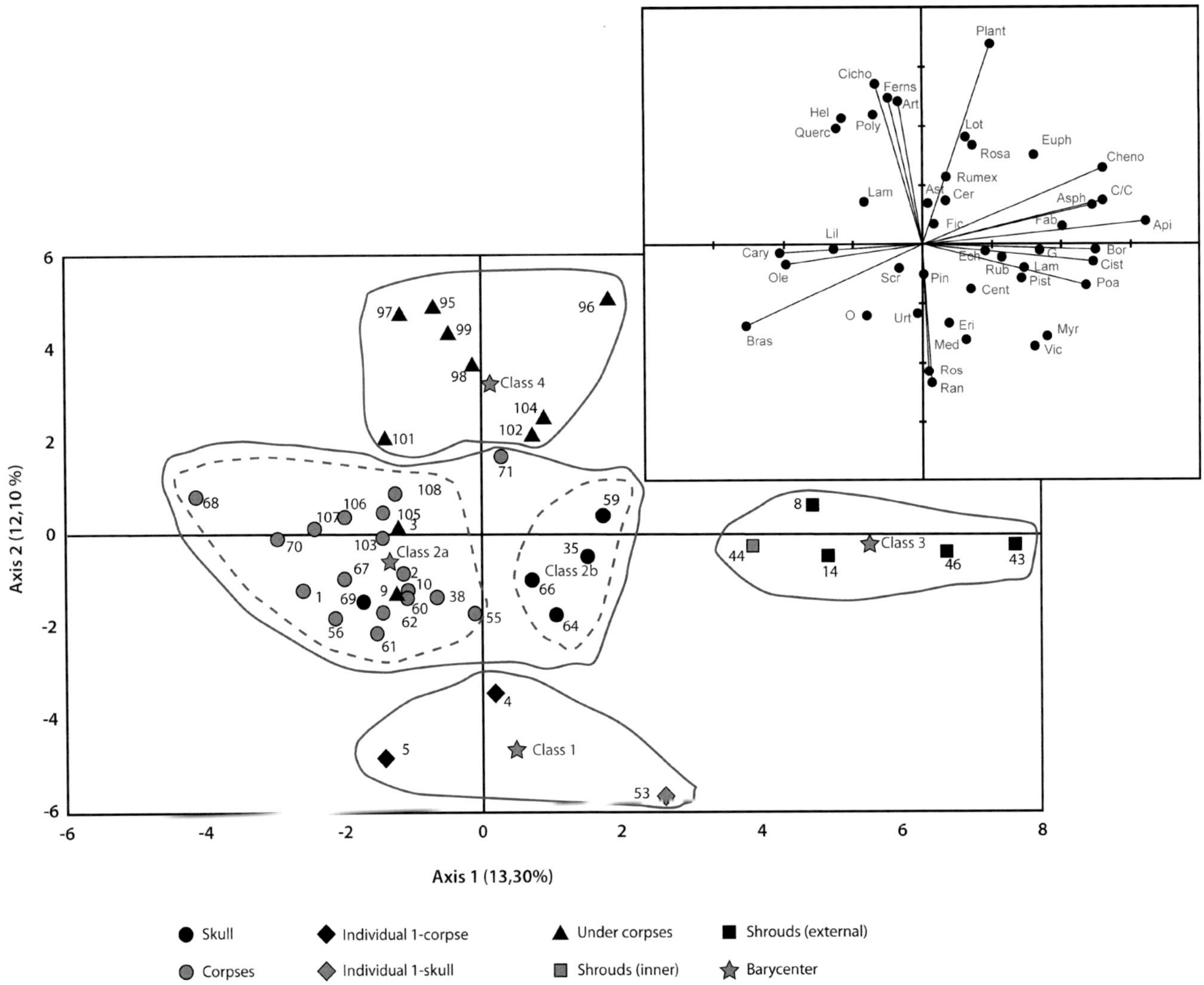

Figure 3.3 Principal Component Analysis (PCA) of pollen assemblages. Upper chart variables: Cicho = Cichorioideae; Art = Artemisia*; Poly =* Polygonum aviculare *type; Hel =* Helianthemum*; Querc =* Quercus*; Lam =* Lamium *type; Lil = Liliaceae; Cary = Caryophyllaceae; Ole =* Olea*; Bras = Brassicaceae; O =* Ononis *type; Urt = Urticaceae; Plant =* Plantago *sum; Lot =* Lotus *type; Ros = Rosaceae; Euph =* Euphorbia*; Cheno = Chenopodiaceae; Ast = Asteroideae; Cer = Cerealia type; Asph =* Asphodelus*; C/C =* Cardus/Cirsium*, Fic =* Ficus*; Fab = Fabaceae undiff.; Api = Apiaceae; Ech =* Echium*; Rub = Rubiaceae; Lam = Lamiaceae undiff.; Bor =* Borago officinalis *type; Pin =* Pinus*; Pist =* Pistacia*; Poa = Poaceae; Cist =* Cistus*; Cent =* Centaurea sosltitialis *type; Eri = Ericaceae; Med =* Medicago cf.*; Vic =* Vicia *type; Myr =* Myrtus*; Ros =* Rosmarinus *type; Ran =* Ranunculus ficaria cf.*; Scr = Scrophullariaceae; Rumex =* Rumex acetosa/acetosella*; Ferns = Fern spores; G =* Genista *type. The lower chart plots samples and the four classes resulted from the cluster analysis (see Figure 3.1).*

external faces. The results suggested that clear differences exist between pollen assemblages from the two sides of the skin shroud that wrapped individuals 33, 37 and 4. High percentages of Brassicaceae were found inside the shroud while its values sharply decreased at the external face of skins, where percentages of 7% and 4% were observed for individuals 37 and 33 (S14 and S8) respectively (Figs. 3.2 and 3.4). In addition, a microtransect was taken on a piece of shroud of individual 4, from the internal face of the shroud to the external rope, including samples S43, S44 and S46 (Figs. 3.2 and 3.3), with the objective to specify the location of Brassicaceae in corpses. The percentages of Brassicaceae decreased from 47% in the corpse of this individual (S9), to 17% in the internal part of the shroud (S44) and to 5% in its external part (S43). A very low value of 2.8% of Brassicaceae was recorded in sample S46, which was located between the external face of the shroud and the rope, a microcontext that remained particularly protected from subsequent pollen inputs. All these data, together with the presence of Brassicaceae pollen clumps in corpses (Fig. 3.2), indicate that these flower offerings were placed in direct contact with the corpse, which was later covered by the shroud. In consequence, in the case of individuals wrapped with shrouds these plants were not visible when the corpse was transported into the cave.

Brassicaceae is a large plant family, widely present in a number of habitats of the island of Minorca. A common feature of most of its species is their yellow or white eye-catching flowers and for this reason, they could have been used as a floral tribute. Offerings of Brassicaceae flowers in Bronze Age tombs were also observed by other pollen studies, such as at Scottish cysts (Clarke 1999; Whittington 1997) and a necropolis of the Paris basin (Girard 1988). In the case of the cysts of Whitsome in Scotland for instance, high percentages (82.5%) and pollen clumps of Brassicaceae were recorded and interpreted as a probable floral tribute (Clarke 1999).

The PCA analyses also underline the significance of Caryophyllaceae and *Olea*, particularly for samples coming from the corpses (Class 2) (Fig. 3.2), suggesting the possible inclusion of other plants as offerings for the thorax part.

Pollen assemblages from the external face of the shrouds were clearly discriminated by PCA analyses in Class 4 (Fig. 3.3), and their distribution can be mainly explained by the presence of Apiaceae, *Carduus/Cirsium*, *Cistus*, Fabaceae (pea family) and *Asphodelus*. These entomophilous taxa reach higher percentages in the external face of the shroud of individuals 4 (sample S46: *Cistus* 1.2%, Apiaceaea, 6.2%, *Carduus/Cirsium*, 3%, *Genista* type 2.8% and occurrence of *Asphodelus* and sample S43: *Cistus* 1.8%, Apiaceae 7.4% and *Asphodelus* 1.2%) and 37 (sample S14: Apiaceae 10% and *Cistus* 1.5%). These are tall herbs, characterised by eye-catching flowers composing a possible 'out-of-shroud' flower tribute that was different to the tribute of Brassicaceae placed inside the mortuary package.

Other taxa such as Cerealia type and *Ficus* (fig) were present in all contexts and individuals but these pollen taxa have not been defined as discriminative by the PCA analyses. Cerealia type is present in most samples with mean percentages of 4.5%, reaching maximum values of 9% in S101 of individual 33 (Fig. 3.4). Pollen clumps of this taxon also occurred regularly in most contexts. Cereals could have thus played a role in mortuary rituality in Cova des Pas but their unclear attribution to specific microcontexts does not allow further insights into their specific use in the preparation of the corpses or as tributes. The presence of cereal pollen in graves has been interpreted as alimentary offerings, as for instance in a Neolithic cyst in Sweden where cereal pollen values up to 26% were recorded from a pot (Lageras 2000). In the study of a Bronze Age Scottish individual cyst, Clarke (1999) also suggested the possible presence of cereal tributes with low values of cereal pollen (4.5%) but this interpretation is precarious. Nevertheless, cereal grains were absent at Cova des Pas. Other studies have suggested that the presence of cereal pollen can result from the use of straw in corpses, as for instance, in the case of certain Roman and Late Roman sarcophagi where high pollen percentages with values ranging from 25% to 57% were recorded (Corbineau 2014). The reduced values of cereal pollen in Cova des Pas (with the highest percentages ranging between 4.5 and 9.7%), however, cannot be interpreted as clear evidence for the presence of cereal straw as a bed or lining material into the cavity.

Pollen and seed evidence also suggests that fruits were offered in the burials of Cova des Pas. *Ficus* pollen grains and clumps were recorded, reaching the highest pollen percentage (21%) in S104 (individual 41). Due to the specific pollination biology of *Ficus* (Valdeyron 1984), the presence of fig tree pollen and clumps must indicate the presence of fig fruits inside the cave. The offering of fig fruits is also supported by the high amounts of fig seeds found in the cave's sediments. Fruit offerings in the cave could have also included other plants, as indicated by the presence of seeds of *Rubus* sp. (bramble). Similarly, large quantities of seeds of these fruits were recovered from the nearby Cova des Càrritx funerary cave (Lull *et al.* 1999).

Vegetal beds and strewn plants

The set of pollen assemblages from the samples located under the corpses (Class 3) are clearly discriminated by the PCA analyses from samples located in other contexts (Fig. 3.3). These samples include mainly high percentages of *Plantago,* reaching maximum pollen values of more than 50% in samples S97 (individual 42), S95 and S96 (individual 37) (Fig. 3.2), and 34% under the corpse of

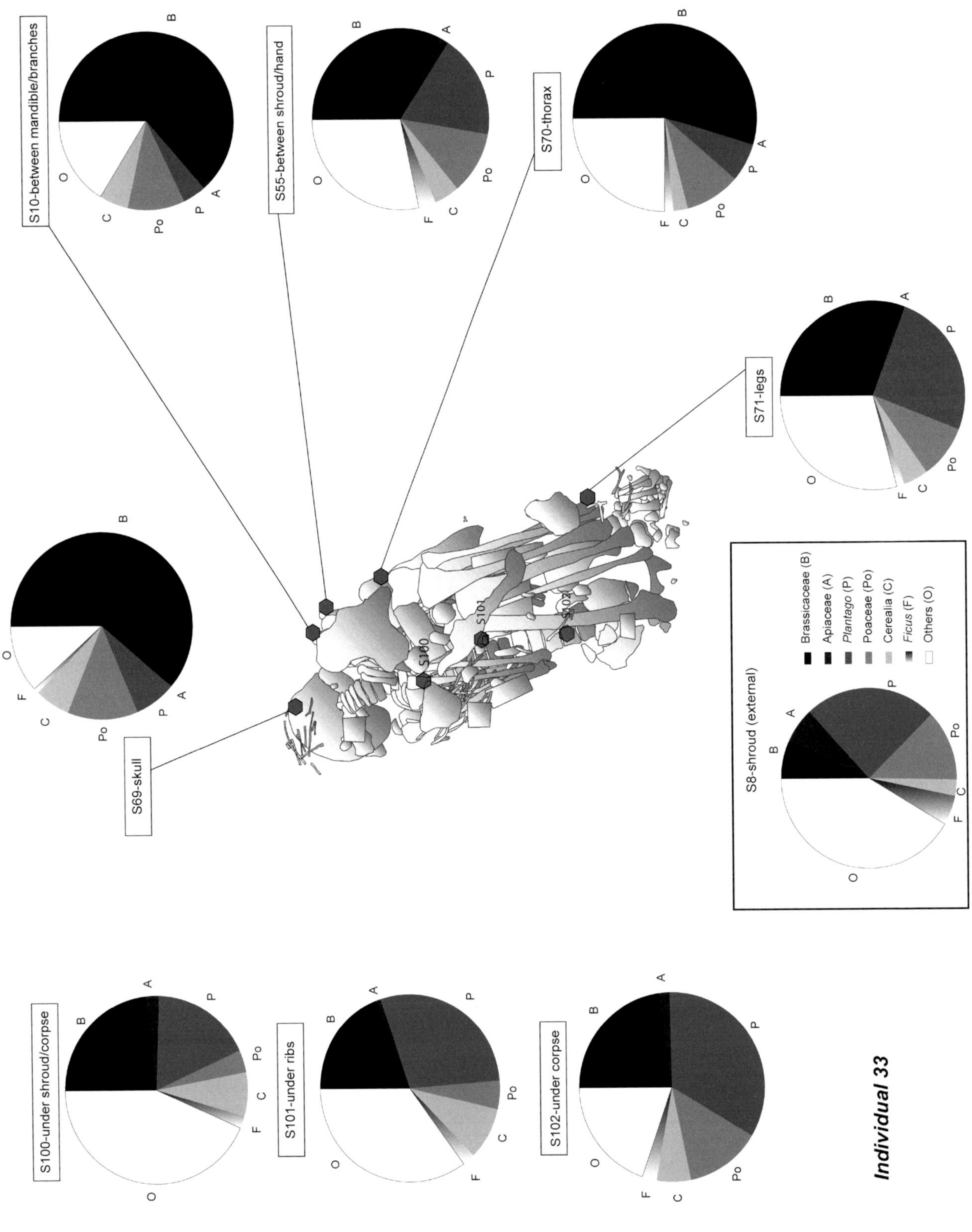

Figure 3.4 Selected pollen taxa from samples located in different parts of the corpse of individual 33. Values in percentages. Drawing of the skeleton by Xavier Esteve.

individual 33 (S102) (Fig. 3.4). The regular occurrence of pollen clumps in these samples probably indicates the presence of plantain plants inside the cave. Class 3 samples are also explained by the presence of Cichorioideae and fern spores (including Trilete and Monolete types). Cichorioideae reached values of *c*. 6% and ferns between 1.5 and 3.3% in S95, S96 and S97 (Fig. 3.2). Furthermore, Apophyte pollen types were mainly present in this context, including *Artemisia*, *Polygonum aviculare* type (knotweeds) and *Lotus* type. The presence of these taxa is corroborated by the seed assemblage, which shows the regular occurrence of wild plants, such as *Polygonum aviculare*, *Lithospermum arvensis* (corn gromwell), *Euphorbia helioscopia* (sun spurge) and *Calendula arvensis* (field marigold). Otherwise, all sampled sediments located under the corpses have the lowest presence of Brassicaceae with minimum values of *c*. 2% in samples S95, S96 and S97 (Fig. 3.2).

The pollen composition of Class 3 is similar to that recorded in the lower sample of the sedimentological profile (Fig. 3.1), suggesting that these taxa were mainly present on the ground of the cave. Due to the anemophilous pollination and high pollen production of plantains, their deliberate enrichment in samples taken from underneath the corpses is unclear. Nevertheless, some pollen studies have suggested that high frequencies of this taxon could indicate its use as lining material in burials (Bunting *et al.* 2001). In Cova des Pas, the presence of pollen clumps and the high percentages of plantains (50%) in comparison with the 20% recorded at the bottom of the pollen profile, reinforce the interpretation that plantain corresponds to pollen enrichments under the corpses due to its use as a component of a plant bed. Deliberate use of plantain has been suggested also for some Bronze Age cysts' floors in Scotland according to the presence of this taxon's pollen and plant fibres (Bunting *et al.* 2001). As plantains do not have particularly distinctive flowers, their presence has been interpreted as part of other aspects of the ritual rather than of decorative significance (Bunting *et al.* 2001). The plant beds prepared at Cova des Pas may have also contained ferns, a plant that has been interpreted as component of strews, pillows and lining in Neolithic (Coûteaux 1962) and Bronze Age graves (Bunting *et al.* 2001; Whittington 1997).

These data could suggest that once the corpses were wrapped with the shroud and introduced into the cave, they were deposited on a vegetal bed. The possibility that the corpse was never laid in direct contact with the floor of the chamber was also suggested by the anthropological study in order to explain the good preservation of biological non-skeletal tissues (Armentano *et al.* 2012).

The arrangement of heads

The detailed study of individuals 22 (Fig. 3.5) and 41 revealed that samples from their head contexts were characterised by pollen assemblages that were also influenced by pollen signatures of the out-of-the-shroud contexts of Class 4 (Fig. 3.3). The specific location of skull samples in the PCA, forming sub-Class 2b that is slightly different from the corpse samples, highlights this particularity of heads with pollen assemblages that share taxa from both the corpses and the external faces of the shrouds. These pollen assemblages from the skulls, therefore, contain taxa that were interpreted as flower offerings deposited in the external face of the shrouds, such as *Cistus* (2.7%), Apiaceae (2.3%) and Fabaceae undiff. (4%) in individual 22 (S66) and Apiaceae (6.5%) and Fabaceae undiff. (2%) in individual 41 (S35) (Fig. 3.2). Furthermore, *Plantago*, a plant interpreted as the main component of the vegetal beds under the corpses, reached higher values near the skulls compared to those from the corpses. This was observed for the heads of individual 22, where *Plantago* percentages reached 18% (S66) and pollen clumps were present (Fig. 3.5), and individual 41, where its values reached 15% in S35. All these suggest that the heads were not covered by the skin shrouds when the corpses were packaged and tied up, and they probably remained uncovered once the corpses were deposited on the vegetal bed inside the cave.

A case of distinctive plant offerings

Although a general pattern of plant uses can be observed for most of the studied individuals, the PCA analyses isolated a group of samples (Class 1) that comes from individual 1 (Fig. 3.3). This implies that different plants were employed for this individual, an adult woman of 25–35 years old (Armentano *et al.* 2012).

In this individual, instead of Brassicaceae other entomophilous taxa, such as *Rosmarinus* type and *Ranunculus ficaria* cf., with values of 14.5% and 31.5% in S5 and 20% and 32% in S4 respectively (Fig. 3.2) were the main tribute at the thorax. The presence of flowers of both taxa is also attested by the occurrence of pollen clumps that suggest plant offerings. The noticeable values of *Vicia* type (vetches) (1.8% in S4) and Ericaceae (heather family) (8.5% in S5) could indicate that these were also part of these floral tributes.

The fact that other adult women (*e.g.* individuals 3, 33 and 41), and men (*e.g.* individuals 13, 22 and 47) of different ages (Table 3.1) received the most common plant offering of Brassicaceae suggests that the distinctive floral tributes of individual 1 were not associated with its age or gender. The different floral treatment of this woman could thus be related to her particular role or social significance within her community. The archaeological record of the cave reinforces this assumption, as this individual was the object of specific material treatment. She had a long plait with a wooden hairpin and a hair slide in pure tin, an exotic

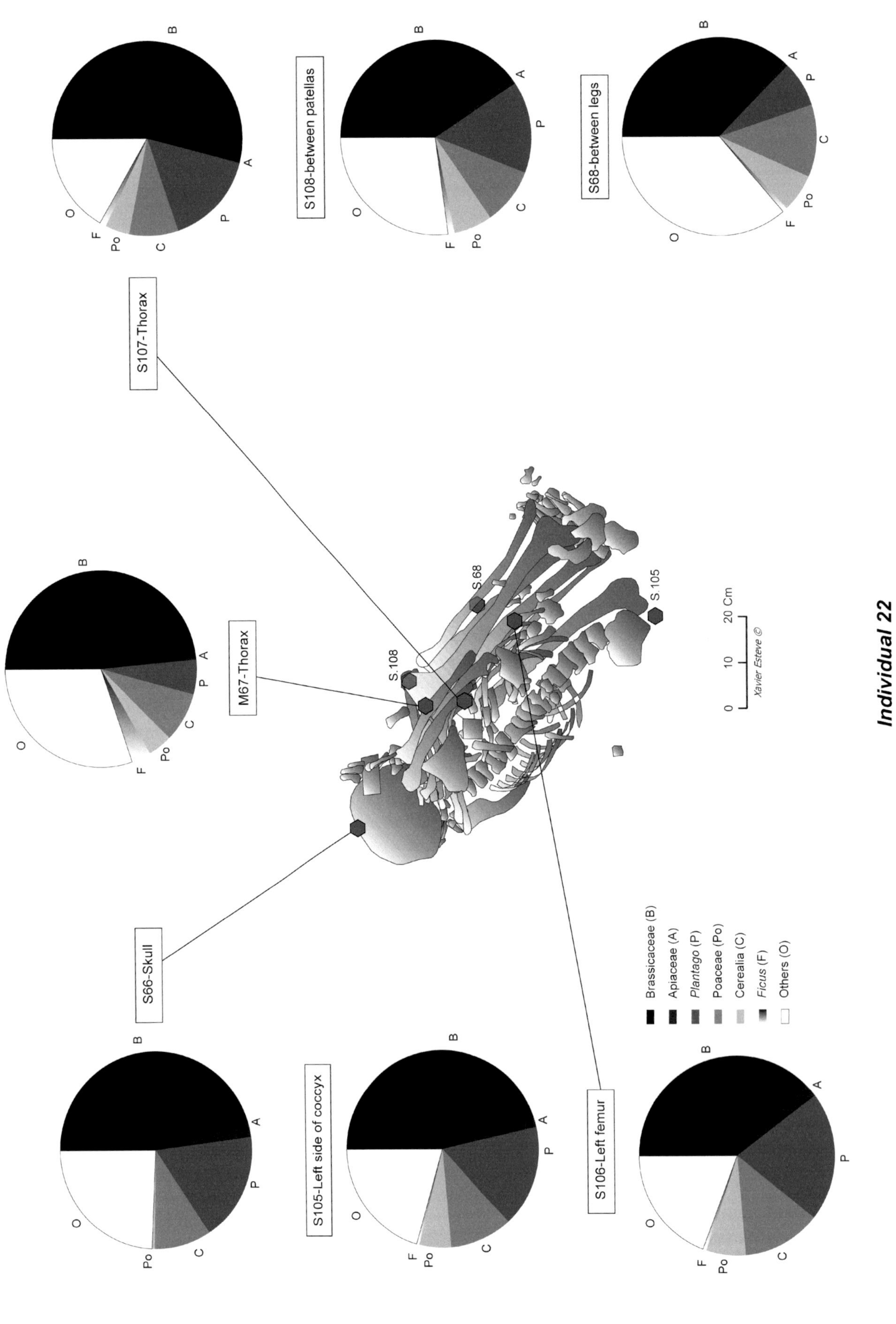

Figure 3.5 Selected pollen taxa from samples located in different parts of the corpse of individual 22. Values in percentages. Drawing of the skeleton by Xavier Esteve.

material in Minorca that was necessarily imported (Guerrero *et al.* 2006). The presence of these artefacts is particularly significant, considering that other material, including ceramics, were very scarce in Cova des Pas.

Conclusions

Pollen analyses carried out at Cova des Pas demonstrate the key role played by plants in a complex funerary ritual, which was only partially visible in the archaeological and anthropological record. Spatial microcontextualisation of pollen assemblages, revealing both specific pollen enrichments and pollen signatures, allowed us to suggest:

(a) the successive steps of the funerary sequence;
(b) the plants used in each step of the corpse preparation; and
(c) the diversity of plants used for sensory purposes, as tributes, and for the preparation of vegetal beds.

The sequence of the ritual actions and the plants involved in each step can be thus described as follows (Plate 7): during the first steps of the ritual, Brassicaceae was the main taxon deposited on the thorax of most individuals, without any distinction of gender or age. After these flowers were offered, limbs were flexed against the trunk and the corpse was wrapped with an animal skin shroud and tied up with plant fibre ropes. As Brassicaceae plants were covered by the shroud, the flower bouquets were not visible once the pack containing the corpse was ready. Once the corpse was covered, it was transported into the cave and deposited on a vegetal bed prepared on the cave's ground. In all likelihood, this bed was mainly made by plantains and ferns. Flower offerings were deposited also on the packaged corpse but it remains unclear in which step of the funerary ceremony these tributes were conducted: before the transport to the cave or when the corpse was already deposited. These flower tributes were composed by tall plants with eye-catching flowers, such as Apiaceae, *Asphodelus* and *Cistus*. In addition to flowers, some fruits were selected for offerings, including figs and berries.

In the reconstruction of the packaged corpse recently proposed by archaeologists, the head was wrapped by the shroud (Solé *et al.* 2016). However, the pollen data suggested that the head of most individuals probably lied in contact with the vegetal bed and with out-of-shroud flower tributes. This could imply that heads were uncovered at least when the corpses were lying on the vegetal bed inside the cave.

Only one adult woman received distinct plant tributes that included aromatic and eye-catching plants and flowers, such as *Rosmarinus* type and *Ranunculus ficaria* cf. (Plate 7). This specific floral treatment, also associated to outstanding objects, suggests that this woman could have had a specific social role in this community.

This study showed that this Late Bronze Age community selected specific plants for use in funerary rituals. These plants could be chosen for their aromatic properties and aesthetic reasons (*e.g.* rosemary and tall plants with eye-catching flowers) while others without any such attribute, like plantains, could have been selected for other symbolic or functional reasons. The fact that specific plants were used in certain steps of the ritual process and were assigned to special individuals allows us to affirm that the use of plants was highly codified by this community. In conclusion, this Minorcan Late Bronze Age society had strong ties with the vegetal world, which were translated into cultural patterns and plant conceptualisations. General patterns were applied to most individuals of the community, while differentiated flower grave-goods were used for those with a key social role.

Acknowledgements

The Cova des Pas archaeological excavations were directed by Dr V. M. Guerrero, Dr M. Calvo, Dr J. M. Fullola and Dr M. A. Petit. We particularly thank Xavier Esteve for the design of the plans and M. Girard for his help with pollen identifications. Also, Nuria Armentano for her help in sampling and for her useful information on taphonomy and the corpses. The Ministry of Science and Innovation of Spanish Government (projects HAR2012-32602 and HAR2015 67211-P), the Minorca Council, Caixa Catalunya and Catalan Government (EXCAVA) provided financial support. Llorenç Picornell-Gelabert was supported by the Balearic Conselleria d'Educació and the European Social Fund and Gabriel Servera-Vives by the Ministry of Economy and Competitiveness of Spanish Government (Juan de La Cierva Programme).

References

Addingsoft (1995–2011) *Data Analysis and Statistical Solutions for MS Excel*. New York, Addingsoft.

Andersen, S. Th. (1979) Identification of wild grass and cereal pollen. *Danmarks Geologiske Undersoegelse, Arbog* 1978, 69–92.

Arnau, P., Gornés, S. and Stika, H. P. (2003) Los hipogeos de s'Alblegall (Ferreries) y la agricultura cerealística a mediados del segundo milenio cal ANE en Menorca. *Trabajos de Prehistoria* 60, 117–30.

Armentano, N., Jordana, X. and Malgosa, A. (2010) Aproximación paleodemográfica a una población protohistórica de las Baleares. El yacimiento de la Cova des Pas. *Revista de Demografía Histórica* XXVIII(1), 91–108.

Armentano, N., Esteve, X., Nociarova, D. and Malgosa, A. (2012) Taphonomical study of the anthropological remains from Cova des Pas (Minorca). *Quaternary International* 275, 112–9.

Bergadà, M. M., Poch, R. M. and Cervelló, J. M. (2015) On the presence of gypsum in the archaeological burial site of Cova des Pas (Menorca, western Mediterranean). *Journal of Archaeological Science* 53, 472–81.

Bertolani Marchetti, D., Forliani, L. and Trevisan Grandi, G. (1983) Analisi polliniche in reperti delle grotte del Lago Copaide (Beozia-Grecia). *Le Grotte d'Italia* 11, 173–82.

Beug, H.-J. (2004) Leitfaden der Pollenbestimmung für Mitteleuropa und angrenzende Gebiete. Stuttgart, Fisher.

Boi, M. and Llorens, L. (2007) *Atlas polínico de las Baleares. Flora endémica*. Palma de Mallorca, Govern de les Illes Balears, Conselleria de Medi Ambient.

Bui Thi Mai and Girard, M. (2003) Pollens, ultimes indices de pratiques funéraires évanouies. In Sens Dessus Dessous. La recherche du sens en Préhistoire, recueil d'études offert à Jean Leclerc et Claude Masset. *Revue Archéologique de Picardie* 21, 127–37.

Bui Thi Mai and Girard, M. (2010) Pollen et archéologie: une alliance passionnante. In C. Delhon, I. Théry-Parisot and S. Thiébault (eds.) *Des hommes et des plantes. Exploitation du milieu et gestion des ressources végétales de la Préhistoire à nos jours. XXX Rencontres internationales d'archéologie et d'histoire d'Antibes. Anthropobotanica* 1(8), 3–23.

Bunting, M. J., Tipping, R. and Downes, J. (2001) 'Anthropogenic' Pollen Assemblages from a Bronze Age Cemetery at Linga Fiold, West Mainland, Orkney. *Journal of Archaeological Science* 28, 487–500.

Cabanes, D. and Albert, R. M. (2011) Microarchaeology of a collective burial: Cova des Pas (Minorca). *Journal of Archaeological Science* 38, 1119–26.

Ciuffarella, L. (1998) Palynological analyses of resinous materials from the roman mummy of Grottarossa, second century A.D.: a new hypothesis about the site of mummification. *Review of Palaeobotany and Palynology* 103, 201–8.

Cho, S. and Chinchilla, J. (2008) La conservació i restauració dels materials arqueològics dela Cova des Pas. *Unicum Revista de l'Escola Superior de Conservació i Restauració de Béns Culturals de Catalunya* 7, 30–54.

Clarke, C. (1999) Palynological Investigations of a Bronze Age Cist Burial from Whitsome, Scottish Borders, Scotland. *Journal of Archaeological Science* 26, 553–60.

Corbineau, R. (2014) Pour une archéobotanique funéraire. Enquêtes interdisciplinaires et analyses polliniques autour de la tombe et du corps mort (ère Chrétienne, France-Italie). Unpublished thesis, Université du Maine.

Coûteaux M. (1962) Analyse pollinique d'une tombelle de la Tène I à Namoussart (commune d'Hamipré). *Ardenne et Famenne* 3, 115–24.

Dickson, J. H. (1978) Bronze Age Mead. *Antiquity* 52, 108–13.

Faegri, K. and Iversen, J. (1989). *Textbook of Pollen Analysis*. 4th edn. Chichester, Wiley.

Franquesa, A., Estradé, S. and Gallofré, A. (2011) *Cartografia Climàtica de Menorca (1950–2008). Aproximació mitjançant eines SIG. Maó*, Menorca. Institut Menorquí d'Estudis (IME), Observatori Socioambiental de Menorca (OBSAM). Available from: http://www.obsam.cat/documents/informes/Cartografia-climatica-Menorca-1950-2008.pdf (accessed 30/9/2017).

Fornaciari, G., Bortolotti, F., Gortenuti, G., Guido, G. C., Marchesini, M., Martinucci, S. and Tagliaro, F. (2008) The mummy of Cangrande della Scala Lord of Verona (1291–329): A case of Medieval acute Digitalis intoxication. In P. A. Peña, C. R. Martin, A. R. Rodriguez (eds.) *Mummies and Science. World Mummies Research. Proceedings of the VI World Congress on Mummy Studies*, 371–7. Lanzarote, Academia Canaria de la Historia.

Fullola, J. M., Guerrero, V. M., Petit, M. A., Calvo, M., Malgosa, A., Armentano, N., Arnau, P., Cho, S., Esteve, X., Fadrique, T., Galtés, I., Garcia, E., Fornés, J., Jordana, X., Pedro, M., Riera, J., Sintes, E. and Zubillaga, M. (2007) La Cova del Pas (Ferreries, Menorca): un avanç. In *L'arqueologia a Menorca: eina per al coneixement del passat. Llibres del Patrimoni Històric i Cultural* 3, 95–110.

Girard, M. (1968) Annexe III: Analyse pollinique. In G. Bailloud and M. Brézillon (eds.) *L'hypogée de l'Homme-Mort à Tinqueux (Marne). Bulletin de la Société Française de Préhistoire* 65, 498–503.

Girard, M. (1988) Analyse pollinique de la nécropole de Buno-Bonnevaux (Essonne). In P. Brun and C. Mordant (eds.) *Le Groupe Rhin–Suisse–France orientale et la notion de civilisation des Champs d'Urnes*, 291–3. Actes du colloque international de Nemours. Nemours, A.P.R.A.I.F.

Girard, M. (2006) La sépulture collective néolithique de la Chaussée-Tirancourt (Somme). Analyse pollinique. *Bulletin de la Société Préhistorique Française* 103, 133–42.

Girard, M. and Maley, J. (1987) Étude palynologique. In L. David and R. Mourer (eds.) *Autopsie d'une momie égyptienne du Muséum de Lyon. Nouvelles archives du Muséum d'histoire naturelle de Lyon* 25, 103–10.

Girard, M. and Maley, J. (1999) La sépulture féminine du cercueil en plomb du quartier Trion-Gerlier de Lyon (IV e siècle ap. J.-C.). Analyse pollinique. *Revue archéologique de l'Est et du Centre* 50, 397–410.

Giuffra, V., Fornaciari, A., Marvelli, S., Marchesini, M., Caramella, D. and Fornaciari, G. (2011a) Embalming methods and plants in Renaissance Italy: two artificial mummies from Siena (central Italy). *Journal of Archaeological Science* 38, 1949–56.

Giuffra, V., Fornaciari, A., Marvelli, S., Marchesini, M., Fornaciari, G., and Vitiello, A. (2011b) The children of the Medici, grand Dukes of Florence: embalming in Renaissance Italy (XVI–XVII century). *Atti della Società Toscana di Scienze Naturali* serie B 118, 81–8.

Guerrero, V. M., Calvo, M.; Gornés, S. (2006) *El poblamiento prehistórico de las islas Baleares. Desde los inicios al fin de la Edad del Bronce*. Historia de las Baleares vol. 1. Palma de Mallorca, Rey Sol S.A.

Hadjouis, D., Corbineau, R., Ruas, M. P. and Verdin, P. (2011) Techniques d'embaumement sur le corps d'un noble anglais mort de la peste (Saint-Maurice, Val-de-Marne). Étude anthropologique et archéobotanique. In P. Charlier (ed.) *Actes du 3 ème Colloque International de Pathographie*, 31–62. Paris, De Boccard.

Heim, J. (1970) Les relations entre les spectres polliniques récents et la végétation actuelle en Europe occidentale. Unpublished thesis, Université de Louvain.

Kvavadze, E., Sagona, A., Martkoplishvili, I., Chichinadze, M., Jalabadze, M. and Koridze, I. (2015) The hidden side of ritual: New palynological data from Early Bronze Age Georgia, the Southern Caucasus. *Journal of Archaeological Science reports* 2, 235–45.

Lagerås, P. (2000) Burial rituals inferred from palynological evidence: results from a late Neolithic stone cist in southern Sweden. *Vegetation History and Archaeobotany* 9, 169–73.

Lull, V., Micó, R., Rihuete, C., Risch, R. (1999) *La Cova des Càrritx y la Cova des Mussol. Ideología y sociedad en la prehistoria de Menorca*. Barcelona, Consell Insular de Menorca.

Marchesini, M. and Marvelli, S. (2009) Il contributo delle indagini palinologiche alla scoperta delle cause di morte di Cangrande della Scala, Signore di Verona (1291–329 d.C.) In A. M. Mercuri and A. Travaglini (eds.) *La palinologia Forense: le metodologie e il futuro. European Journal of Aerobiology and Environmental medicine* V(2), 49–53.

Mercuri, A. M., Massamba N'Siala, I. and Olmi, L. (2009) Polline giallo: la palinologia applicata alle scienze forensi. In A. M. Mercuri, A. Travaglini (eds.) *La palinologia Forense: le metodologie e il futuro. European Journal of Aerobiology and Environmental Medicine* V(2), 11–9.

Moore, P. D., Webb, J. A. and Collinson, M. E. (1991) *Pollen Analysis*, 2nd edn. Oxford, Blackwell.

Picornell-Gelabert, Ll. (2012) Paisaje vegetal y comunidades prehistóricas y protohistóricas en Mallorca y Menorca (Illes Balears): una aproximación desde la antracología. Unpublished thesis, Universitat de Barcelona.

Pons, G. X., Gómez-Pujol, L. (2003) Introducció al medi físic de Menorca (Illes Balears). In V. M., Rosselló, J. J. Fornós and L. Góméz-Pujol (eds.) *Introducción a la Geografía Física de Menorca, 1–30. Monografies de la Societat d'Història Natural de les Balears*, 10. Ciutat de Mallorca, Societat d'Història Natural de les Balears, Universitat de València-Universitat de les Illes Balears.

Prats-Muñoz, G., Galtés, I., Armentano, N., Cases, S., Fernández, P. L. and Malgosa, A. (2013) Human softtissue preservation in the Cova des Pas site (Minorca Bronze Age). *Journal of Archaeological Science* 40, 4701–10.

Reille, M. (1992–1998) *Pollen et spores d'Europe et d'Afrique du nord*. Marseille, Laboratoire de Botanique historique et Palynologie, Université d'Aix-Marseille III.

Roux, I. (1967) Analyses polliniques. In P. Horemans, F. Poplin, Poulain-Josien, T., and I. Roux (eds.) *Les sépultures collectives de Marolles-sur-Seine (Seine-et-Marne) IV, sépulture II: le milieu. Gallia Préhistoire* 10(1), 156–67.

Solé, A., Picornell-Gelabert, Ll., Allué, E. and Fullola, J. M. (2016) How did the dead turn up to the burial? A technological and experimental approach to the late Bronze Age wooden biers from Cova des Pas (Minorca, Balearic Islands). *Archaeological and Anthropological Sciences* 8(4), 849–65.

Stockmarr, J. (1971) Tablets with Spores used in Absolute Pollen Analysis. *Pollen and Spores* 13, 615–21.

Ter Braak, C. J. F. (1987) Ordination. In: C. J. F. ter Braak and O. F. R. van Torgeren (eds.) *Data Analysis in Community Ecology*, 91–173. Wageningen, Pudoc.

Tipping, R. (1994) Ritual Floral Tribute in the Scottish Bronze Age. Palynological evidence. *Journal of Archaeological Science* 21, 133–9.

Valdeyron, G. (1984) La pollinisation du figuier. In P. Pesson and J. Louveaux (eds.), *Pollinisation et productions végétales*, 393–408. Paris, Institut National de la Recherche Agronomique.

Van Strydonck, M., Boudin, M., Ayuso, V. M. G., Calvo, M., Fullola, J. M. and Petit, M. A. (2010) The necessity of sample quality assessment in C-14 AMS dating: the case of Cova des Pas (Menorca e Spain). *Nuclear Instruments and Methods in Physics Research*, section B 268, 990–4.

Vermeeren, C. and van Haaster, H. (2002) The embalming of ancestors of the Dutch royal family. *Vegetation History and Archaeobotany* 11, 121–6.

Whittington, G. (1993) Palynological investigations at two Bronze Age burial sites in Fife. *Proceedings of the Society of Antiquaries of Scotland* 123, 211–3.

Whittington, G. (1997) Upper Kenly farm, Belliston farm and Dalgety bay: 'The Pollen evidence'. *Taysideand Fife Archaeological Journal* 3, 1–21.

Wiltshire, P. E. J. (2006a) Forensic palynology: Why do it and how it works. *Forensic Science International* 163, 163–72.

Wiltshire, P. E. J. (2006b) Consideration of some taphonomic variables of relevance to forensic palynological investigation in the United Kingdom. *Forensic Science International* 163, 173–82.

Wiltshire, P. E. J. (2006c) Hair as a source of forensic evidence in murder investigations. *Forensic Science International* 163, 241–8.

Chapter 4

The Final Masquerade: Resinous Substances and Roman Mortuary Rites

Rhea Brettell, Eline Schotsmans, William Martin, Ben Stern and Carl Heron

---

Introduction

Ironically, while studying the manner in which the Romans treated the bodies of their dead the myriad practical and symbolic facets entwined in mortuary practices were quite literally 'brought home' as a result of the death of my mother, Gloria June Brettell. As so powerfully demonstrated by Rosaldo (2004), this intermingling of real life and academic research provided a different perspective from which to consider the nature of humanity's responses to corporeal remains. My experience revealed that conflicts may exist between the wishes of the deceased, the needs of the survivors (immediate family and friends) and the demands of society and that these tend to focus around material aspects of the treatment of the body. A similar discrepancy between cultural expectations and individual understandings was encountered by Erasmo during his mother's funeral (2012, xi–xii). In both cases, social norms were victorious resulting in rites that ascribed to dominant regional practices to the emotional discomfort of the living. These observations are, however, of considerable interest as a starting point for investigating the materiality of archaeological mortuary contexts.

Our last performance in this world, the period between death and the concluding funerary rites, affords the final sensory impact of the deceased upon the living. Mortuary rituals performed during this liminal phase must deal with the biological reality of the decomposing body but they have also been shown to be crucial in the renegotiation of the structure of society. Through enactment of the correct, culturally-situated, processes the transfer of the spirit of the deceased to the otherworld is navigated so that the survivors may return to their transformed lives within the re-balanced social order (Hertz 1907; van Gennep 1909). Thus, although diverse in nature, the treatment of the dead involves a series of considered, materially manifested and socially significant decisions brimming with 'patterned behaviours [and] symbolic action' (Pader 1980, 143). As a result, burial contexts are one of the most structured deposits encountered by archaeologists although the full potential of this repository of material evidence has rarely been explored (Shanks and Tilley 1982). This is particularly true with regards to amorphous organic substances applied to or deposited with the body as these may go unnoticed during excavation or persist only as invisible comminuted traces among the grave deposits (Brettell *et al.* 2015a; Evershed *et al.* 1997).

The importance of considering the active agency of materialities (*i.e.* the social significance of materials and their capacity to act within a given environment) in relation to mortuary practices has recently been highlighted (Fahlander and Oestigaard 2008). Application of this theoretical approach has shown that asking questions about which objects or substances were selected, why this was and how they were employed can be highly rewarding in terms of re-constituting social relations (Hallam and Hockey 2001). More concrete methods such as those advocated by Duday (2009) have also demonstrated that through deconstruction of the material components present in burial contexts and the tracing of taphonomic processes, event sequences may be recreated and funerary practices reconstructed. In order to address questions about the treatment of the dead within a specific cultural milieu, a combination of these approaches should, therefore, permit the substances present to be identified and hypotheses relating to their role as part of socially-motivated action to be tested. Even the physical and emotional impact on the survivors of this transformative process should be considered, in particular, the role of smell in the maintenance of memory (Graham 2011).

Returning to my mother's funeral, one question that is rarely raised in relation to the manner in which the body is prepared is: at whose instigation? As frequently observed, 'the dead do not bury themselves' although they may continue to act through the agency of others (Erasmo 2012, 61–104). We should, therefore, also reflect on how the discourse revealed by the materials identified might relate to the personal desires of the deceased or to choices made by the survivors or whether the entire process was scripted by social mores. These issues will be explored below in relation to recent chemical evidence for the use of resinous plant exudates in the treatment of the body as part of Roman mortuary rites.

Tantalising hints: the primary sources

Traditionally the domain of women, although increasingly placed in the hands of funerary professionals (*e.g. libitinarii* and *pollinctores*), the finer details regarding the treatment of the corporeal remains of the dead during the Roman Empire are difficult to ascertain (Hope 2009, 69–70; Noy 2011). Some references do appear in the literature, as discussed below, but due to poetic license or the fact that these ritual acts were deemed common knowledge (Hope 2007, 99, 108), specific information regarding the substances employed is rarely provided. It is also possible that this aspect of mortuary practices may have been perceived as an indelicate or even taboo subject. Contact with the dead was certainly proscribed as social restrictions were placed on those who handled them and on the immediate household of the deceased, with cleansing rituals required to enable the survivors to re-enter the community of the living (Graham 2011; Toynbee 1971, 50–1).

Nonetheless, there is evidence to suggest that aromatic plant products played an important role during the Roman *funus* (the period between death and interment). Soon after death, the body of the deceased was 'washed [and] anointed with...choicest unguents' (Lucian, *Of Mourning*, 11–2) while incense was burnt around the funerary couch and, probably, during the procession to the pyre or tomb (Hope 2009, 74–5). Scented substances were piled around the body as it was carried through the streets with excess offerings transported on additional biers, by incense bearers (alight?) or even fashioned into effigies (Plutarch, *Sulla*, 38.2). If the individual was cremated, these perfumed products were burnt along with the body or were added to the 'ashes' after collection so that the dead should not be consigned 'unscented to the urn' (Persius, *Satire 6*, 34–6). Once inhumation became the prevalent method of body disposal the fate of these 'fragrant forests of incense'(Statius, *Silvae*, 2.6:85–8) is obscure with only Lucian's satirical account *Of Mourning* (11–8) and Statius's description of the funeral of Priscilla hinting that they may have passed into the tomb with the dead: 'age will no longer wither you, nor will the effects of the years do you harm: such care is taken of your body, such are the riches that the venerable marble breathes' (*Silvae*, 5.1:228–31).What is known about early Christian burial practices, which were influenced by both Roman and Jewish rites, may support this hypothesis as reflected in complaints voiced by the Doctors of the Church in the late 4th century AD: 'If a rich man is buried with perfumes he may delay bodily corruption but will he not still decay?' (St Augustine, *Sermon* 177:7, author's translation). Indeed, the sacred texts of the Jews stipulate that the body of the deceased should be purified and anointed using water and oil then sprinkled with perfumes, funerary spices and/or dried herbs. Aromatic substances also appear to have been burnt as incense during Jewish funerary processions, scattered over the bier and deposited within the tomb, with *unguentaria* (ceramic or glass vessels) common finds in these rock-cut burial chambers (Green 2008; Hachlili 2005, 376–85). As early Christian practices seem to have been conducted along similar lines (with inhumation the preferred method of disposal; Hope 2009: 82), in terms of their religious identity, the dead were generally interred side by side with archaeologically indistinguishable rites in the late Roman Empire (Harries 1992; Toynbee 1996, 234–44).

Classical authors also provide information regarding the nature of the substances selected. Nine fragrant botanical products are named with varying degrees of specificity (*amomum*, balsam, cassia, cherry, cinnamon, frankincense, myrrh, nard and saffron) alongside generic references to the produce (crops, harvests, scents) associated with various ethnic groups or geographical regions (Table 4.1). In addition, the Gospels refer to the use of nard, myrrh and aloes as part of 1st century AD burial customs (SEV 2009, Mark 14:3–8, John 19:29–40). Of these, cherry, a gum unlikely to be chemically identifiable in the archaeological record, is mentioned primarily as a literary device, 'he will stint the funeral feast, and will commit your bones unscented to the urn, not caring to enquire whether the cinnamon has lost its fragrance or the cassia has been adulterated with cherry' (Persius, *Satire 6*, 34–6). According to sources such as Pliny the Elder (*Naturalis Historia* (*NH*)) and Dioscorides (*De Materia Medica*), the others represent some of the most highly desirable spices, plant exudates and unguents employed in the ancient world with most obtained by long-distance trade from southern Arabia, India and eastern Africa, as recounted by the anonymous author of the *Periplus Maris Erythraei* (*PME*).

Thus, the earliest references alongside Ausonius' 4th-century AD evocation of past practices (see Table 4.1 for details of relevant passages) are to nard (spikenard). Probably obtained from the leaves of *Nardostachys jatamansi* from the Gangetic region of India, a number of other varieties of valerian from Syria, Gaul and Crete were also exploited in the Roman period (Pliny, *NH*, 12:26). In addition, the term nard was used to denote an unguent, described as holding 'a foremost place among perfumes' (Pliny, *NH*, 12:26, 42–3), which was made from an elaborate mixture of plant extracts (Groom 1997, 315). Given the context in which nard appears in these texts (*e.g.* Petronius, *Satyricon*, 77–8), it seems that it was, on occasion, used to anoint the deceased: 'a woman came with an alabaster flask of ointment of pure nard, very costly, and she broke the flask and poured it over his head ... Jesus said ... she has done a beautiful thing to me ... she has anointed my body beforehand for burial' (SEV 2009, Mark 14:3–8). Similarly, aloes, the leaf extract of members of the genus *Aloe* from Arabia and eastern Africa (Groom 1997, 5), and the unidentifiable *amomum* from Assyria, seem to have been linked to the preparation of the body with the powerful scent of the latter strongly associated with mortuary imagery: 'reeking ... with [amomum] enough to out-scent two funerals' (Juvenal, *Satire 4*, 108).

The products most frequently mentioned are, however, cassia (today the bark of *Cinnamomum* spp. shrubs) and the closely related cinnamon (correctly the inner bark of *C. verum*). According to Pliny, such spices could be purchased in the ports of Ethiopia (Pliny, *NH*, 12:41–3) with 'true cinnamon' presumably obtained from the Indian subcontinent (Anon, PME). Moreover, well-attested vagaries in ascribing botanical sources (Groom 1997, 55–6, 65–6) mean that the bark or resinous exudates of other species which contain cinnamates, such as *Styrax officinalis* (styrax), might have been included under this heading. Even more highly prized were frankincense and myrrh, the exudates of *Boswellia* spp. and *Commiphora* spp. respectively (Groom 1981, 55–76; Pliny, *NH*, 12:30–5; Theophrastus, 9:4, 1–10). Traded by sea through Alexandria from Somali ports and southern Arabia (the 'Frankincense Kingdom') or transported overland to Persia and Judaea, frankincense, in particular, seems to have been in demand for temple worship and funerary rites (Anon, PME; Groom 1981, 13–20; Pliny, *NH*, 12:26, 41). Myrrh, however, was more commonly used in unguents and perfumes as a stabilising agent (Pliny, *NH*, 13:2). The reason that southern Arabia gained the epithets '*felix ac beatae*' (happy and blessed; Pliny, *NH*, 12:30), both of these gum-resins would have readily sprung to mind when mention was made of the riches of Arabia and Sabaea.

Other products such as saffron (the stigmas of *Crocus sativus*) could be obtained closer to Rome as this spice was closely associated with Corycia in Cilicia (southern Turkey) and Sicania (Sicily) (Pliny, *NH*, 13:2). Phrases such as 'Cilician harvests' are, therefore, generally interpreted as referring to saffron, although the balsamic exudates of *Liquidambar orientalis* (storax) also came from this region (Pliny, *NH*, 12:55). Likewise, the Levant (*Judaea/ Syria Palastina*) was renowned for its aromatic produce with the most desirable being balsam (Safrai 1994, 83–7). Generally thought to have been extracted from members of the family Burseraceae (*e.g. Commiphora opobalsamum, C. gileadensis* and/or related species) (Howes 1949, 153; Langenheim 2003, 356–7), even in antiquity there was considerable uncertainty as to the precise botanical source of this exudate which seems to have been particularly difficult to extract on a commercial scale (Pliny, *NH*, 12:54–5). Referred to under a variety of names (*e.g.* Balsam of Gilead, Judaea, Mecca or Syria), even its geographical origins were debated which suggests that more than one botanical source may have been involved (Groom 1981, 126–31; Theophrastus, 9:6, 1–4). Indeed, as a range of plant exudates were harvested from the Levant, references to Palestinian incense and Hebrew perfumes may also include styrax (*S. officinalis*), storax (*L. orientalis*) and terebinth, the resin secreted by *Pistacia terebinthus* or closely related species (Mills and White 1989) with the latter possibly the source of balsam of Syria (Theophrastus, 3:2, 6).

Despite the fascinating hints provided by the sources (see Table 4.1), it must be stressed that the vast majority of these texts were written by and for educated, elite males working in and around Rome during the 1st century BC–1st century AD. Thus, their relevance as guides to wider Roman rites is questionable. The situation, particularly in the later Empire, remains unclear as does the extent to which such practices were adopted in the northern provinces far beyond the geographical range of the botanical sources mentioned.

Archaeological traces: speculation about embalming

Reports regarding the discovery of Roman period burials (outside of Egypt) containing the remains of unusually well-preserved individuals span the 15th to 20th centuries. Although most derive from Italy, references to mummified or embalmed individuals also relate to other regions of the Empire (Chioffi 1998; Reifarth 2013, 492–511). In the 1970s, similar observations regarding the presence of sticky substances in connection with late Roman plaster burials where individuals interred in stone sarcophagi and/ or lead-lined coffins had additionally been encased in lime or gypsum (although subsequent research has indicated that plaster may also be present in more normative inhumations in wooden coffins, *e.g.* Barber and Bowsher 2000, 101–3), led to a long-running debate about the significance of this 'package' seemingly aimed at some degree of body preservation (Philpott 1991; Ramm 1971; Sparey Green 1977). The majority of these finds are, however, no longer extant due to reburial soon after excavation or subsequent misadventure (*e.g.* theft, fire, loss).

This is unfortunate and raises the fraught issue of terminology. Inconsistencies abound, particularly in the antiquarian reports, with individuals wrapped in textile bandages, accompanied by dark residues or plant remains (*e.g.* leaves, flowers, twigs), coated

Table 4.1 Fragrant plant products mentioned in the primary sources by name or geographical region in relation to Roman mortuary practices.

Date	*Original term/phrase*	*Translation (by author)*	*Source*
1st century BC	ampullam nardi	Jar of nard	Petronius *Satyricon, 77–78*
	Nardo	Nard	Propertius *Elegies 4.7*
	Assyrios...odores	Assyrian scents	Tibullus *Elegies 1.3*
	dives Panchaia merces Eoique Arabes, dives et Assyria	Riches and gifts of Panchaia, the East, Arabia, and Assyria	Pseudo-Tibullus *Elegies 3.2*
1st century AD	ναρδου σζμυρνης και αλοης	Nard Myrrh and aloes	*Mark 14:3–8* *John 19:29–40*
	ἀρωμάτων λιβανωτοῦ κινναμώμου	Perfumes Frankincense Cinnamon	Plutarch *Sulla*
	Cinnama Casiae Ceraso	Cinnamon Cassia Cherry	Persius *Satires* *6, 34–36* *3, 104–105*
	crassisque lutatus amomis	thick and [?] amomum	
Late 1st century AD	Arabum Cilicumque fluit floresque Sabaei Indorumque arsura seges... tura Palaestinis, simul Hebraeique liquores Coryciaeque comae Cinyreaque germina	Produce of Arabia and Cilicia Sabaean flowers Indian crops for burning Palestinian incense Hebrew liquids (perfumes) Corycian threads (saffron) Seeds of Cinyreia (myrrh)	Statius *Silvae* *5.1, 281–287*
	Cilicum flores graminis Indi Arabes Phariique Palaestinique liquores	Cilician flowers Indian spices Arabian, Egyptian, Palestinian perfumes	*2.1, 214–227*
	Assyrio amomo Arabum respirant gramine Sicaniisque crocis	Assyrian *amomo* Aromatic crops of Arabia Sicanian saffron	*2.4, 43–46*
	odoriferous...Sabaeos Cilicum messes Pharian cinnama Assyrio manantes gramine sucos	Scented produce…of Sabaea Cilician harvests Egyptian cinnamon Juices that flow from Assyria	*2.6, 85–88*
	Eoa germina messes Cilicumque Arabumque	Produce of the East Harvests of Cilicia and Arabia	*3.3, 42–48*
	Sicanii...croci, Sabaei cinnama odoratas Arabs...aristas	Sicanian saffron Sabaean cinnamon Scented Arabian grains	*5.3, 53–59*
	Casias Murram Incense Cinnama	Cassia Myrrh, Incense (?frankincense) Cinnamon	Martial *Epigrams* *11.54*
	Murram Casias	Myrrh Cassia	*10.97*
	Amomum	Amomum	*5.64*
	Amomum	Amomum	Juvenal *Satires 4.108*
4th century AD	unguine nardi balsama	Nard Balsam	Ausonius *Epitaphs 6.31*

with traces of plaster or just elaborately interred (*e.g.* with a box containing ointment bottles) interpreted as 'mummified' or 'embalmed' (*cf.* Chioffi 1998; Reifarth 2013). Indeed, re-evaluation of the remains of an infant from a sarcophagus with inner lead coffin found in the Fin Renard necropolis, Bourges, France, has indicated that remnants of soft tissue were preserved due to natural processes acting within this protected microenvironment rather than being the result of artificial mummification (Thillaud 2006). Similar issues of misidentification have been noted with respect to the presence of white residues in burial contexts with inwash or infill from the surrounding geology, for example, mistakenly reported as evidence of the deliberate inclusion of plaster (Schotsmans 2013, 31–7; Schotsmans *et al.* 2014; in press). Confusion between different aspects of body treatment also occurs. For example, individuals from Pannonia who were originally recorded as 'desiccated with saline' appear, in fact, to have been interred with plaster (Reifarth 2013, 501–2). Thus, as proposed by Chioffi (1998, 24), care should be taken to distinguish between artificial mummification (desiccation and the application of often heated mixtures aimed at long-term preservation) and other forms of body treatment, with the term embalming reserved for the presence of resinous substances (as perfumed oils/unguents or solid fragments) in close association with the body or textile wrappings.

All of this does not, of course, preclude the presence of resinous substances within these burial contexts since chemical analysis was rarely undertaken, with one notable exception. The earliest account of an embalmed individual from Rome dates to 1485 and describes the recovery of the remains of a 12–15-year-old girl from a marble sarcophagus found beside the via Appia Antica (Chioffi 1998, 66–7). The corpse was covered in a thick casing of transparent substances which reportedly comprised a gum-resin (with myrrh, frankincense or aloes, actually an essential oil, suggested) mixed with plant oils (cedar and olive). Clearly a body treatment in this instance, the purpose of other plant exudates mentioned in these records remains speculative as they may have been included as offerings rather than used to prepare the dead. Reports of 'frankincense and herbs', in the sarcophagus of a young adult female from L'Abbaye Saint Victor, Marseilles (Reifarth 2013, 506) and of residues posited to be frankincense or myrrh in association with elaborate cremation burials from Bartlow Hills, Cambridgeshire (Gage 1834) and Weston Turville, Buckinghamshire, UK (found in 1855, Waugh 1962), reinforce this issue. In addition, scepticism about such identifications has arisen due to the Christian milieu of the excavators, date of examination and/or a lack of information about the methods employed.

Chemical confirmation: resinous plant exudates

Since the 1990s, research using modern analytical techniques (principally gas chromatography-mass spectrometry) has begun to address questions regarding the use of resinous products in antiquity (Colombini and Modugno 2009; Evershed 2008). Inevitably, when considering mortuary practices, initial studies focused on Ancient Egyptian mummification processes and their continuance during the Ptolemaic and Roman periods (*e.g.* Buckley and Evershed 2001; Colombini *et al.* 2000; Koller *et al.* 1998; Tchapla *et al.* 2004). Even in these complex mixtures, the characterisation of natural plant exudates, the sticky scented substances secreted when the bark of certain trees or shrubs are 'wounded' (Langenheim 2003, 23–6), proved a particularly fruitful area of inquiry. Such studies have shown that the higher molecular mass terpenic compounds (di- and triterpenoids) present in the resin fraction are both diagnostic and relatively resistant to decay (Pollard and Heron 2008, 235–69). Their presence can, therefore, permit the identification of specific resins and gum-resins, generally to the level of genus.

Such findings sparked interest in the analysis of visible residues from burials elsewhere in the Roman Empire and began with assessment of the well-preserved remains of a girl *c.* eight years old curated in the National Museum (Palazzo Massimo), Rome (Ascenzi *et al.* 1993). Samples of organic substances associated with the skin, silk textiles and linen bandages revealed the presence of sesquiterpenes and abietic acid derivatives indicative of a conifer resin. Amorphous masses and adhering residues from sarcophagus burials around Rome (Ghini *et al.* 2005; Mitschke and gen. Schieck 2012) and an exceptional mass grave in the catacombs of Santi Marcellino e Pietro, also in Rome (Devièse 2008; Devièse *et al.* 2010) subsequently provided indications of the inclusion of plant exudates. In addition, from further afield, chemical confirmation of the use of resins and gum-resins was obtained from elaborate sarcophagus burials found in Milan, Italy (Bruni and Guglielmi 2005; 2014), Thessaloniki, Greece (Papageorgopoulou *et al.* 2009), Naintré, near Poitiers and Anché in the Loire valley, France (Devièse 2008; Devièse *et al.* 2011) and Trier, Germany, with additional finds from Iovia, Hungary and Palmyra, Syria (Reifarth 2009; 2013; Plate 8; Table 4.2).

Although full details have yet to be published about some of these discoveries, others were found to contain suites of biomarkers indicating the presence of diterpenoid (conifer) resins and/or triterpenoid (angiosperm) resins and gum-resins. The botanical sources identified encompass members of the Pinaceae, a family of the conifers which includes pines, cedars and larches (Langenheim 2003, 35–7). Of widespread occurrence across Europe, it seems that conifers had a special significance in Roman eschatology, being grown on graves with their woods 'placed at the doors

Table 4.2 Results of the chemical analysis of samples from Roman period inhumations from continental Europe, compiled from the literature.

Find	*Publications*	*Date AD*	*Brief details*	*Results of chemical analysis*
Via Appia Antica Rome	Cited in Chioffi 1998	2nd–3rd century	12–15 yr old female in sarcophagus with marble lid.	Gum-resins (*e.g.* frankincense or myrrh). Aloes also suggested. Cedar oil & olive oil
Via Latina Grottaferrata, Rome	Ghini *et al.* 2005	2nd century	18 yr old, Carvilius Gemellus, in marble sarcophagus accompanied by well preserved floral offerings.	Traces of ?myrrh & pine reported (method used unclear, so may not be directly indicative of presence of plant exudates)
San Sebastian Via Appia Antica Rome	Mitschke & gen. Schieck 2012	2nd–3rd century	Adult ?male in marble sarcophagus	Cinnamic acid + *p*-hydroxy cinnamic acid – balsamic resin suggested (although no terpenic compounds mentioned)
Via Cassia Grottarossa, Rome	Ascenzi *et al.* 1993	2nd century	*ca.* 8 yr old girl possibly interred in decorated marble sarcophagus	Sesquiterpenes & abietic acid derivatives indicative of presence of conifer resin. Egyptian-style mummification
Catacombs of *Santi Marcellino e Pietro*, Rome	Deviese 2008 Deviese *et al.* 2010	2nd–3rd century	Multiple burials within pits, wrapped in shrouds, gypsum present	Diterpenoids indicative of presence of conifer resin, sandarac
Necropolis of the *Università Cattolica,* Milan	Bruni & Guglielmi 2005 Bruni & Guglielmi 2014	3rd–4th century	'Lady of the sarcophagus', young adult female in gneiss sarcophagus	Triterpenoids characteristic of *Pistacia* spp. resin
Basilica of *St. Ambrogio*, Milan	Bruni & Guglielmi 2014	4th century	Remains of *SS. Ambrogio, Gervasio e Protasio* in porphyry sarcophagus	Triterpenoids characteristic of *Pistacia* spp. resin
Eastern cemetery, Thessaloniki, Greece	Papageorgopoulou *et al.* 2009	3rd century	Mature adult female in marble sarcophagus with inner lead coffin	Range of fatty acids + sesqui-, di- & triterpenoids indicative of scented unguent or oil
Naintré, Gaul (Poitou-Charentes, France)	Devièse 2008 Devièse *et al.* 2011	3rd century	Adult female & 12 yr old child in sarcophagi with 2 vaulted tombs	*Boswellia* spp. (frankincense) & *Pistacia* spp. (mastic) on & around both bodies. Black substance provisionally interpreted as *Boswellia* spp. bark located above the remains
Anché, Gaul (Indre-et-Loire, France)	Devièse 2008	2nd–4th century	Adult female & 12 yr old child in stone sarcophagi	Pinaceae and frankincense with the adult.*Pistacia* spp. resin with the child
Trier, Rhineland (Germany)	Reifarth 2013	4th century	Individuals of all ages & both sexes in stone sarcophagi in crypt	Pinaceae, Cupressaceae, *Pistacia* spp., balsamic resin (GC-MS); gums & gum-resins (FTIR)
Iovia, Pannonia (Hungary)	Reifarth 2013	2nd–4th century	Sarcophagus from Roman period SE cemetery	'Thermoanalysis' indicated pine & frankincense. GC-MS of double burial confirmed Pinaceae & *Pistacia* spp. resins
Palmyra, Syria	Reifarth 2013	1st–3rd century	Sarcophagus burial from the tomb of Atenatan	Mixture of gum & ?balsam – cinnamic acid + related compounds

of houses as a token of bereavement', and their exudates used 'for torches at religious ceremonies' (Pliny, *NH*, 16:16-22). Indeed, pinecones, as symbols of immortality or mourning, are often found as finials and carvings on funerary monuments (Alcock 1980; see also Caracuta and Fiorentino, this volume). In addition, the range of diterpenoids present in red crystalline fragments recovered from the remarkable multiple burials in the catacomb of Santi Marcellino e Pietro indicate that sandarac (*Tetraclinis articulata*, Cupressaceae) was deposited with these individuals (Devièse *et al.* 2010).

Characteristic biomarkers for the use of the pale-yellow resin harvested from the genus *Pistacia* (family Anacardiaceae) have also been documented in sarcophagus burials from Italy, France and Germany (Bruni and Guglielmi 2005; 2014; Devièse 2008; Reifarth 2013). Previously found in a number of Ancient Egyptian contexts including as part of mummification balms (*e.g.* Buckley and Evershed 2001; Colombini *et al.* 2000; Serpico 2000), in the Roman period these highly prized exudates were obtained from species native to the Mediterranean and the Levant (Pliny, *NH*, 14:25). In addition, evidence for the use of gum-resins of the genus *Boswellia*, frankincense, was recovered from both sarcophagus burials at Naintré, and that of the adult from Anché, France (Devièse 2008). This is a rare find in the archaeological record despite the apparent widespread exploitation of frankincense in antiquity (Groom 1981) which was in such demand that two harvests a year were needed to meet the requirements of increasingly lavish Roman funerary displays (Pliny, *NH*, 12:41). Evidence for a triterpenoid-containing balsamic resin, possibly derived from *Liquidambar orientalis* (storax) although seemingly modified by interactions in the burial environment (Townley *et al.* 2015; ongoing research), has also been found with a number of individuals from Trier (Reifarth 2013, 91–114).

The most recent research projects have, moreover, revealed additional information about the treatment of the body (Devièse 2008; Reifarth 2013). For example, exposed areas of skin (Naintré adult and both Anché individuals) appeared to have been coated with a layer of clay whose iron content resulted in a pink colouration (Devièse *et al.* 2011). Likewise, two of the sarcophagus burials from below St Maximin's, Trier, provided evidence of iron-containing pigments in close association with the remains of two infants and a young adult (Reifarth 2013, 112–3). Encased in plaster or surrounded by fir-wood shavings (*Abies* spp.), the presence of gums or other gum-resins (*e.g.* myrrh) as part of these complex mixtures was also indicated (Reifarth 2013, 110–2). Predominantly composed of polysaccharides such evidence is generally lost in the archaeological record due to their solubility in water and the inevitability of leaching (Colombini and Modugno 2009). Likewise, evaporation of the more volatile fraction of perfumed oils and unguents may leave only traces of a degraded generic oil/fat and so what is recoverable may often be the remnants of the resin base, added 'in order to retain the scent in the solid part' (Pliny, *NH*, 13:2, 7).

These discoveries provided confirmation for the use of resinous plant exudates in continental Europe. Nonetheless, most came from isolated finds and derived from burial grounds associated with major cities of the Empire while all relied on initial visual recognition of organic residues. A province-wide evaluation of such evidence and the possibility that invisible chemical traces might remain in less favourable burial environments had not been tested. A systematic study of extant examples of sarcophagi and/or lead-lined coffin burials from late Roman Britain was, therefore, proposed (Brettell 2016). Analysis of samples from 39 such 'package' burials alongside 31 more normative inhumations (interred in a wooden coffin and/or shroud but where plaster or unusual residues were also indicated) dated between the 2nd and 4th centuries AD provided evidence of suites of terpenic compounds characteristic of resinous exudates in sixteen cases (Brettell *et al.* 2014; 2015a; Table 4.3).

These finds were associated with individuals of all ages and both sexes, with a slight preponderance of females (male:female ratio, 0.7:1), all of whom had been interred in sarcophagi and/or lead-lined coffins, and were located in both urban and rural burial grounds across the province. Notably, the exudates identified comprised that same limited range of natural products discussed above in relation to the continental finds. Thus, Pinaceae and *Pistacia* spp. resins, a balsamic resin (possibly *Liquidambar orientalis*) and even *Boswellia* spp. gum-resins can now be shown to have been transported all the way to the remote province of *Britannia* and used as part of mortuary practices (Plate 9; Brettell 2016). In addition, amorphous masses from two multi-container cremation burials interred at Purton, Wiltshire and in the centre of the Mersea Island barrow, Essex, contained indicators of frankincense (Brettell *et al.* 2015b). Possible traces of *Pistacia* spp. resin were observed in the former while the latter was mixed or, perhaps, 'cut' with a small amount of Pinaceae resin since the practice of adulterating more valuable exudates with cheaper products was seemingly as rife in the ancient world as it is today:

> It [a member of the Pinaceae] gives out considerable quantities of resin … so similar in appearance to frankincense, that when mixed, it is impossible to distinguish them; hence the adulterations we find practiced in the Seplasia [an area of perfumer's shops in Capua, Italy] (Pliny, *NH*, 16:18, 40–1).

In the find from Purton, the indicators comprised characteristic degradation products of the boswellic acids (biomarkers for frankincense) together with oleanonic acid (found in *Pistacia* spp. resins). These were incorporated within an abundance of fatty matter, probably representative of a plant oil based on the presence of phytosterols and absence of cholesterol (a marker of mammalian fats). The material from the Mersea Island barrow, however, provided even more exceptional results. The vast quantity (over 90g) of material recovered and reduced oxygen conditions (within a glass vessel, originally part-filled with a liquid, inside a covered lead ossuary in a tile-built chamber at the centre of the barrow) had facilitated the survival not only of di- and triterpenoids but also of volatile

Table 4.3 Burials from Roman Britain found to contain resinous substances.

Location	*Date (AD)*	*Age*	*Sex*	*Body position*	*Container(s) + fill*	*Associated finds*	*Nature of sample*	*Resinous exudate(s) present*
Wraggs Farm, Arrington, Cambridgeshire	2nd-3rd century	Infant	–	extended, supine, head to W	wood with lead-liner *traces of plaster (LSC)*	Pipeclay figurines (in box on foot end; dyed-wool frags; hair	Amorphous orange frags	*Pistacia* spp.
'Spitalfields Lady' SK15903 280 Bishopsgate, London, E1	4th century	YA	F	extended, supine, head to W	limestone sarcophagus decorated inner lead coffin	Textiles (wool, silk); gold thread; 'pillow' of bay leaves; jet & glass items	Residue adhering to hyoid & grave deposits	*Pistacia* spp. + Pinaceae
Eagle Hotel site, Andover Road, Winchester, Burial G336	4th century	MA	M	extended, supine, NS aligned	wood with lead-liner	Mineralised textiles; coin of Constantine	Grave deposits from base of lead-liner	?*L. orientalis* + Pinaceae
Northview Hospital, Purton, Wiltshire, Grave 1	4th century	YA	F	extended, supine	stone sarcophagus undecorated lead-liner *traces of plaster?*	Glass and ceramic vessels; shale bracelet; animal bones; wool/dyed-border	Residues/debris associated with skeletal elements & lead frags	*Pistacia* spp. + Pinaceae
Northview Hospital, Purton, Wiltshire, Grave 2	?	MA	?F	cremation burial	limestone ossuary cylindrical lead urn blue-green glass vessel	Cremated bird & animal bones; frags burnt ceramic; charcoal	Amorphous residue from surface of liquid within glass cremation vessel	*Boswellia* spp. + ?*Pistacia* spp.
Poundbury, Dorchester, Dorset: R2 mausoleum, Grave 8	4th century	–	–	extended, supine, head to W	limestone sarcophagus *gypsum body-casing*	Textile impressions; fragments of bone comb	Residues adhering to inner surface of body-casing	Pinaceae
Poundbury, Dorchester, Dorset: Site E, Grave 127	4th century	SA	F	extended, supine headto W	wood with lead-liner *traces of plaster (LSC)*	Textile impressions	Residues adhering to mineralised textiles	Pinaceae
Poundbury, Dorchester, Dorset: R10 mausoleum, Grave 517	4th century	MA	F	extended, supine head to W	limestone sarcophagus *gypsum body-casing*	Textile impressions; bone comb; copper alloy ring	Residues adhering to inner surface of body-casing	Pinaceae
Poundbury, Dorchester, Dorset: R9 mausoleum, Grave 529	4th century	MA	F	extended, supine head to W	wood with lead-liner *plaster body-casing*	Textile impressions; hair	Residues adhering to inner surface of body-casing	Pinaceae
Poundbury, Dorchester, Dorset: R9 mausoleum, Grave 530	4th century	MA	M	extended, supine head to W	wood with lead-liner *gypsum body-casing*	Textile impressions; hair [Wool band, with organic residue; Reifarth 2013]	Residues adhering to inner surface of body-casing	Pinaceae [?Balsam]
Poundbury, Dorchester, Dorset: Site E, Grave 892	4th century	MA	M	extended, supine	wood with lead-liner *traces of plaster*	Textile impressions	Residues adhering to mineralised textiles	Pinaceae

(Continued)

Table 4.3 Burials from Roman Britain found to contain resinous substances. (Continued)

Location	*Date (AD)*	*Age*	*Sex*	*Body position*	*Container(s) + fill*	*Associated finds*	*Nature of sample*	*Resinous exudate(s) present*
Poundbury, Dorchester, Dorset: Site E, Grave 1040	4th century	MA	M	extended, supine	wood with lead-liner *plaster (LSC) body-casing*	Textile impressions	Residue adhering to mineralised textiles	Pinaceae
Alington Avenue, Dorchester, Dorset, Burial 3664, SF 1075	3rd century	OA	F	extended, supine, head to SW	wood, lid of sandstone *chalk rubble packing*	Hobnailed footwear; glass vessels; ?food offerings	Grave deposits associated with cranium	*Boswellia* spp.
Alington Avenue, Dorchester, Dorset, Burial 4378, SF 1169	3rd century	Child	?	extended, supine, head to SE	wood with lead-liner *chalk environmental ingress*	Ceramic jar; coin; iron rod; garment, purple dyed *clavi*	Grave deposits & adhering residues	*Boswellia* spp.
Railway excavations, York YORYM: 2007.6206/11931	–	–	–	–	stone cist/ cedarwood coffin *plaster body casing*	Frags coarse textile	Residue adhering to inner surface of body-casing	*?Boswellia* spp.
Mill Mount, York YORYM: 2007.6205i/11928	–	–	–	–	– *plaster body casing*	Textile impressions	Residue adhering to inner surface of body-casing	*?Boswellia* spp.
Railway excavations, York YORYM: 2010.1219	–	A	–	–	stone sarcophagus *plaster body casing*	–	Grave deposits from base of sarcophagus	*Pistacia* spp.
Mersea Island barrow (Mersea Mount), Mersea Island, Essex	2nd century	MA	M	cremation burial	tile-built burial chamber square lead ossuary green glass vessel	–	Amorphous yellow-white masses enclosing orange frags	*Boswellia* spp. + Pinaceae

Age determinations: infant <12 months; child 1<11 years; sub-adult 12–20 years; young adult 20–35 years; middle adult 36–50 years; older adult 50+ years. LSC = lead substituted carbonate. References: Arrington (Taylor 1993); Spitalfields Lady (Museum of London 1999); Eagle Hotel (Richards 1999); Northview Hospital (Chandler 1994); Poundbury (Farwell and Molleson 1993); Alington Avenue (Davies *et al.* 2002); York (Ramm 1971; RCHME 1962); Mersea Island barrow (Hazzeldine Warren 1913).

components (mono- and sesquiterpenes) rarely observed in the archaeological record (Plate 10). This confirmed that an unburnt offering had been deposited in the cinerary urn prior to the tomb being sealed (Brettell *et al.* 2015b).

Ritual reconstruction: moving towards meaning

Now that this corpus of research has provided definitive evidence for the use of resinous plant exudates in mortuary contexts, what can be said about this aspect of Roman ritual? Dealing with death requires culturally-situated strategies whose material manifestations perform a significant role in mediating relationships between the living and the dead. Through the agency of materials, the socially disruptive emotions created by loss can be transformed into lasting counter-memories through sensory experiences (Hallam and Hockey 2001). Fortunately for us, in the Roman cultural milieu, materiality appears to have been key in negotiating human-divine boundaries and during rites of passage. This was a corollary of the fact that the core of Roman eschatology 'lay in ritual, not in belief' (Rives 2000, 251). Based around interpreting signs in order to take the correct steps to gain divine support, materialised actions were all important in establishing positive relationships and ensuring a successful outcome (Ando 2003). Thus, as death approached, the focus fell on material rather than spiritual preparations centred on an individual's *fama* (reputation) (Graham 2011).

The text of Roman death was designed, therefore, to establish the credentials of the deceased in this world and create an idealised and enduring image in order to negotiate entrance to the next and maintain memory (Noy 2011). During the preparation of the body, its display in the atrium and highly visible journey through the streets, the material aspects of ritual actions would have diverted attention from the reality of the decomposing corpse towards remembrance of the beautiful and fragrant dead (Hallam and Hockey 2001, 14–5). Visually, audibly and olfactorily stimulating, the pageantry and substances employed signified the status of the deceased and the respect in which the survivors held their loved one while providing a collective social memory of the event. Similarly, continued interaction through memorial feasts and the provision of offerings, principally at the tomb, sustained the dead in the afterlife and perpetuated their memory, both privately and publically (Gee 2008). As roles in these rituals (both for the living and the dead) were assigned according to age, gender and, above all, status, the treatment of the body mirrored, reinforced and re-negotiated fundamental aspects of Roman social structure (Hope 2009, 93–6).

Thus, as part of a 'package' of materials employed in the Roman mortuary sphere, aromatic substances played an important role in the transformation of the dead with males and females of all ages accorded these rites. What is particularly striking about the findings from both British and continental contexts, are the commonalities between these burials and their remarkable correspondence with literary descriptions of rites conducted by the Roman elite. With the exception of the multiple inhumations in the catacombs of SS. Marcellino e Pietro, Rome (although these also contain related materials; Blanchard *et al.* 2007; Devièse *et al.* 2010), all of those interred with resinous exudates had been placed in stone sarcophagi and/or lead-lined coffins often located within mausolea, vaulted tombs or earthen mounds. Most were accompanied by jewellery and other high quality grave goods, including hair nets of twisted gold threads, glass vessels and items made of amber and ivory. In addition, where taphonomic conditions have allowed, a range of textiles have been recovered comprising gender and status appropriate garments (*e.g.* tunics, dresses, cloaks) made from wool, linen and damask silk with decorative elements provided by murex ('Tyrian') purple-dyed *clavi* (stripes) and gold tapestry-work alongside bandages, face-covers and shrouds (Mitschke and gen. Schieck 2012; Reifarth 2013, 47–90; Wild 2013).

Other elements such as the inclusion of wood shavings, the presence of plaster (lime or gypsum) and the application of pigments to the skin are also of considerable interest. Only the latter are mentioned in the literary sources with the *pollinctores* so named because they used a fine powder to cover the progressive discolouration of the corpse (Graham 2011). Indeed, the inclusion of fir-bark turnings appears to be restricted to Trier although other plant materials including crowns, pillows, beds/biers or garlands of leaves and flowers have been recovered elsewhere (*e.g.* Devièse 2008, 65–8; Ghini *et al.* 2005; Papageorgopoulou *et al.* 2009). For example, in Britain, bay leaves were found around the cranium of the Spitalfields Lady, London (Thomas 1999). The use of plaster (lime or gypsum) in late Roman inhumations has, however, been more widely observed (*e.g.* Ramm 1971; Reifarth 2013, 433–77; Sparey Green 1977). Seemingly added at the graveside (Barber and Bowsher 2000, 101–3), calcitic materials have been shown to retard decay by absorbing body fluids and restricting microbial action (Schotsmans *et al.* 2012). Their application to the body can, therefore, be rationalised when sarcophagi were to be placed within vaults or mausolea that would be regularly revisited for acts of remembrance and subsequent burials. What remains inexplicable, however, is the inclusion of plaster in otherwise normative soil-cut graves and so their potential relationship with other facets of identity, for example an association with Christianity, remains a matter for debate (Sparey Green 1977). What is key in terms of this research agenda is, that all of these components appear, primarily, to signify attempts to disguise the visual and olfactory impact of corporeal decay although symbolic connotations cannot be ruled out as the ascription of meaning to both plant species and sensory effects, such as colour, are attested in Roman eschatology (Bradley 2011; Giesecke 2014).

With regards to the resinous plant exudates identified, both the literary and archaeological evidence suggests that a limited palette of substances was deemed appropriate for use in the mortuary sphere. There is, however, little direct correspondence between the two sources as conifer and *Pistacia* spp. resins appear to have played a significant, but unmentioned, role in the later Empire, if not before. This may be due to the generic, poetic terminology employed or to changing tastes as Pliny refers both to the concept of 'branding' by place of origin in his discussion on perfumes and to changes in fashionable preferences for certain products (Pliny, *NH*, 13:2). Only frankincense definitively appears on both lists reinforcing the well-attested and long-standing ritual importance of *Boswellia* spp. gum-resins (Groom 1981). Thus, on a practical level, in response to the biological reality of the decomposing corpse, it appears that these 'choicest unguents [were indeed used] to arrest the progress of decay' (Lucian, *Of Mourning*, 11–2). This is supported by an absence of evidence for extensive heating, if any, prior to their application to the body (Brettell *et al.* 2015a) and makes sense in chemical terms as only the steady release of volatile components would act to mask the odour of putrefaction and slow invasion of the corpse by decomposer organisms. Likewise, the anti-microbial properties of the substances selected would have temporarily retarded soft-tissue decay. Such considerations would have been particularly pressing during the extended funerary rites accorded the social elite (Hope 2009, 71–4).

Nonetheless, it must have been their ritual significance that warranted transportation across the length and breadth of the Empire. This is confirmed by the abundant, unburnt offering of frankincense found in the Mersea Island barrow cremation urn (Brettell *et al.* 2015b) which could have served no practical purpose. The precise method (possibly in *amphorae*, *unguentaria* or *pyxides* depending on the type of exudate and whether it could be shipped as a viscous liquid, solid fragments or formed part of a scented oil/unguent) and extent of this trade, however, remains to be ascertained. Likewise, the issue of ethnicity requires investigation. Were all of those interred in this manner, migrants, as in the case of the Spitalfields Lady, London, who appears to have been a recent arrival from the vicinity of Rome (Montgomery *et al.* 2010)? Were they members of the native elite keen to display their *romanitas*, as may have been the case with the mature male interred with composite pre-Roman and Roman rites in the Mersea Island barrow? Or, by the late Roman period, were these practices so embedded in Britain that these scented substances could simply be selected, alongside other material manifestations of wealth and status, from a range of options supplied by the local *libitinarii* (undertakers)? It is hoped that isotope studies will provide further answers in this regard.

To return to focus of this paper, the additional level of information gained through organic residue analysis and its ability to enhance our understanding of funerary practices, we now need to combine scientific fact with mortuary theory. So, how do these findings fit with the widely observed tripartite structure of the final rite of passage (van Gennep 1909)? During the Roman *funus*, initial separation was marked by the application of oils/unguents, possibly scented, to purify and anoint the body. Evidence for this would only be recoverable in exceptional circumstances, for example, where natural preservation of the soft-tissue, including the skin, had occurred. In the liminal stage when the deceased was prepared for display, however, the dressing of the dead appears to have been accompanied by substantial amounts of resinous substances placed around the body: 'perfumes … given to the gods a grain at a time … are piled up in heaps to the honour of dead bodies!' (Pliny, *NH*, 12:41, 83). Viewed both as gifts from the gods and gifts acceptable to the gods, these aromatic exudates would have provided an odour of sanctity and are, as shown in the images from Trier (Plate 8), those most readily accessed in the archaeological record. This intimate association with the body and textile wrappings is also found in many of the British 'resin burials' (*e.g.* Arrington, Poundbury, York). For example, the multiple samples collected from within the lead coffin of the Spitalfields Lady, London, showed that *Pistacia* spp. and Pinaceae resins were in high abundance in the mid-line but in low abundance or absent from samples collected beyond the outline of the skeletal remains (*i.e.* between the arm bones and the side of the lead coffin) (Brettell *et al.* 2015a). Finally, during the rites marking the final transformation, offerings were made at the tomb with water, wine, milk, grain and scented substances deemed appropriate tributes to accompany the dead (Gee 2008; Hope 2009: 85–8). This aspect is clearly demonstrated by the find from the Mersea Island barrow (Brettell *et al.* 2015b) but may be missed when amorphous masses are no longer visible. The recovery of such evidence would, therefore, be facilitated by the collection of samples not only from the interior but also the exterior of the container, when the individual had been inhumed, in order to distinguish such deposits from the use of resinous substances in the treatment of the body.

Ultimately, can anything be said about who instigated these actions? Just as with modern mortuary practices and so personally (and uncomfortably) experienced during preparations for the funeral of my mother, the choices made appear, predominantly, to be driven by social mores. Thus, as this research has revealed, members of the Roman elite over a broad chronological period were provided with a very similar material package, modified with respect to the dominant method of disposal. Although acting to disguise the smells arising from corpse, the primary role of these costly plant products, through their conspicuous consumption on the pyre or deposition in the tomb, was to signify wealth and social status. Heaped around the body, they could also be viewed as offerings made as a mark of respect:

> the women contributed such a vast quantity of spices … that, apart from what was carried on 210 litters, a large image of Sulla himself … was moulded out of costly frankincense and cinnamon. (Plutarch, *Sulla*, 38.2)

As illustrated in this account, it was the survivors (interestingly, in this case, women) who chose to follow this path. Thus, although the deceased might have requested certain rites including the choice of perfumed products, as implied in the Satyricon by Trimalchio asking to be anointed with nard (Petronius, *Satyricon*, 77–8), it was the need of the survivors to honour their dead and ensure a successful transition (both for the spirit of the deceased and for themselves as the living) that were addressed by meeting the expectations of society. They could, therefore, gain solace in having 'done the right thing' even if they had, perhaps, had to put aside their own conflicting personal preferences.

Finally, the socially embedded desire for immortality through remembrance appears to have been of considerable importance in the Roman world (Graham 2011). If a 'good death' could be ensured, then the ubiquitous reality of mortality was not something to be avoided but rather embraced in the knowledge that some form of continued existence would follow. Part of the iterative spectacle surrounding death, the strongly scented substances used in the treatment of the body would have enabled the power of smell to trigger recall:

> Sprinkle my ashes with pure wine and fragrant oil of spikenard; bring balsam too ... Unending spring pervades my tearless urn: I have but changed my state, and have not died. I have not lost a single joy of my old life, whether you think that I remember all or none. (Ausonius, *Epitaph* 6.31)

Designed to engage the survivors in individual and collective commemorative acts (*e.g.* conspicuous funerary processions; official festivals for feasting with the dead), these social strategies transformed death into a celebration, with the greater the sensory impact on the survivors, the more lasting the embedded memory. Thus, adorned to meet their maker in a manner signifying their idealised status in this life (as inversions must not be forgotten), the materiality of resinous substances would have served to imprint a final image of the deceased in the minds of the living and help play out this final masquerade.

References

Alcock, J. P. (1980) Classical religious belief and burial practice in Roman Britain. *The Archaeological Journal* 137, 50–85.

Ando, C. (2003) *Roman Religion*. Edinburgh, Edinburgh University Press.

Anon (nd) *The Periplus Maris Erythraei (PME)*. Transl. L. Casson [1989] Princeton NY, Princeton University Press.

Ascenzi, A., Bianco, P., Nicoletti, R., Ceccarini, G., Fornaseri, M., Graziani, G., Giuliani, M. R., Rosicarello, R., Ciuffarella, L. and Granger-Taylor, H. (1993) The Roman mummy of Grottarossa. In K. Spindler, H. Wilfing, E. Rastbichler-Zissernig, D. zur Nedden, and H. Nothdurfter (eds.) *Human Mummies: A Global Study of their Status and the Techniques of Conservation*, 205–17. New York, SpringerWein.

Ausonius (nd) *Epitaphs* [In *Ausonius*]. Transl. H. G. Evelyn White [1919]. London–New York, Heinemann.

Barber, B. and Bowsher, D. (2000) *The Eastern Cemetery of Roman London*. MoLAS Monograph 4. London, Museum of London Archaeology Service.

Bradley, M. (2011) *Colour and Meaning in Ancient Rome*. Cambridge, Cambridge University Press.

Blanchard, P., Castex, D., Coquerelle, M., Giuliani, R. and Ricciardi, M. (2007) A mass grave from the catacomb of Saints Peter and Marcellinus in Rome, second–third century AD. *Antiquity* 81, 989–99.

Brettell, R. 2016. The Final Masquerade: A Molecular-based Approach to the Identification of Resinous Plant Exudates in Roman Mortuary Contexts in Britain and Evaluation of their Significance. Unpublished PhD thesis, University of Bradford.

Brettell, R., Stern, B., Reifarth, N. and Heron, C. (2014) The '*semblance of immortality*'? Resinous materials and mortuary rites in Roman Britain. *Archaeometry* 56, 444–59.

Brettell, R., Schotsmans, E. M. J., Walton Rogers, P., Reifarth, N., Redfern, R. C., Stern, B. and Heron, C. P. (2015a) '*Choicest unguents*': molecular evidence for the use of resinous plant exudates in late Roman mortuary rites in Britain. *Journal of Archaeological Science* 53, 639–48.

Brettell, R., Stern, B. and Heron, C. P. (2015b) Mersea Island barrow: molecular evidence for frankincense. *Essex Society for Archaeology and History* 4(2013), 81–7.

Bruni, S. and Guglielmi, V. (2005) Le analisi chimiche. In M. P. Rossignani, M. Sannazaro, and G. Legrottaglie (eds.) *La signora del sarcofago: una sepoltura di rango nella necropoli dell'Università Cattolica*, 131–6. Milan, Vita e Pensiero.

Bruni, S. and Guglielmi, V. (2014) Identification of archaeological triterpenic resins by the non-separative techniques FTIR and 13C NMR: the case of *Pistacia* (mastic) in comparison with frankincense. *Spectrochimica Acta Part A: Molecular and Bimolecular Spectroscopy* 121, 613–22.

Buckley, S. A. and Evershed, R. P. (2001) Organic chemistry of embalming agents in Pharaonic and Græco-Roman mummies. *Nature* 413, 837–41.

Chandler, C. J. (1994) Excavations at the Romano-British Walled Cemetery, Northview Hospital, Purton. Unpublished site monograph (second draft), WILT MC890036, Purton 806. Swindon, UK: Archives c/o Swindon Museum and Art Gallery.

Chioffi, L. (1998) *Mummificazione e imbalsamazione a Roma ed in altri luoghi del mondo Romano, Opuscula epigraphica 8*. Rome, Edizioni Quasar.

Colombini, M. P. and Modugno, F. (2009) Organic materials in art and archaeology. In M. P. Colombini and F. Modugno (eds.) *Organic Mass Spectrometry in Art and Archaeology*, 1–36. Chichester, Wiley.

Colombini, M. P., Modugno, F., Silvano, F. and Onor, M. (2000) Characterization of the balm of an Egyptian mummy from the seventh century B.C. *Studies in Conservation* 45, 19–29.

Davies, S. M., Bellamy, P. A., Heaton, M. J. and Woodward, P. J. (2002) *Excavations at Alington Avenue, Fordington, Dorchester, Dorset, 1984–87*. Dorchester, DNHAS.

Devièse, T. (2008) Elucidating Funeral Rituals in Burials from the End of the Roman Empire: Development of a multi-analytical approach. Unpublished PhD thesis, University of Pisa.

Devièse, T., Vanhove, C., Blanchard, P., Colombini, M. P., Regert, M. and Castex, D. (2010) Détermination et function des substances organiques et des matières minérals exploitées dans les rites funéraires de la catacombe des Saints Pierre-et-Marcellin à Rome (1er–IIIe siècle). In I. Cartron, D. Castex, P. Georges, M. Vivas and M. Charageat (eds.) *De corps en corps: traitement et devenir du cadaver*, 115–39. Pessac (Aquitaine), Maison des Sciences de l'Homme d'Aquitaine.

Devièse, T., Ribechini, E., Baraldi, P., Farago-Szekeres, B., Duday, H., Regert, M. and Colombini, M. P. (2011) First chemical evidence of royal purple as a material used for funeral treatment discovered in a Gallo-Roman burial (Naintré, France, third century AD). *Analytical and Bioanalytical Chemistry* 401, 1739–48.

Dioscorides (nd) *De Materia Medica [The Greek herbal of Dioscorides: englished by John Goodyer A.D. 1655]*. Transl. R. T. Gunther [1959]. New York, Hafner Publishing.

Duday, H. (2009) *The Archaeology of the Dead: Lectures in Archaeothanatology*. Oxford, Oxbow Books.

Erasmo, M. (2012) *Death: Antiquity and its Legacy*. London and New York, I. B. Tauris.

Evershed, R. P., van Bergen, P. F., Peakman, T. M., Leigh-Firbank, E. C., Horton, M. C., Edwards, D., Biddle, M., Kjølbye-Biddle, B. and Rowley-Conwy, P. A. (1997) Archaeological frankincense. *Nature* 390, 667–8.

Evershed, R. P. (2008) Organic residue analysis in archaeology: the archaeological biomarker revolution. *Archaeometry* 50, 895–924.

Fahlander, F. and Oestigaard, T. (2008) The materiality of death: bodies, burials and beliefs. In F. Fahlander and T. Oestigaard (eds.) *The Materiality of Death: Bodies, Burials, Beliefs*, BAR International Series 1768, 1–16. Oxford, Archaeopress.

Farwell, D. E. and Molleson, T. I. (1993) *Excavations at Poundbury 1966–80. Volume 2: the cemeteries*. Dorchester, DNHAS.

Gage, J. (1834) A plan of barrows called the Bartlow Hills in the parish of Ashdown in Essex with an account of Roman sepulchral relics recently discovered in the lesser barrows. *Archaeologia* 25, 1–23.

Gee, R. (2008) From corpse to ancestor: the role of tombside dining in the transformation of the body in Ancient Rome. In F. Fahlander and T. Oestigaard (eds.) *The Materiality of Death: Bodies, Burials, Beliefs*, BAR International Series 1768, 59–68. Oxford, Archaeopress.

Ghini, G., Cecere, M. G. G., Rubini, M. and Arietti, F. (2005) L'ipogeo delle Ghirlande a Grottaferrata (Roma): una storia vissuta 2000 anni fa. In P. Attema, A. Nijboer, A. Zifferero, O. Satijn, L. Alessandri, M. Bierma and E. Bolhuis (eds.) *Communities and Settlements from the Neolithic to the Early Medieval Period, Papers in Italian Archaeology VI*, BAR International Series 1452, 246–57. Oxford, Archaeopress.

Giesecke, A. (2014) *The Mythology of Plants: Botanical Lore from Ancient Greece and Rome*. New Haven CT, Yale University Press.

Graham, E.-J. (2011) Memory and materiality: re-embodying the Roman funeral. In V. M. Hope, and J. Huskinson (eds.) *Memory and Mourning: Studies on Roman Death*, 21–39. Oxford, Oxbow Books.

Green, D. (2008) Sweet spices in the tomb: an initial study on the use of perfume in Jewish burials. In O. P. Laurie Brink and D. Green (eds.) *Commemorating the Dead: Texts and Artifacts in Context. Studies of Roman, Jewish and Christian Burials*, 145–73. Berlin–New York, Walter de Gruyter.

Groom, N. (1981) *Frankincense and Myrrh: A Study of the Arabian Incense Trade*. Harlow, Longman.

Groom, N. (1997) *The New Perfume Handbook*. 2nd edn. London, Blackie Academic and Professional.

Hachlili, R. (2005) *Jewish Funerary Customs, Practices and Rites in the Second Temple Period*. Leiden–Boston MA, Brill.

Hallam, E. and Hockey, J. (2001) *Death, Memory and Material Culture*. Oxford, Berg.

Harries, J. (1992) Death and the dead in the late Roman West. In S. Bassett (ed.) *Death in Towns: Urban Responses to the Dying and the Dead, 100–1600*, 56–67. Leicester, Leicester University Press.

Hazzledine Warren, S. (1913) The opening of the Romano-British barrow on Mersea Island, Essex. *Transactions of the Essex Archaeological Society* 13, 114–41.

Hertz, R. (1907) *Death and the right hand*. Transl. R. Needham and C. Needham [2004, reprint]. Abingdon, Routledge.

Hope, V. M. (2007) *Death in Ancient Rome: A Sourcebook*. London–New York, Routledge.

Hope, V. M. (2009) *Roman Death: The Dying and the Dead in Ancient Rome*. London, Continuum UK.

Howes, F. N. (1949) *Vegetable Gums and Resins*. Waltham MA, Chronica Botanica.

Juvenal (nd) *Satires* [In *Juvenal and Persius*]. Transl. G. G. Ramsey [1918]. London–New York, William Heinemann.

Koller, J., Baumer, U., Kaup, Y. Etspüler, H. and Weser, U. (1998) Embalming was used in Old Kingdom. *Nature* 391, 343–4.

Langenheim, J. H. (2003) *Plant Resins: Chemistry, Evolution, Ecology and Ethnobotany*. Portland OR, Timber Press.

Lucian (nd) *Of Mourning [The works of Lucian of Samosata* (reprint)]. Transl. H. W. Fowler and F. G. Fowler [2007, reprint]. Charleston SC, Forgotten Books.

Mills, J. S. and White, R. (1989) The identity of the resins from the Late Bronze Age shipwreck at Ulu Burun (Kaş). *Archaeometry* 31, 37–44.

Mitschke, S., gen. Schieck, A. P. (2012) Dressing the dead in the city of Rome: burial customs according to textiles. In M. Carroll and J. P. Wild (eds.) *Dressing the Dead in Classical Antiquity*, 115–33. Stroud, Amberley Publishing.

Montgomery, J., Evans, J., Chenery, S., Pashley, V. and Killgrove, K. (2010) 'Gleaming, white and deadly': using lead to track human exposure and geographic origins in the Roman period in Britain. In H. Eckardt (ed.) *Roman Diasporas: Archaeological Approaches to Mobility and Diversity in the Roman Empire*, Journal of Roman Archaeology, Supplementary Series 78, 199–226. Portsmouth RI, JRA.

Museum of London (1999) *The Spitalfields Roman*. London, Museum of London.

Noy, D. (2011) 'Goodbye Livia': dying in the Roman home. In V. M. Hope and J. Huskinson (eds.) *Memory and Mourning: Studies on Roman Death*, 1–20. Oxford, Oxbow Books.

Pader, E.-J. (1980) Material symbolism and social relations in mortuary studies. In P. A. Rahtz, T. M. Dickinson and L. Watts (eds.) *Anglo-Saxon cemeteries 1979*, BAR British Series 82, 143–59. Oxford, Archaeopress.

Papageorgopoulou, C., Xirotiris, N. I., Iten, P. X., Baumgartner, M. R., Schmid, M. and Rühli, F. (2009) Indications of embalming in Roman Greece by physical, chemical and histological analysis. *Journal of Archaeological Science* 36, 35–42.

Persius (nd) *Satires* [In *Juvenal and Persius*]. Transl. G. G. Ramsey [1918]. London–New York, William Heinemann.

Petronius (nd) *Satyricon* [In *The Satyricon/Seneca/The Apocolocyntosis*]. Transl. J. P. Sullivan [1977]. Harmondsworth, Penguin.

Philpott, R. (1991) *Burial Practices in Roman Britain: A Survey of Grave Treatment and Furnishing AD 43–410*, BAR British Series 219. Oxford, Archaeopress.

Pliny the Elder (nd) *Naturalis Historia*. Transl. H. Rackham [1968, Loeb Classical Library, revised edition]. Cambridge MA, Harvard University Press.

Plutarch (nd) *Sulla* [In *Plutarch's Lives*]. Transl. B. Perrin [1916]. London, William Heinemann Ltd.

Pollard, M. A. and Heron, C. 2008. *Archaeological Chemistry*. 2nd edn. Cambridge, Royal Society of Chemistry.

Ramm, H. G. (1971) The end of Roman York. In R. M. Butler (ed.) *Soldier and Civilian in Roman Yorkshire*, 179–99. Leicester, Leicester University Press.

Reifarth, N. (2009) Textiles in their scientifical context: interdisciplinary cooperation during the evaluation of burial textiles. In C. Alfaro, J.-P. Brun, P. Borgard and R. Pierobon Benoit (eds.) *Textiles y tintes en la ciudad Antigua. Purpureae vestes 3: actas del symposium internacional sobre textiles y tintes del Mediterráneo en el mundo antiguo, Naples*, 101–7. València, Universitat de València.

Reifarth, N. (2013) *Zur Ausstattung spätantiker Elitegräber aus St. Maximin in Trier: Purpur, Seide, Gold und Harze, Internationale Archäologie 124*. Rahden, Westfalen, Verlag Marie Leidorf.

Richards, J. (1999) *Meet the Ancestors: Unearthing the Evidence that Brings Us Face to Face with the Past*. London, BBC Worldwide Ltd.

Rives, J. (2000) Religion in the Roman empire. In J. Huskinson (ed.) *Experiencing Rome: Culture, Identity and Power in the Roman Empire*, 245–75. London, Routledge.

Rosaldo, R. (2004) Grief and a headhunter's rage. In A. C. G. M. Robben (ed.) *Death, Mourning and Burial: a Cross-Cultural Reader*, 167–78. Malden MA, Blackwell Publishing.

Royal Commission on the Historic Monuments of England (RCHME) (1962) *Roman York (Eburacum)*. Leicester, RCHME.

Safrai, Z. (1994) *Economy of Roman Palestine*. Oxford, Routledge.

Schotsmans, E. M. J. (2013) The Effects of Lime on the Decomposition of Buried Human Remains. Unpublished PhD thesis, University of Bradford.

Schotsmans, E. M. J., Denton, J., Dekeirsschieter, J., Ivaneanu, T., Leentjes, S., Janaway, R. C. and Wilson, A. S. (2012) Effects of hydrated lime and quicklime on the decay of buried human remains using pig cadavers as human body analogues. *Forensic Science International* 217, 50–9.

Schotsmans, E. M. J., Wilson, A. S., Brettell, R., Munshi, T. and Edwards, H. G. M. (2014) Raman spectroscopy as a non-destructive screening technique for studying white substances from archaeological and forensic burial contexts. *Journal of Raman Spectroscopy* 45, 1301–8.

Schotsmans, E. M. J., Brettell, R., Becker, K., Chapoulie, R. and Castex, D. (in press) L'étude des matières blanches dans des sépultures: intérêt d'une démarche scientifique raisonnée. In S. de Larminat, R. Corbineau, A. Corrochano, Y. Gleize, and J. Soulat (eds.) *Nouvelles approches de l'archéologie funéraire, Thanat'Os 4*, xxx. Pessac (Aquitaine), Maison de Sciences de l'Homme d'Aquitaine.

Serpico, M. (2000) Resins, amber and bitumen. In P.T. Nicholson and I. Shaw (eds.) *Ancient Egyptian Materials and Technology*, 430–74. Cambridge, Cambridge University Press.

Shanks, M. and Tilley, C. (1982) Ideology, symbolic power and ritual communication: a reinterpretation of Neolithic mortuary practices. In I. Hodder (ed.) *Symbolic and Structural Archaeology*, 129–54. Cambridge, Cambridge University Press.

Sparey Green, C. J. (1977) The significance of plaster burials for the recognition of Christian cemeteries. In R. Reece (ed.) *Burial in the Roman World*, CBA Research Report 22, 46–53. London, Council for British Archaeology.

Standard English Version (SEV) (2009). *Bible with Apocrypha, Standard English Version*. New York, Oxford University Press.

Statius (nd) *Silvae [The Silvae of Statius]*. Transl. B. R. Nagle [2004]. Bloomington IN, Indiana University Press.

Taylor, A. (1993) A Roman lead coffin with pipeclay figurines from Arrington, Cambridgeshire. *Britannia* 24, 191–225.

Tchapla, A., Méjanelle, P., Bleton, J. and Goursaud, S. (2004) Characterisation of embalming materials of a mummy of the Ptolemaic era. Comparison with balms from mummies of different eras. *Journal of Separation Science* 27, 217–34.

Theophrastus (nd) *Historia Plantarum [Enquiry into plants and minor works on odours and weather signs, Volume 2* (reprint)]. Transl. A. F. Hort [1980]. London, Heinemann.

Thillaud, P. L., Glon, Y., Charlier, P., Vignal, J.-N. (2006) Autopsy of the mummy of 'Fin Renard', Bourges, France. *Journal of Biological Research* 80, 96–8.

Thomas, C. (1999) Laid to rest on pillow of bay leaves. *British Archaeology* 50.

Townley, C., Brettell, R. C., Bowen, R. D., Gallagher, R. T., Martin, W. H. C. (2015). The application of positive mode atmospheric chemical ionisation to distinguish epimeric oleanolic and ursolic acids. *European Journal of Mass Spectrometry* 21, 433–42.

Toynbee, J. M. C. (1971) *Death and Burial in the Roman World*. London, Thames and Hudson.

Van Gennep, A. (1909) *The Rites of Passage*. Transl. M. B. Vizedom and G. L. Caffee [1960]. Chicago IL, University of Chicago Press.

Waugh, H. (1962) The Romano-British burial at Weston Turville. *Records of Buckinghamshire* 17, 107–14.

Wild, J. P. (2013) Luxury? The north-west end of the silk-purple-and-gold horizon. In C. A. Giner, J. O. García and J. J. M. García (eds.) *Luxury and Dress: Political Power and Appearance in the Roman Empire and its Provinces*, 169–80. València, Universitat de València.

Chapter 5

Plant Rituals and Fuel in Roman Cemeteries of Apulia (SE Italy)

Valentina Caracuta and Girolamo Fiorentino

Introduction

During the Roman Republic and the early centuries of the Empire, from about the 5th century BC to the 2nd century AD, cremation was the primary form of burial on the Italian peninsula. The Roman cult of the departed, whether public or private, had a double purpose to preserve the memory of the dead and provide subsidence for the deceased after life (Toynbee 1971). The corpse, together with the wooden stretcher on which it laid, was burnt either at the place where the ashes were to be buried (*bustum*) or at a place specially constructed and reserved for cremations (*ustrinum*). The pyre (*rogus*) was a rectangular pile of wood, sometimes mixed with residues of the funerary meal consumed by those present at the burial, and the food offered to the dead. Vergil (*Aeneid* 6, 215–23) mentions:

> Some heat water, setting cauldrons bubbling on the flames, and wash and anoint the cold body. Loud is the wailing; then, their weeping done, they lay his limbs upon the couch, and over them cast purple robes, the familiar dress. Some shouldered the heavy bier – sad ministry – and in ancestral fashion, with averted eyes, held the torch below. The gifts were piled up in the blaze – frankincense, viands, and bowls of flowing oil. After the ashes fell in and the flame died away, they washed with wine the remnant of thirsty dust, and Corynaeus, gathering the bones, hid them in a bronze urn.

Food played an important role in the ritual and, despite the scarce information in the Roman written sources, it is possible to distinguish three principal phases of its involvement in the funerary process: the libation, the funeral meal and the offering to the dead (André 2001). Ovid (*Fasti* 2, 533–41) refers to the subject:

> And the grave must be honored. Appease your fathers' Spirits, and bring little gifts to the tombs you built. Their shades ask little, piety they prefer to costly offerings: no greedy deities haunt the depths. A tile wreathed round with garlands offered is enough, a scattering of meal, and a few grains of salt, and bread soaked in wine, and loose violets.

Since residues of the meal were often discarded into the pyre, it is difficult to distinguish between what was offered to the dead during the cremation of the body and what could be related to the funerary meal shared by the family members. To date no systematic study exists in the Italian peninsula that has identified specific patterns regarding which plant offerings were related to particular graves, burial types or individuals. Furthermore, when archaeobotanical studies are included in the investigation of cremations, they usually involve the study of seeds and fruits, while neglecting the fuel of the pyre (Castiglioni *et al.* 2003). This chapter examines both the remains of edible plants and the fuel used in the pyre and contributes new primary data to the scant knowledge of the role of plants into funerary rituals in south Italy. Whereas several studies have investigated the Roman burial customs in central Europe and northern Italy (Bouby and Marinval 2004; Castiglioni and Rottoli 2007; Cooremans 2008; Giorgi 2000; Marinval 2001; Šoštaric *et al.* 2006; Zach 2002), the only other analysis available for the south part of the Italian peninsula is that of the necropolis of Porta Nocera in Pompei (Matterne and Derreumaux 2008). The data collected thus from two necropolises, Via Ciotta and Masseria Amendola, in north Apulia offer an unprecedented opportunity to shed light on the funerary use of plants in a core area of the Roman Empire.

Plants and rituals in the Italian peninsula

In recent years plant remains have been found in many Roman cemeteries in Italy, mainly in the north part of the peninsula (Castelletti 1999; Castiglioni 1996; Castiglioni *et al.* 1998; 1999; Castiglioni and Rottoli 2000; 2006; 2007; Cottini and Rottoli 2000; Forlani and Bandini Mozzanti 1984; Marchesini and Marvelli, 2006; 2007; Motella 1990). The investigation of these northern cemeteries showed a predominance of fruit remains among the plant offerings, followed by cereals and pulses. A good portion of these remains are fruits and nuts that could have been grown locally, such as grape (*Vitis vinifera*), hazelnut (*Corylus avellana*), and, from the 1st century BC onwards, also walnut (*Juglans regia*) and peach (*Prunus persica*). Remains of pine nut (*Pinus pinea*) are also plentiful and indicate trade with the coastal areas of the peninsula while those of dates (*Phoenix dactylifera*) are indicative of imports from other areas of the Mediterranean basin (Rottoli and Castiglioni 2011).

The rituals observed suggest some continuity of traditions from the Iron Age, when grapes and hazelnuts have been found associated with burials together with uncultivated plants, such as acorn (*Quercus* spp.), bramble (*Rubus* spp.), dogwood (*Cornus sanguinea*), rose (*Rosa* sp.) and strawberry (*Fragraria vesca*), which are also occasionally attested in Roman cemeteries (Castiglioni *et al.* 1992; Rottoli and Castiglioni 2011). The presence of exotic fruits, such as dates, has been interpreted in central Europe as sign of Romanization by some authors (Bouby and Marinval 2004; Preiss *et al.* 2005), while others attribute the diffusion of dates partly to the spread of oriental rituals and mystic cults (Livarda 2013). In the north Italian necropolises, dates are usually found in high-rank burials, suggesting they are an expression of the wealth of the deceased (Rottoli and Castiglioni 2011). Fewer studies have been carried out on material from cemeteries in the south and, besides the cemetery of Porta Nocera in Pompei (Matterne and Derreumaux 2008), all others are pre-Roman, such as those of Lylibaeum in Sicily (Madella 1999) and Piazzetta Cervi, Epulione, Caserma Roasio and Via Perella in Apulia (Colaianni 2007/2008). The excavation of the necropolis of Porta Nocera, which is dated to the 1st century AD, indicated that the ceremonial offerings on the graves included figs (*Ficus carica*), grapes, apples (*Malus* spp.), dates and chestnuts (*Castanea sativa*), which were deposited as complete fruits on the pyre. Shells of walnuts and hazelnuts and olive stones (*Olea europaea*) were interpreted, instead, as possible residues of the funerary meals. A much wider spectrum of food plants was found inside the graves, including a greater variety of fruits, but also cereals, pulses and wild taxa (Matterne and Derreumaux 2008). The species recovered from the cemetery, whether deposited inside the graves or offered above the grave, were those commonly used in everyday life in Pompei (Ciarallo 2000; Jashemski *et al.* 2002) and, apart from dates, all could have been locally grown (*ibid.*).

In Lylibaeum an incineration necropolis that included tombs dated to between the 3rd and the 1st century BC was discovered (Madella 1999). The study of the plant offerings from four tombs revealed the presence of pine cones and nuts, and shells and nuts of walnut and hazelnut. The analysis of the vegetal material in the necropolis of Lylibaeum included also wood tissue, interpreted as probably burnt as fuel on the pyre. The tree species identified were olive, cork oak (*Quercus suber*), strawberry tree (*Arbutus unedo*) and fan palm, all of which belong to the maquis forest that still grows around the necropolis (*ibid.*). Two out of four tombs were associated to individuals with specific characters. The first, tomb 4, that included the remains of a four-year-old child, contained wood of olive and evergreen oak (*Quercus ilex*) but no offerings of seeds or fruits. The second, tomb 52, belonged to a female individual, who was cremated on a pyre made by olive tree and evergreen oak wood and had plant offerings of pine cones and nuts, and shells and nuts of walnut and hazelnut (*ibid.*). No further information was provided on the possible associations between gender or other social aspect of the deceased and plant offerings and the fuel used for the pyre.

Pre-Roman funeral rituals in Apulia, dated to between the 4th/3rd and the 2nd century BC, are characterised by the practice of inhumation rather than incineration. In the burials discovered in Piazzetta Cervi (Cavallino-LE), Epulione (Lecce) and Caserma Roasio (Lecce) plants had been burnt on fire and offered to the deceased as part of the ritual. The majority of these were cereals, legumes, such as peas, and in one case, figs (Colaianni 2007/2008). Charcoal of local maquis taxa was found associated with the bodies, but it is unclear if they were used as fuel for the fire where the seeds were burnt or if the wood itself was meant as funeral offering (*ibid.*).

The cemeteries of Via Ciotta and Masseria Amendola

Via Ciotta

This necropolis was discovered in 2009 during a salvage excavation conducted at the village of Ascoli Satriano (Fig. 5.1) (Corrente 2011). The site is located at *c.* 48km from the coast, at the foothill of the Sub-Apennine, a system of hills that ranges from 400 to 700m.a.s.l. and has annual rainfall 700–900mm per year. While the largest part of the area around Ascoli Satriano is nowadays devoted to cereal cultivation and olive orchards, remains of the local woodland survive in patches, the most common plants of which are deciduous and semi-deciduous oak (*Quercus cerris, Q. robur*), hazelnut and maple (*Acer campestre*) (Fenaroli 1966; Campanile and Cocca 2005).

The excavation of the necropolis brought to light three *ustrina* coded 2, 3 and 4, which included grave goods representative of the gender of the dead. The presence of three rings and three mirrors in *ustrinum* 2 and a mirror and a miniature ointment jar in *ustrinum* 3 suggest that two female individuals were cremated in the urns. The *strigil*, a metal tool used by athletes to scrape off the dirt after the training, was found in *ustrinum* 4, and consequently, the individual buried there was identified as a male (De Venuto *et al.* 2012a). In the proximity of the *ustrina*, another burial was found. Remains of a cremated body were discovered in a large pit within a rectangular enclosure. The traces of thermal alteration on the clay of the walls of the pit were interpreted as evidence of *in situ* combustion of the body (*bustum*). Along with the ashes and the calcined bones, the burial contained goods of female connotation, which allowed dating the funeral event to the 1st century AD. These funerary goods were thirty-three miniature ointment jars, a set for make-up, which included a bronze spatula and powder compact, a metal hairpin and fragments of oil lamps. The discovery of some iron nails, especially along the edges of the walls of the pit, was considered proof that the body was placed in a wooden coffin (*ibid.*).

Masseria Amendola

This Roman cemetery was discovered in 2011 few kilometres east of the necropolis of Via Ciotta (Fig. 5.1). Despite the proximity of the two sites, Masseria Amedola is located at lower altitude (about 250m.a.s.l.) at a heavily cultivated part of the Tavoliere plain (De Venuto *et al.* 2012b). The proximity of the site to the coast affects the rate of rainfall that is below 500mm per year. In turn, the rainfall influences the natural vegetation of the area, which is composed mostly of thermo-Mediterranean species, such as olive, mastic-tree (*Pistacia lentiscus*) and juniper (*Juniperus oxycedrus*) (Macchia *et al.* 2000).

The necropolis of Masseria Amendola has 41 burials, including both cremations and inhumations, although the former outnumber the latter by far. Most of the burials had been damaged by ploughing, and therefore, plant remains could be collected only from ten of the original burials. Nine of those burials were *ustrina* containing, besides the urns, a few grave goods (*ustrina* 2, 13, 20–3, 26 and 36). One of these, *ustrinum* 13, contained the remains of a female individual. The tenth burial was a grave within a rectangular enclosure (*bustum* 37), where the remains of a male individual were laid on a large stock of glass and metal objects. On the basis of the material associated with the graves, two different phases of occupation were identified: one dated to the first half of the 1st century AD (*ustrina* 13, 20, 21, 26 and 36) and a later one, dated to between the second half of the 1st century and the 2nd century AD (*ustrina* 2, 22, 23, 25 and *bustum* 37).

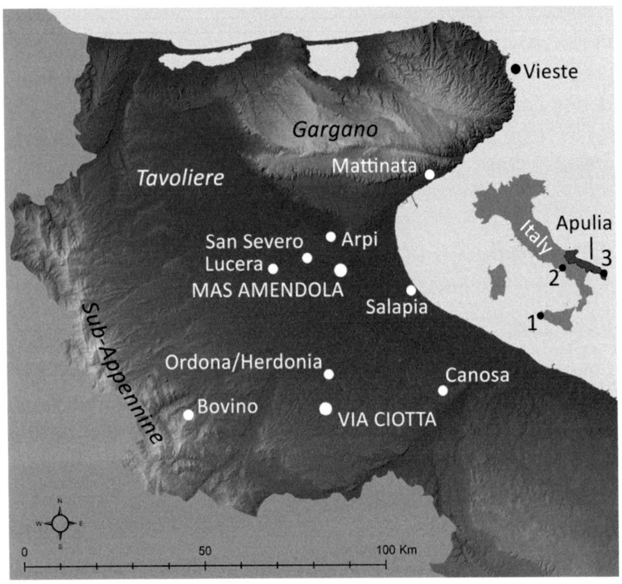

Figure 5.1 a) The location of the necropolises mentioned in the text in Italy: 1. Lylibaem; 2. Porta Nocera; 3. Piazzetta Cervi, Epulione and Caserma Roasio. b) Position of Via Ciotta and Masseria Amendola in north Apulia and other key archaeological sites mentioned in the text. The three major morphological units of north Apulia, the Gargano Promontory, the Tavoliere Plain and the Sub-Appennine hills, are included to define the geographical setting of the area.

Materials and methods

A total of 38 and 51Lt of sediment were collected from the necropolis of Via Ciotta and Masseria Amendola respectively and were wet-sieved, using sieves with apertures 5mm and 1mm. Due to the sieving procedure it is possible that seeds smaller than 1mm were lost and this can account, for instance, for the absence of smaller seeds and other plant parts (see results below). The sediment obtained was left to dry and was subsequently submitted to the archaeobotanists for analysis at the Laboratory of Archaeobotany and Paleoecology of the University of Salento. Plant remains were sorted and identified using a stereomicroscope, while the microscopic features of the anthracological remains were investigated by means of a metallographic microscope. Reference plant material and atlases were used to refine the identification of each specimen (Bojňanský and Fargašová 2007; Schweingruber 2007). The minimum number of individuals in regards to nuts was assessed by dividing the total number of medium size nuts, such as hazelnut, by three, and those of large size nuts, such as walnut, by seven, following observations recorded during experimental cracking of nuts.

Results

The analyses of the plant material of Via Ciotta led to the identification of 1267 archaeobotanical remains that include shell fragments of walnut and hazelnut, pine cones

Table 5.1 Results of the plant macroremains' analysis in the necropolis of Via Ciotta, 1st century AD. MNI = Minimum Number of Individuals.

Date	*Context*	*Gender*	*Archaeobotanical remains*				*Charcoal remains*									
					Pinus pinea (pine)											
			Corylus avellana (hazelnut fragments)	*Juglans regia (walnut fragments)*	*cone*	*scale*	*Abies sp. (fir)*	*Acer sp. (maple)*	*Buxus sempervirens (boxwood)*	*Corylus avellana (hazel)*	*Erica arborea (heather)*	*Olea europaea (olive)*	*Ostrya carpinifolia (hornbeam)*	*Quercus pubescens (deciduous oak)*	*Volume (Lt)*
1st century AD	*bustum* 1	female	0	179	0	0	1	0	6	0	0	1	0	33	19.5
	ustrinum 3	female	474	0	0	0	0	0	10	0	2	0	0	45	7
	ustrinum 2	female	567	0	0	0	0	1	5	1	12	1	3	125	5.5
	ustrinum 4	male	0	0	44	3	0	0	0	0	0	110	0	0	6
Total			1041 (350 MNI)	179 (40 MNI)	44	3	1	1	21	1	14	112	3	203	38

Table 5.2. Results of the archaebotanical analysis of the necropolis of Masseria Amendola, 1st–2nd century AD.

Chronology	*Context*	*Gender*	*Archaeobotanical remains*		*Charcoal remains*								
			Triticum aestivum/ compactum (free-threshing wheat)	*Olea europaea (olive)*	*Juglans regia (walnut)*	*Pistacia lentiscus (mastic-tree)*	*Prunus persica/dulcis (peach/almond)*	*Prunus avium (cherry-tree)*	*Pyrus sp. (pear)*	*Olea europaea (olive)*	*Juniperus oxycedrus (juniper)*	*Unidentified conifer wood*	*Volume (Lt)*
1st century AD	*ustrinum* 13	Female	0	0	0	0	0	0	3	143	0	1	5
	ustrinum 20	Unknown	0	0	0	0	16	0	0	0	0	0	3
	ustrinum 21	Unknown	0	0	4	0	0	0	10	34	0	0	4.5
	ustrinum 26	Unknown	0	1	0	0	0	0	0	127	0	0	3
	ustrinum 36	Unknown	0	0	0	0	0	0	0	316	0	0	6
1st–2nd century AD	*ustrinum* 2	Unknown	0	0	0	0	1	0	0	11	8	0	3.5
	ustrinum 22	Unknown	0	0	0	0	0	0	0	101	10	0	5.5
	ustrinum 23	Unknown	0	0	0	6	0	0	2	4	0	0	3
	ustrinum 25	Unknown	1	0	0	0	6	12	82	35	0	0	3.5
	bustum 37	Male	0	0	0	0	137	0	0	0	0	0	12
Total			1	1	4	6	161	12	97	776	18	1	55.5

and scale fragments, and 356 specimens of woody tissue that belong to eight taxa. The most common wood taxa in descending order are deciduous oak (*Quercus pubescens*), olive, boxwood (*Buxus sempervirens*), heather (*Erica arborea*), hornbeam (*Ostrya carpinifolia*), maple (*Acer* sp.), hazel and fir (*Abies* sp.) (Table 5.1; Caracuta and Fiorentino 2011).

The necropolis of Masseria Amendola yielded 1069 plant remains and these were mostly fragments of wood that could have been used to ignite the funeral pyre or to build wooden cases for the deceased. Only two archaeobotanical remains (other than charcoal) were found: one grain of free-threshing wheat (*Triticum aestivo-compactum/durum*) and one olive stone. Among the wood, olive was the most abundant species, followed by almond/peach (*Prunus dulcis/persica*) and pear-like tree (*Pyrus* sp.). Few fragments of juniper, mastic-tree, cherry tree (*Prunus avium*) and walnut were also found (Table 5.2).

The use of plants in funerary rituals between the 1st and the 2nd century AD in Apulia

Analysis of the plant remains in these funerary contexts indicated variations between the different graves. Chronological differences, the composition of the local vegetation, religious beliefs and gender-orientated choices might have influenced the selection of the plants offered or used in the funerary pyre and these are explored below. Remains of edible plants were found almost exclusively in Via Ciotta, although we cannot exclude that remains with sizes less than 1mm were lost during the sieving process, leaving thus open the possibility that small seeds or other plant parts could have been present in both necropolises.

Via Ciotta

In the necropolis of Via Ciotta the distribution of archaeobotanical material in the burials seems to follow a pattern that could be related partly to the gender of the individual. The remains recovered from the male burial (*ustrinum* 4) differ markedly from those found in the female burials, which, in turn, are characterised by different plants depending on whether the individual was buried in a funerary enclosure or in simple pits (*bustum* 1 and *ustrina* 2–3 respectively).

Broken pieces of pine cones and scales were found associated only to the male individual. Offerings of pines were very common in Roman cemeteries both in the north and the south part of the peninsula (Castiglioni and Rottoli 2006; 2007; Forlani and Bandini Mozzanti 1984; Madella 1999; Marchesini and Marvelli 2007; Motella 1990). The use of pine in funerary rituals is observed also in Roman necropolises elsewhere in the Empire, such as in England (*e.g.* Giorgi 2000), France (*e.g.* Marinval 1993) and Algeria (*e.g.* Lancel 1970). Its use in rituals, however, seems to extend far beyond the funeral, since numerous remains were found in contexts linked to the cult of oriental deities, such as Mithra (Kislev 1988) and Isis (Lignereux *et al.* 1997; Megaloudi 2005; Zach 2002). In Rome, the use of pine seems to increase with the rise of the Phrygian cult of Attis, son (or lover) of Cybele at least since the 1st century BC, mentioned, for instance, by Ovid (*Metamorphoses* X, 103–5): 'the shaggy-topped pine tree, armed with needles, sacred to Cybele, mother of the gods, since Attis exchanged his human form for you, and hardened in your trunk'. Attis appears to have been a god of vegetation, and his death and resurrection were annually mourned and rejoiced over at a festival in spring (Brosse 1989; Frazer 1922). On the 22nd day of March a pine-tree was cut in the woods and brought into the sanctuary of Cybele, where it was treated as a great divinity. The duty of carrying the sacred tree was entrusted to a guild of tree-bearers. The trunk was swathed like a corpse with woollen bands and carried in procession (see also Koch *et al.* this volume). In the night between the 24th and 25th, dirges accompanied the mourning for the death of Attis, while the next day, the day of the equinox, they sang songs of joy to celebrate the resurrection of the son (or lover) of Cybele, which was an allegory for the rebirth of vegetation (Brosse 1989; Frazer 1922). Julian the Apostate (*Oration upon the Mother of the Gods*, 266–7) wrote:

> This great god of ours is Attis; this is the meaning of the 'Flight of King Attis' that we have just been lamenting; his 'Concealments,' his 'Vanishings,' his 'Descents into the Cave.' Let my evidence be the time of year when all these ceremonies take place; for it is said that the Sacred Tree – Pine – is cut down at the moment when the Sun arrives at the extreme point of the equinoctial arc: next in order follows the sounding of the trumpets, and lastly is cut down the sacred and ineffable harvest of the god Gallos: after these come, as they say, the Hilaria and festivities.

Several other studies show that the offering of stone pines in the funeral ritual was a rather common practice. During rituals, pine cones could have been burnt on temple altars, for instance, for their aromatic properties (Megaloudi 2005). The analysis of domestic contexts in Pompeii indicates that pine was also part of everyday activities (Jashemski *et al.* 2002). It is worth noting here that pine was part of the funerary ritual since at least the 3rd–2nd century BC, as shown by the finding of pine stones in two burials, a female and a non-gender-specific individual, in the necropolis of Lylibaeum (Madella 1999). The findings of pine in the male cremation at Via Ciotta suggests that the species had some symbolic meaning and one possibility, although without any conclusive evidence, is that the use of pine might have been related to the cult of Attis and/or the idea of rebirth.

On the other hand, the nutshells recovered from the female burials could be related to symbolism associated with the universe of marriage. Roman authors, such as Pliny (*NH* 1–7, 15, 86), attributed sacral importance to walnut in weddings, comparing the protectiveness of the nutshell to the strength of marriages. Pliny (*ibid.*) also mentions that nuts were thrown during the wedding procession when the guests recited the licentious verses known as *Fescennini*:

> The walnut has won from the service-berry in point of size the place that it has yielded to it in popularity, although the walnut also accompanies the Fescennine songs sung at weddings. The walnut has a distinction of structure that is peculiar to it, in that it is protected by a double covering, consisting first of a cushion-shaped cup and then of a woody shell. This is the reason why walnuts have become emblems consecrated to

weddings [possibly as a fertility charm; these were thrown by the bridegroom to the boys carrying the torches], because their progeny is protected in so many ways—a more likely explanation of the custom than that it is due to the rattling rebound which it makes when it falls on the floor.

Remains of walnut (shells and nuts) have been found in several cemeteries in northern Italy (Castiglioni and Rottoli 2003; Castigloni *et al.* 2007; Marchesini and Mavelli 2007; Rottoli and Castiglioni 2011). In the south, the discovery of this taxon is reported at the necropolis of Lylibaeum (Madella 1999) and the Pompeian Porta Nocera (Matterne and Derreumaux 2008). In the former case due to the nature of the evidence (cremations) determination of the gender was possible only in a few instances but it is noticeable that the only female burial identified included walnut remains. In the latter case, walnut remains were recovered from graves enclosed in funerary walls of the same type of *bustum* 1 as in Via Ciotta but no information on the gender of the deceased is available.

Hazelnut, the other food-plant remains found in female *ustrina* of Via Ciotta (Fig. 5.2), must have also had symbolic value. The fruit of hazel is, for instance, symbolically tied to the virgin goddess Artemis/Diana and a sacred grove of hazelnut trees was dedicated to Artemis at Karyai in Lakonia. According to Servius (*Commentary on the Eclogues of Virgil* 8, 29–30) Karya was a Lakonian maiden loved by the god Dionysus. When her two sisters tried to prevent the liaison, the pair was driven mad and went to Mount Taygetos. Karya meanwhile died and was transformed into a hazelnut tree. The goddess Artemis informed Karya's father, Dion, of the affair and commanded that he found a sanctuary in honour of Artemis Karyatis. Remains of hazelnuts have been found in many Roman cemeteries in Emilia-Romagna (Marchesini and Marvelli 2007), Lombardia (Castiglioni *et al.* 2007) and in Pompeii (Matterne and Derreumaux 2008). Unfortunately, these studies do not provide information on the gender of the burials and the only comparison available comes from the single female burial of the pre-Roman necropolis of Lylibaeum as mentioned above (Madella 1999). No hazelnuts have been found so far in any pre-Roman necropolis of Apulia (Colaianni 2007/2008) and their ritual meaning is far from fully understood. Pollen of hazelnut was found in the Iron Age site of Arpi and the Roman site of Herdonia, near Ascoli Satriano, suggesting that this species grew in the area already in the 4th century BC (Antonacci Sanpaolo 1995; Heim 1995). Its use as food is attested in Pompeii, where hazelnut remains were found in the garden of the House of the Ship (Jashemski 1979, 241), while fragments of its wood were found in the House of the Vestals (Veal 2014). Evidence for the consumption of this nut also exists in the *villa rustica* at Oplontis-Campania, where imprints of hazelnuts that had been stored were found perfectly preserved. Three hazelnuts can also be identified on small fragments of a wall painting in the Museum at Castellamare di Stabia-Campania (Jashemski *et al.* 2002).

Another aspect that distinguishes the male cremation at Via Ciotta is the selection of wood taxa as fuel for the funerary pyre. The olive tree is the only arboreal taxon found in *ustrinum* 4 and the selection of a single species to fuel the pyre reflects a different pathway compared to those of the other incinerations of the necropolis. The wood used in the funerary pyres of the three female burials included a mixture of taxa (Table 5.1), most of which were growing in the proximity of the necropolis (Caracuta and Fiorentino 2009; Caracuta 2011). Deciduous oak, heather, maple and hornbeam were definitely part of the local meso-Mediterranean vegetation (Antonacci San Paolo 1995; Heim 1995), and therefore, could have been easily collected as fuel for the pyre or to build the wooden coffin. In contrast, boxwood was not locally available. The branches recovered of this species were very small (less than 1cm of diameter), which suggests its possible use to ignite the pyre. In addition, in several myths related to Athena and Apollon boxwood was often mentioned as material for musical instruments. According to Ovid (*Fasti* 6, 69) 'I [Athene] first enabled the long flute to produce notes through spaced holes in perforated boxwood'; Statius (*Silvae* 5, 3, 87) writes 'He [Marsyas] who dared make music against Phoebus [Apollon], while Pallas [Athene] rejoiced that the boxwood-pipe deceived him'; and Philostratus the Elder (*Imagines* 1, 10) mentions 'All the wood required for the lyre is of boxwood'. If a functional or sensorial value is excluded, then it is possible that burning boxwood had a symbolic meaning, which on the basis of this evidence still remains elusive (see also Bilimoff 2003; 2006).

Overall, the analysis of plant remains from the necropolis of Via Ciotta provided new information on the funerary ritual of a Roman community in north Apulia in the early centuries of the Imperial Age (1st–2nd century AD). The presence of non-edible parts of archaeobotanical material, such as various nutshells, suggests that those were remains of the funeral meals, discarded in the pyre after the participants of the ritual had consumed the edible part. The selection of the plants for the funerary meal could be correlated to the gender of the dead for which the ritual was performed but also to other aspects as the differences in the type of plants for instance according to burial type indicates. The study of the necropolis of Via Ciotta offers a glimpse into the universe of possible meanings that plants assumed in the funeral but it should be kept in mind that any pattern observed may be specific to that particular group of people and does not necessary reflect beliefs shared by other groups. It is only by building up the corpus of funerary archaeobotanical data from across the peninsula that a pattern can emerge.

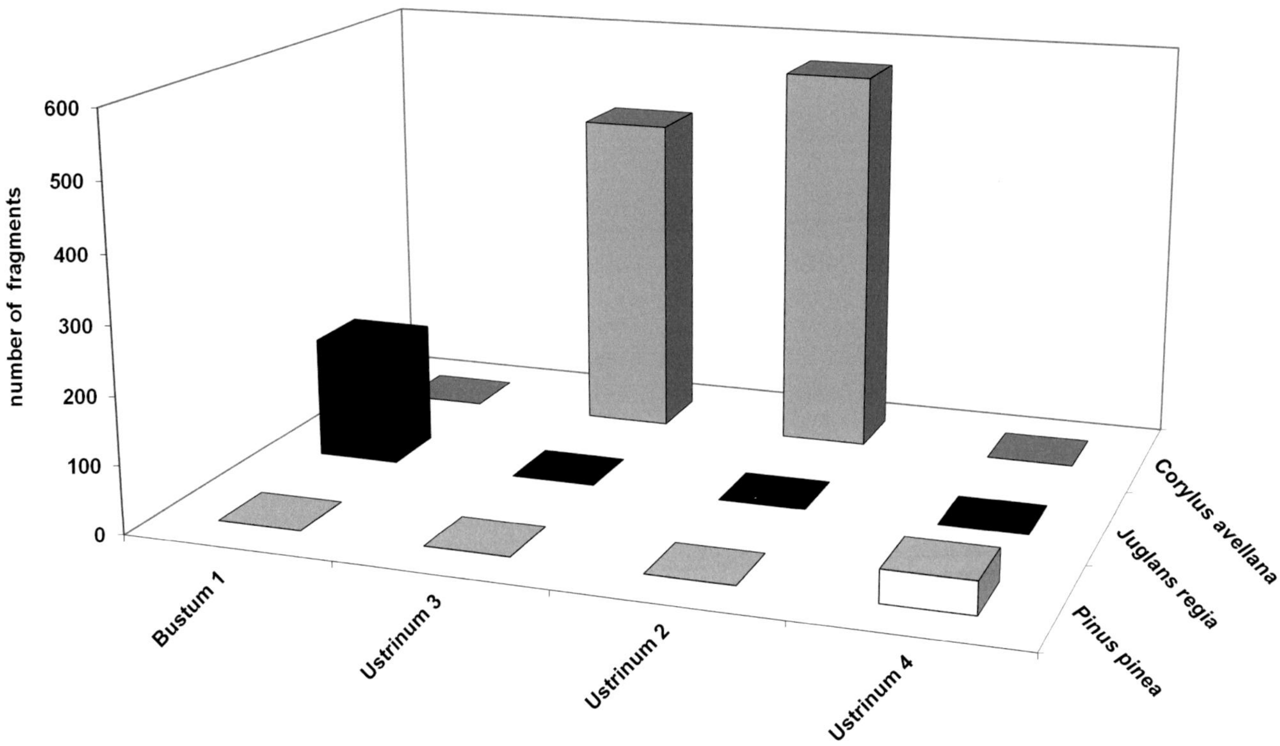

Figure 5.2 Histogram of the archaeobotanical remains found in the burials of the necropolis of Via Ciotta.

Masseria Amendola

Roman funerary rituals were in fact far from uniform even within one area. The evidence from the necropolis of Masseria Amendola highlights this variation in north Apulia. At Masseria Amendola almost no archaeobotanical remains were found. A grain of free-threshing wheat was recovered from *ustrinum* 25, but this may have been residual or intrusive, while the single other plant remains, an olive stone, could have accidentally ended up in the burial as part of the branch of an olive tree, the presence of which is attested by the study of charcoal. Smaller plant macroremains might have been present but if they did they were not retrieved, due to the sieving method.

Almost all plant remains found in these burials were chunks of charred wood, with minimum length 5cm. Most of the wood came from fruit trees, including olive, almond/peach, pear, cherry and walnut. This wood was used in all likelihood to fuel the pyre, but it is also possible that wooden coffins/stretchers or other wooden objects were burnt during the cremation of the body, and therefore, the origin of the wood is difficult to securely identify. When, however, only a single charcoal taxon is present in a grave, it is then safe to assume that this type of wood was also used to fuel the pyre. At Masseria Amendola four out of the ten burials studied included just one type of wood: olive in *ustrina* 26 and 36, and peach/almond in *ustrinum* 20 and *bustum* 37. *Ustrinum* 25 had the highest number of wood taxa, but also the only possible food remains found at the necropolis. Although no conclusions can be drawn on the presence of a single free-threshing wheat grain in a grave of this necropolis, the possibility that wheat was part of the offerings and/or the funerary meal cannot be excluded. Castiglioni and Rottoli (2011) reported that free-threshing wheat was one of the most common offerings in Roman cemeteries between the 1st century BC and the 3rd century AD, while large quantities of free-threshing wheat were found in the pre-Roman graves of Apulia, suggesting that offering of cereals was part of the funeral traditions, at least, before the Romans settled in Apulia (Colaianni 2007/2008).

The type of wood recovered from the burials at Masseria Amendola was mainly olive and it is possible that this may relate to its availability in the area. During the early centuries of the *imperium*, southern Italy underwent a series of administrative transformations, which led to the creation of Regio II (*Apulia et Calabria*) and the foundation of new colonies in Apulia and adjacent regions (Pliny, NH 3, 11, 103). The land was divided in small and medium size estates and was given to the new settlers to be cultivated (Sirago and Volpe 1993). Archaeological discoveries of olive presses and storage systems for olive oil in the Roman farms of Luceria, Ordona, Salapia, Canosa, Mattinata, Bovino, Vieste and San Severo and the epigraph of a *collegium* of olive oil refiners in the area of Ascoli Satrinao dated to the 1st and

2nd century AD attest to intensive activities related to the production of olive oil in north Apulia (Brun 2003; Gasperini 1971; Volpe 1990). The archaeobotanical data from Masseria Amendola confirm that olive was available around the site and provide an invaluable source of information about the landscape around the necropolis, considering also that archaeobotanical studies in north Apulia are only limited to later periods (Caracuta and Fiorentino 2011; 2012).

Most other wood taxa recovered in this necropolis seem to differ according to the chronology of the graves. For instance, walnut was found only in one of the earliest burials (*ustrinum* 21), while cherry-tree, juniper and mastic-tree wood were recovered only from the burials of the later period (*ustrina* 2, 22, 25). Whereas the latter three species could have grown in the Tavoliere plain, walnut must have come from higher altitudes, such as the hills of the Sub-Apennine. In this case it is hard to infer any ritual pattern based only on the wood remains, especially since most of the burials have no gender connotation and the grave goods mark the status of just one individual (*bustum* 37), who was buried with a sword (De Venuto 2012b). The only certain female burial is *ustrinum* 13, where olive was mixed with pear wood, and the only certain male individual was buried in a grave within a rectangular enclosure (*bustum* 37) over a layer of charred chunks of peach/almond wood. Such associations can only be used as a base upon which future studies can build, but at this stage, they cannot allow any explicit link between gender, burial type and wood use.

Conclusion

The practice of cremation deters the preservation of human bones, making the identification of the sex and age of the deceased difficult. One way to infer information about the gender is by looking at the funerary goods associated with the cremation, but even so, the information about the gender is rarely reported in the publication. To our knowledge, the only necropolises with incinerations for which some information about the gender are available, are that of Lylibaeum and Via Ciotta. In other necropolises of Apulia, such as those of Piazzetta Cervi, Epulione and Caserma Roasio, the body was not cremated and the information about the gender was obtained directly from the osteological analysis. Both the necropolis of Lylibaeum and the Apulian cemeteries, however, are dated to before the Roman occupation, and therefore, their funeral rituals might reflect traditions no longer followed during the Roman period. In this regard, the new data collected from the cremation cemeteries of Via Ciotta and Masseria Amendola added original information on the funerary rituals in a core area of the Roman Empire.

The necropolises of Via Ciotta and Masseria Amendola, although contemporaneous, exhibited differences in the use of plants. Gender-related grave goods, so common in the first necropolis, were scant in the latter, where only two burials included commodities related to a specific gender. The presence of remains of edible plants in the burials of Via Ciotta also contrasts with the scarcity of those in the necropolis of Masseria Amendola. The archaeobotanical remains recovered in Via Ciotta were in the form of the non-edible parts of the plant, and therefore, it is safe to assume that they were residues of the funeral meals rather than offerings of food to the dead. When the macrobotanical assemblages recovered from Via Ciotta and Masseria Amendola are compared to those of other Roman cemeteries of the north and south part of the Peninsula, it is clear that these included fewer fruit species, and that imported plants, such as dates, were absent from the funeral ritual. The rituals observed in Via Ciotta and Masseria Amendola also differ from those observed in the pre-Roman cemeteries of Apulia in regards to the way of corpse disposal (cremation *versus* inhumation), to the selection of the plants and their use. Whereas unconsumed cereals and legumes characterised the plant offerings in the pre-Roman phases studied, nutshells, pine cones and scales were thrown in the pyre as residues of funerary meals and possibly for other sensory purposes in the graves of Via Ciotta.

The selection of wood for the pyre is another element that distinguishes the necropolis of Via Ciotta to that of Masseria Amendola. Whereas species of the meso-Mediterranean forest dominated at Via Ciotta, thermo-Mediterranean species were mostly found at Masseria Amendola. In both cases, woods were selected among the local natural vegetation, but some species, such as olive and plums, could have also come from cultivated fields.

One thing to bear in mind is that any pattern expressed in the rituals of these two necropolises could have been specific of these particular groups of people and it does not necessary reflect shared beliefs. Wherever we look at graves with religious, economic or social questions in mind, the analysis of burials is ultimately the analysis of symbolic actions. In rituals people use symbols to make explicit the social structure and provide interpretations of the meaning of daily life (Morris 1992), as understood within specific groups of people. The study of plants thus holds great potential towards unravelling past societies.

Acknowledgements

The present research was possible thanks to Marisa Corrente, inspector of the Apulian Antiquity Authority (Soprintendenza per i Beni Culturali della Puglia). Many thanks go to the archaeologists of the 'Archeologica' SrL who excavated the necropolises and studied the material: Giovanni De Venuto, Antonella Napoletano, Andrea Fratta, Cristiano Moscaritolo, Alessandra De Stefano. The archaeobotanical and anthracological analysis was carried out by Valentina Caracuta during her PhD research, under the direction of Prof. G. Volpe, and was funded by the Department of Human Science of the University of Foggia, Italy.

References

André, J. (2001) *Les offrandes alimentaires dans le culte des morts à Rome*. In P. Marinval (ed.) *Historie d'Hommes, Histoires des Plantes*, 215–21 Toulouse, Mergoil.

Antonacci San Paolo, E. (1995) Ricerche archeoambientali nella Daunia antica. Paesaggio vegetale e allevamento tra documentazione archeologico-letteraria ed analisi dei reperti naturalistici. In L. Quilici and S. Quilici Gigli (eds.) *Atlante Tematico di Topografia Supplemento*, 73–102. Rome, Bretschneider.

Bilimoff, M. (2003) *Enquête sur le Plantes Magiques*. Rennes, Ouest-France.

Bilimoff, M. (2006) *Les Plantes les Hommes les Dieux*. Rennes, Ouest-France.

Bojňanský, V., Fargašová, A. (2007) *Atlas of Seeds and Fruits of Central and East-European Flora*. Dordrecht, Springer.

Bouby, L. and Marinval, P. (2004) Fruits and seeds from Roman cremations in Limagne (Massif Central) and spatial variability of plant offerings in France. *Journal of Archaeological Science* 31, 77–86.

Brosse, J. (1989) *Mythologie des Arbres*. Paris, Plon.

Brun, J. P. (2003) *Le vin et l'Huile dans la Méditerranée Antique: Viticulture, Oléiculture et Procédés de Fabrication*. Paris, Errance.

Campanile, G. and Cocca, C. (2005) I boschi della Puglia: caratteristiche e problematiche. *Forest@* 2(2), 172–7.

Caracuta, V. (2011) Ambiente naturale e strategie agroalimentari in Puglia settentrionale tra Tardoantico ed Altomedioevo: indagini archeoambientali e analisi del δ13C sui resti vegetali di Faragola (FG). *Post-Classical Archaeology* 1, 275–95.

Caracuta, V. and Fiorentino, G. (2009) L'analisi archeobotanica nell'insediamento di Faragola (FG): il paesaggio vegetale tra spinte antropiche e caratteristiche ambientali, tra tardoantico e altomedioevo. *Atti del V Congresso Nazionale di Archeologia Medievale* (V SAMI), 717–26. Florence, All'insegna del Giglio.

Caracuta, V. and Fiorentino, G. (2011) Riti funerari e paesaggio vegetale nel territorio di Ausculum: il contributo dell'analisi archeobotanica per lo studio della necropoli di via Giuseppe Ciotta. In M. Corrente (ed.) *Lo Spreco Necessario. Il Lusso nelle Tombe di Ascoli Satriano*, 160–4. Foggia, Claudio Grenzi Editore.

Caracuta, V. and Fiorentino, G. (2012) Ambiente e strategie produttive nei siti di San Lorenzo in Carminiano e Pantano (FG) tra XIII e XIV secolo. In P. Favia, H. Houben and K. Tomaspooeg (eds.) *Federico II e i Cavalieri Teutonici in Capitanata. Recenti Ricerche Storiche e Archeologiche*, 317–32. Galatina, Congedo Editore.

Castelletti, L. (1985) *Resti vegetali macroscopici e resti di cibo nella necropoli romana di Angera* (Varese). In G. Sena Chiesa (ed.) *Angera Romana. Scavi della Necropoli*, 591–5. Rome, L'Erma di Bretschneider.

Castelletti, L. (1999) Storie naturali ed umane della città dei morti. In G. Frigerio (ed.) *Storia di Mariano Comense, dalla Preistoria all'Altomedioevo* 1, 9–18. Mariano Comense (CO), Comune di Mariano Comense.

Castiglioni, E, Motella, S and Rottoli, M. (1992) Il combustibile nelle cremazioni dell'Italia Nord-orientale. In J. L. Vernet (ed.) *Les Charbons de Bois, les Anciens Écosystèmes et le Rôle de l'Homme. Bulletin de la Société botanique de France* 139, 297–309.

Castiglioni, E. (1996) Le analisi i macroresti vegetali. In M. A. Binaghi Leva (ed.) *Antichi silenzi. La Necropoli Romana di San Lorenzo in Parabiago. Comune di Legnano*, 76–80. Legnano (MI), Comune di Legnano.

Castiglioni, E. and Rottoli, M. (2006) Analisi archeobotaniche. In D. Caporusso (ed.) *Extra Moenia 2. Gli scavi di Via Benzi. I Reperti. Rivista Archeologica dell'Antica Provincia e Diocesi di Como* fascicolo 187, 371–83.

Castiglioni, E. and Rottoli, M. (2006). Analisi archeobotaniche. In D. Caporusso (ed.) Extra Moenia 2. Gli Scavi di Via Benzi. I Reperti. *Rivista Archeologica dell'Antica Provincia e Diocesi di Como* 187, 371–83.

Castiglioni, E. and Rottoli, M. (2007) Resti botanici dagli strati tardoantichi e altomedievali. In E. Roffia (ed.) *Dalla Villa Romana all'Abitato Altomedievale. Scavi Archeologici in Località Faustinella-S. Cipriano a Desenzano*, 79–85. Milano, Et.

Castiglioni, E., Cottini M. and Rottoli, M. (2003). Resti botanici. In M. Fortunati, L. Pagani and R. Poggiani Keller (eds.) *Verdello, dalle Origini all'Alto Medioevo. Ricerche archeologiche e storiche*, 225–8. Verdello (BG), Comune di Verdello

Castiglioni, E., Cottini, M. and Rottoli, M. (1998) I macroresti vegetali. In M. Fortunati Zuccala (ed.) *Verdello (BG) Località Colabiolo, via Garibaldi. Necropoli di Età Romana*, 74–5. Milan, Soprintendenza Archeologica della Lombardia.

Castiglioni, E., Cottini, M. and Rottoli, M. (2007) Ambiente e alimentazione dal I millennio a. C. al medioevo. In M. Fortunati and R. Poggiani Keller (ed.) *I Primi Millenni. Dalla Preistoria al Medioevo. Storia Economica e Sociale di Bergamo*, 277–85. Cenate Sotto (BG), Fondazione per la storia economica e sociale di Bergamo, Istituto di Studi e Ricerche.

Castiglioni, E., Cottini, M. and Rottoli, M. (1999) Mariano, via T. Grossi: analisi archeobotaniche. In G. Frigerio (ed.) *Storia di Mariano Comense, dalla Preistoria all'Altomedioevo* 1, 107– 12. Mariano Comense (CO), Comune di Mariano Comense.

Ciarallo, A. M. (2000) *Gardens of Pompeii*. Rome, L'Erma di Bretschneider.

Colaianni, G. (2007/2008). Analisi archeobotaniche nel Salento tra il IX sec. a.C. e IV sec. d.C. Ricostruzione funzionale e implicazioni vegetali. Unpublished thesis, University of Salento-Lecce.

Cooremans, B. (2008) The Roman cemeteries of Tienen Tongeren: results from the Archaeobotanical analysis of the cremation graves. *Vegetation History and Archaeobotany* 17, 3–13.

Corrente, M. (2011) Lo spreco necessario: il lusso nelle tombe di Ascoli Satriano. Foggia, Claudio Grenzi Editore.

Cottini, M. and Rottoli, M. (2000) L'archeobiologia. Le analisi archeobotaniche. In M. A. Binaghi Leva, P. Cattaneo and R. Volonte (eds.) *Duemila Anni fa a Inveruno. La Necropoli Romana Ritrovata*, 113–4. Inveruno (MI), Villa Tanzi Mira.

De Venuto, G., De Stefano, A., Maruotti, M. (2012a). Nuovi rituali nella Ascoli romana. La necropoli di via Giuseppe Ciotta. In

M. Corrente (ed.) *Lo spreco necessario. Il lusso nelle tombe di Ascoli Satriano*, 155–9. Foggia, Claudio Grenzi Editore.

De Venuto, G., Napoletano, A., Fratta, A. and Moscaritolo, C. (2012b) La necropoli di prima età romano-imperiale (I-II secolo) in località Piano D'Amendola, Ascoli Satriano (FG). Relazione sulle attività di scavo, analisi archeoantropologica, reperti mobili (Unpublished report).

Fenaroli, L. (1966) Il Gargano nei suoi aspetti vegetazionali e floristici. *Annali Accademia Scienze Forestali Italiana* 15, 107–35.

Forlani, L. and Bandini Mozzanti, M. (1984) Indagini paletnobotaniche. In M. Bandini Mazzanti (ed.) *Voghenza, una Necropoli di Età Romana nel Territorio Ferrarese*, 315–9. Ferrara, Belriguardo.

Frazer, J. G. (1922) *The Golden Bough*. New York, Macmillan.

Gasperini, L. (1971) Il municipio tarentino. Ricerche epigrafiche. In G. Barbieri, P. Cavuoto and G. Garofalo Zappa (eds.) *Terza Miscellanea Greca e Romana*, 143–209. Rome, Istituto Italiano Per la Storia Antica.

Giorgi, J. (2000) The plant remains: a summary. In A. Mackinder (ed.) *A Romano-British Cemetery on Watling Street. Excavation at 165 Great Dover Street, Southwark, London*. London, MoLAS Archaeology Studies Series 4.

Heim, J. (1995) *Il paesaggio vegetativo*. In J. Mertens (ed.) *Herdonia, Scoperta di una Città*, 321–4. Bari, Edipuglia.

Jashemski, W. (1979) *The Gardens of Pompeii, Herculaneum and the Villas Destroyed by Vesuvius*. New York, Caratzas Brothers.

Jashemski, W., Meyer F. G. and Ricciardi M. (2002) Plants: evidence from wall paintings, mosaics, sculpture, plant remains, graffiti, inscriptions, and ancient authors. In W. Jashemski and F. G. Meyer (eds.) *The Natural History of Pompeii*, 80–180. Cambridge, Cambridge University Press.

Julian the Apostate, 'Julian the Emperor'. *Oration upon the Mother of the Gods*. Transl. William Heinemann [1913–23]. Cambridge MA, Harvard University Press.

Kislev, M. E. (1988) *Pinus pinea* in agriculture, culture and cult. In H. J. Küster (ed.) *Der Prähistorische Menschund seine Umwelt*, 73–9. Stuttgart, Theiss.

Lancel, S. (1970) La nécropole romaine occidentale de la porte de Cesarée. *Bulletin d'Archéologie Algérienne* 4, 149–266.

Lignereux, Y., Peters, J., Ruas, M. P. and Sillières, P. (1997) Le culte d'Isis en Bétique. *L'Archéologue, Archéologie Nouvelle* 28, 41–5.

Livarda, A. (2013) Date, Rituals and Socio-Cultural Identity in the North-Western Roman Province. *Oxford Journal of Archaeology* 32(1), 101–17.

Macchia, F., Cavallaro, V., Forte, L. and Terzi, M. (2000) Vegetazione e clima della Puglia. *Cahiers Options Méditerranéennes* 53, 28–49.

Madella, M. (1999) Analisi dei resti vegetali carbonizzati provenienti dalle tombe a cremazione della necropoli di Lilibeo a Marsala. In B. Bechtold (ed.) *La Necropoli di Lylibaeum*. Rome, L'Erma di Bretschneider.

Marchesini, M. and Marvelli, S. (2006) Rituale funebre, paesaggio vegetale e ambiente nella necropoli. In F. Berti (ed.) *Mors Inmatura – I Fadieni e il loro Sepolcreto. Quaderni di Archeologia dell'Emilia Romagna* 16, 173–93. Florence, All'insegna del Giglio.

Marchesini, M. and Marvelli, S. (2007) L'alimentazione nell'oltretomba: le offerte votive vegetali nelle necropoli romane dell'Emilia Romagna. *Atti della Società dei Naturalisti e Matematici di Modena* 137, 331–42.

Marinval, P. (1993) Etudes carpologiques d'offrandes alimentaires végétales dans les sépultures gallo-romaines: réflexions préliminaires. In A. Ferdière (ed.) *Monde des morts, monde des vivants en Gaule rurale*, 45–65. 6e supplément à la Revue Archéologique du Centre de la France. Tours, Feracf.

Marinval, P. (2001) Offrandes alimentaires d'origine végétale en contexte funéraire gallo-romaine. In L. Baray (ed.) *Archéologie des Pratique Funéraires. Collection Bibracte 9*, 197–206. Glux en Glenn, Bibracte.

Matterne, V. and Derreumaux, M. (2008) A Franco-Italian investigation of funerary rituals in the Roman world, 'les rites et la mort à Pompéi', the plant part: a preliminary report. *Vegetation History and Archaeobotany* 17, 105–12.

Mégaloudi, F. (2005) Burnt sacrificial offerings at Hellenistic Times: an Archaeobotanical case study from Messene, Peloponnese, Greece. *Vegetation History and Archaeobotany* 14, 329–40.

Monger, G. (2013) *Marriage Customs of the World: An Encyclopedia of Dating Customs and wedding traditions*. Santa Barbara CA, ABC-CLIO.

Morris, I. (1992) *Death-Ritual and Social Structure in Classical Antiquity*. Cambridge, Cambridge University Press.

Motella, S. (1990) I resti vegetali. In S. Pozzi, M. A. Binaghi Leva (eds.) *Arsago. Nullus in Insubria Pagus* Vetustior, Studi in Memoria di Silvio Pozzi, 78–80. Varese, Lativa.

Ovidio. *Fasti*. Transl. J. G. Frazer [1931, Loeb Classical Library No. 253]. Cambridge MA, Harvard University Press.

Philostratus the Elder. *Imagines*. Transl. A. Fairbanks [1931, Loeb Classical Library No. 256]. Cambridge MA, Harvard University Press.

Pliny. *Natural History*. Transl. H. Rackham, W. H. S. Jones, and D. E. Eichholz [1949–54, Loeb Classical Library No. 352]. Cambridge MA, Harvard University Press.

Preiss, S., Matterne, V. and Latron, F. (2005) An approach to funerary rituals in the Roman provinces: plant remains from a Gallo-Roman cemetery at Faulquemont (Moselle, France). *Vegetation History and Archaeobotany* 14, 362–72.

Rettore, E. and Rottoli, M. (2006) *I reperti organici connessi agli oggetti metallici nella necropoli di Bernate Ticino (MI)*, 246–8. Milan, Notiziario della Soprintendenza per i Beni Archeologici della Lombardia.

Rottoli, M. and Castiglioni, E. (2011) Plant offerings from Roman cremations in northern Italy: a review. *Vegetation History and Archaeobotany* 20, 495–506.

Schweingruber, F. H. (1990) *Mikroskopische Holzanatomie*. Bern, WSL.

Servius. *Commentary to the Eclogues of Vergil*. Georgius Thilo Hermannus Hagen. Leipzig. Teubner. 1881. English version available at http://www.perseus.tufts.edu/hopper/text?doc=Perseus:text:1999.02.0053 [accessed 30/10/2017]

Sirago, A., Volpe, G. (1993) *Puglia Romana*. Bari, Edipuglia.

Šoštaric, R., Dizdar, M., Kušan, D., Hršak, V. and Mareković, S. (2006) Comparative analysis of plant finds from Early Roman graves in Ilok (*Cuccium*) and Šćitarjevo (*Andautonia*), Croatia – a contribution to understanding burial rites in southern Pannonia. *Collegium Antropologicum* 30(2), 429–36.

Statius. *Silvae*. Transl. D.R. Shackleton Bailey [2015, Loeb Classical Library No. 206]. Cambridge MA, Harvard University Press.

Strabo. *Geography*. Transl. H. Jones [1912–7, Loeb Classical Library No. 8]. Cambridge MA, Harvard University Press.

Titus Livius. *The History of Rome*. Transl. B. O. Foster [1922, Loeb Classical Library No. 133]. Cambridge, MA, Harvard University Press.

Toynbee, J. M. C. (1971) *Death and Burial in the Roman World*. London, Thames and Hudson.

Veal, R. (2014) Pompeii and its hinterland connection: the fuel consumption of the house of the Vestals, *c*. third century BC to AD 79. *European Journal of Archaeology* 17(1), 1–18.

Vergil. *Eclogues, Georgic, Aeneid*. Transl. H. R. Fairclough [1916, Loeb Classical Library No. 63–64]. Cambridge, MA, Harvard University Press.

Volpe, G. (1990) *La Daunia nell'Età della Romanizzazione, Paesaggio Agrario, Produzione e Scambi*. Bari, Edipuglia.

Zach, B. (2002) Vegetable offerings on the Roman sacrificial site in Mainz, Germany – short report on the first results. *Vegetation History and Archaeobotany* 11, 101–6.

Chapter 6

Feasting in a Sacred Grove: A Multidisciplinary Study of the Gallo-Roman Sanctuary of Kempraten, Switzerland

Pirmin Koch, Örni Akeret, Sabine Deschler-Erb, Heide Hüster-Plogmann, Christine Pümpin and Lucia Wick

Introduction

Between 2009 and 2013 the Archaeology Department of Canton St. Gallen (KASG) carried out a rescue excavation that unearthed the remains of a Gallo-Roman sanctuary on the western periphery of the Roman *vicus* at Kempraten (Rapperswil-Jona/SG, CH). The temple precinct at Kempraten was almost completely excavated using modern methods including extensive sampling for bioarchaeological and geoarchaeological analyses. The Roman-period deposits were preserved over an extended area and, besides oxic preservation, some of the sunken features (pits and ditches) had waterlogged preservation. These ideal starting conditions prompted the KASG, together with the IPAS/University of Basel, to launch a multidisciplinary project involving general archaeology, epigraphy, numismatics, zooarchaeology, archaeobotany, palynology and geoarchaeology. The project's multidisciplinary approach has given us an insight into Roman ritual practices with a broadness that is otherwise rarely possible. The results of the study are presented here for the first time with a special bioarchaeological and geoarchaeological focus.

The aims of this study are to investigate:

1. the natural environment of the sanctuary and its role in the overall worshipers' experience;
2. how the sanctuary was used;
3. the role of burnt offerings;
4. aspects of food consumption in the sanctuary; and
5. the bioarchaeological data within the framework of other sacred and secular contexts across the Gallic and Germanic provinces.

Roman sacrifice and feasting

Roman religious practice was based on the correct execution of the prescribed rituals (orthopraxy), which facilitated the communication between humans and gods (Scheid 2003, 18–20). Rituals and material culture were closely linked, since the former were only possible and perceptible thanks to the latter. Material culture was, therefore, of fundamental importance in practising and experiencing religion and ritual and can thus serve as marker for religious communities and their beliefs (*e.g.* Arweck and Keenan 2006). However, the archaeological interpretation of material culture has its own limitations depending on the archaeological record, the state of research, the quality of the written sources and so on.

Sacrifice was the central religious act. Written records contain detailed information and rules with regard to sacrificial acts and the selection of sacrificial offerings (Rüpke 2001, 137–50; Scheid 2003, 79–110). According to these, a sacrifice was embedded in a two-part system: first the gods received their share, followed by the preparation of the remnants for consumption as part of a religious feast. In Scheid's view, Roman sacrificial acts were primarily banquets. Offering a sacrifice meant to feast with the gods (2003, 94). Such feasting is part of religious practice in many cultures. Feasts can be defined as 'communal food consumption events that differ in some way from everyday practice' (Dietler 1996, 89). This distinction from everyday practice can be expressed by the number and type of persons involved, the amount and type of food consumed and its paraphernalia, the way in which the food is prepared and served, or by the duration, occasion and setting of the feast (Dietler 1996, 89). Some of the characteristic features are thus the consumption of large amounts of food and drink, special meals that are only eaten at feasts because they are complex or difficult to prepare, and special vessels used to cook and serve the food (Hayden 1996, 137–40). The participants at religious feasts are often transported into a world of food symbolism, which is distinct from their everyday food consumption (Douglas 1975).

The examination of Roman religion and ritual has been traditionally dominated by the study of ancient written records, although these mainly reflected the circumstances south of the Alps. In our study area, *i.e.* the Gallic and Germanic provinces, written records are largely absent, with the exception of inscriptions, and we are thus forced to draw heavily on archaeological evidence. The research on Gallo-Roman sanctuaries and in particular their architectural remains, inscriptions and votive offerings, can now look back on a long-standing tradition (Fauduet 2010, 22–8). Bioarchaeological analyses, however, still remain an exception. Only a large number of published studies on large animal bone assemblages from Gallo-Roman temple precincts are now available (*e.g.* Lepetz and Van Andringa 2008; Olive 1989; Rehazek and Nussbaumer 2009), and thanks to this fundamental research it has been possible to put forward the first supra-regional syntheses (Deschler-Erb 2015). The small animal remains, however, have only been examined in a limited number of recent excavations where suitable samples were collected (*e.g.* Hochmuth *et al.* 2005; Hüster Plogmann 2011). Summarised studies on particular research questions, such as the importance of the cock in the cult of Mithras (Lentacker *et al.* 2004), have been published, as have some botanical analyses (*e.g.* Vandorpe and Jacomet 2011; Zach 2002), although due to the preservation conditions the situation is not as favourable with regard to botanical evidence. Some studies also exist on the importance of certain plants in religious practice and on their symbolic or ceremonial meaning (*cf.* Kislev 1988; Livarda 2013; Lodwick 2015). Thanks to waterlogged preservation, pollen analyses have been carried out in a few cases (Groenman-van Waateringe and Pals 1994; Schlumbaum *et al.* 2011). The fact that geoarchaeological analyses are still quite rarely undertaken at Roman sanctuaries (Rentzel 2011) renders our Kempraten study very important. Overall, there is still a general lack of multidisciplinary studies with a similarly broad scope, both in the region and further afield (*cf.* Biesheim/F: Reddé 2011).

Vicus at Kempraten

The site is located at the eastern end of Lake Zurich (Fig. 6.1). Although it has been known since the 19th century, it was only in the 1940s that its character as a *vicus* was established. Large-scale archaeological excavations and systematic monitoring of the intensive construction work by the KASG were carried out in the past ten years and led to a revision of our image of the Roman *vicus*, which we now know covered an area of at least 11 hectares (Ackermann 2013, 17–8).

The *vicus* was established around AD 30/40 and therefore belongs to the final phase of Roman settlement expansion on the Swiss Plateau (Ackermann 2013, 217; Schucany 2010, 108). The settlement was located in an area where the waterway that came from the Alpine passes and led on via the Rivers Limmat and Aare to the Rhine crossed a northern branch. The installation of the road towards the north was probably the reason why a settlement was founded here (Ackermann 2013, 217). During its heyday from *c.* AD 120 onwards, its earlier wooden buildings were replaced by stone-built houses. A monumental, Mediterranean-style public building stood in the centre of the settlement. The residential buildings varied from luxurious courtyard houses with smaller buildings in their backyards in the centre of the settlement to strip houses at its periphery. The current state of research suggests decreased settlement activity in the 4th century AD (Ackermann 2013, 220).

Layout of the sanctuary

The sanctuary was enclosed by a ditch/wall and measured approximately 900 square metres (Fig. 6.1). Two temples were excavated in this area. Both were Gallo-Roman ambulatory temples of the *fanum* type. The temples consisted of an inner, tower-like *cella*, which was surrounded by an open or closed ambulatory/portico (Fauduet 2010, 99–107). This type of temple was particularly common in the Gallic and Germanic provinces and in southern Britain (Fauduet 2010, 12). Apart from two simple post constructions identified as chapels, no other building remains were uncovered within the temple precinct and its southern half appears to have been devoid of any constructions. The construction features of the temple precinct can be divided into two main phases (I and II) (Fig. 6.1).

During phase I preparatory work was carried out in advance of the construction of the temple precinct (phase Ia, second quarter of the 2nd century AD), including digging of two ditches to drain the terrain (Pos. 449 and 1188). The earlier temple complex (phase Ib, second quarter of the 2nd century until 160/180 AD) was surrounded by a ditch. Temple A stood in its northern half and can probably be dated to the second quarter of the 2nd century based on a coin offering retrieved from the *cella* wall. Whilst temple B already existed in this phase, its remains have only survived in fragments due to later alterations. A stone-lined hearth (Pos. 139) was located south of temple B probably from as early as phase Ib. Other concentrations of charcoal may represent simple hearths or may have been redeposited material (*e.g.* Pos. 332 and 997). Two pits that reached down to the groundwater level (Pos. 719 and 1334) were probably open well shafts. This is confirmed by evidence of reinforced walls, like the remains of a perforated barrel in pit 719.

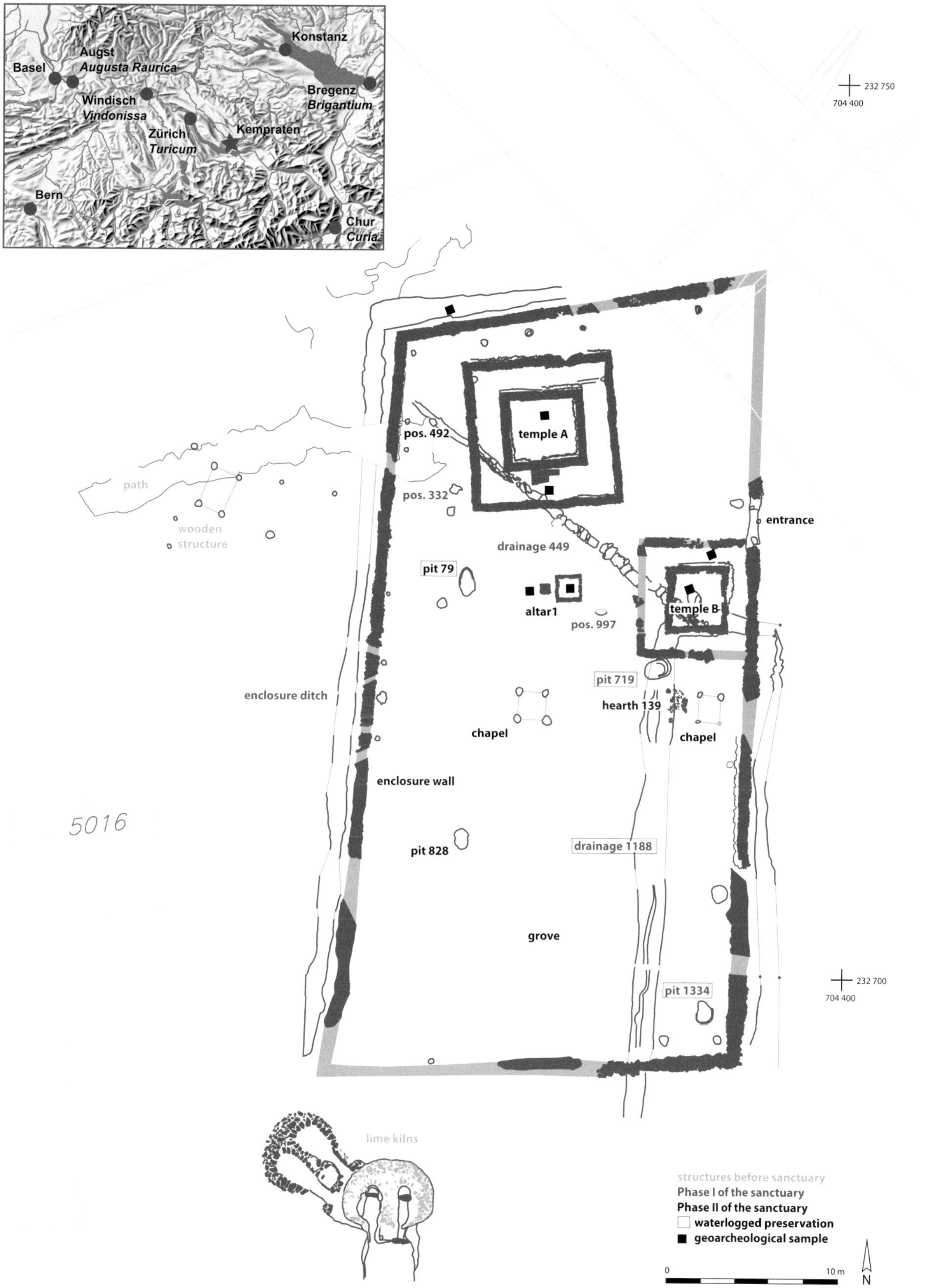

Figure 6.1 Rapperswil-Jona, Kempraten Seewiese. The sanctuary with its main Roman-period structures. Scale 1:250. Top left: Location of Kempraten and Roman settlements in the wider area.

Phase II (160/180 AD until the late 3rd century) saw the enclosure ditch replaced by a wall. This measure was probably accompanied by the renovation of temple A and the reconstruction of temple B, which was now built onto the perimeter wall. A rectangular altar (Pos. 1) bordered with stones was installed on the ground in front of the two temples and used for burnt offerings. Also, pit 828 (probably from phase II) served as a place of sacrifice where burnt offerings were made. As in phase I some concentrations of charcoal were documented (*e.g.* Pos. 492). Another pit (Pos. 79) was installed, probably to replace pit Pos. 719.

According to its layout and location, the sanctuary of Kempraten can be considered representative of sacred districts situated on the periphery of Roman cities and towns in the Gallic and Germanic provinces, although the site itself was relatively small. Quite a large number of such Gallo-Roman sanctuaries are known today in the wider region (*cf.* Lausanne/CH, Schleitheim/CH, Vindonissa/CH, Augusta Raurica/CH, Avenches/CH, Studen-Petinesca/CH; *cf.* Fauduet 2010; Flutsch *et al.* 2002, 310–5). However, most of these sites were excavated a long time ago or have been only partially explored.

Magna Mater

Amongst numerous votive artefacts (small altars, terracotta figurines, miniature axes, jewellery and coins) four lead curse tablets attest to the cult of Magna Mater (in one case she is addressed as Mater Deum) being practised at the sanctuary (Frei-Stolba *et al.* 2015). Magna Mater, also known as Cybele, was officially transferred (*evocatio*) from Asia Minor to Rome in the shape of a meteorite in 204 BC and a temple was devoted to her in a prominent location on the Palatine Hill (Vermaseren 1977, 38–43). She was therefore originally a ‘foreign’ goddess in the Roman pantheon. A feature that the group of ‘foreign’ deities, including Isis and Mithras, had in common, was that, in contrast to the Roman-Greek gods, they had a direct link to death and rebirth (Alvar 2008, 33–4). Due to their origins, earlier scholars often termed them ‘oriental’ deities (for a discussion about the meaning of their oriental roots see for example Bonnet *et al.* 2006). However, having been in Rome for a period of almost 300 years, Magna Mater could hardly be viewed as an oriental goddess in respect of the Imperial period. The cult of Magna Mater was widespread throughout the Empire and, judging by inscriptions, also existed in the Gallic and Germanic provinces (Vermaseren 1977, 131–40). In the wider region, her cult is mainly attested to urban places (*cf.* Avenches/CH and Augusta Raurica/CH; Flutsch *et al.* 2002, 326). The discovery of a double sanctuary devoted to Isis and Magna Mater in Mainz/D, which was the provincial capital of Germania superior, provided unexpected insights into the different types of cult in the northern provinces and how far they had spread throughout the region (Blänsdorf 2012; Witteyer 2013). The bioarchaeological analyses of animal and plant remains from the Mainz sanctuary have so far remained the only ones in respect of this particular cult, and therefore, serve as a reference point (Hochmuth *et al.* 2005; Zach 2002). Whilst no such analyses have yet been undertaken in the cult’s Mediterranean region of origin, the bioarchaeological evidence retrieved from sanctuaries of other ‘foreign’ cults – such as that of Isis (Lignereux *et al.* 1995; Zach 2002) and Mithras in particular (for a summarised account see Hüster Plogmann 2011; Lentacker *et al.* 2004; Olive 2008; Von den Driesch and Pöllath 2000) – have been studied in more detail.

Written records give accounts of two feasts linked with Magna Mater. The Maegalesia began on the 4th of April and celebrated the goddess’s arrival in Rome on that day, whilst the Hilaria was celebrated between 15th and 27th March and also served the worship of Attis, the mythical consort of Magna Mater (Vermaseren 1977). The celebrations included the trunk of a pine tree being carried into the sanctuary and displayed there on the 22nd of March (*arbor intrat*) (*ibid.*). The college of tree-bearers (*dendrofori*), who carried the tree in this procession, is also attested to the northwestern provinces by inscriptions (*cf.* in Avenches/CH, CIL XIII, 5153, found in Amsoldingen/CH, and Heddernheim/D, AE 2001, 1543).

The presence of the urban Roman Cult of Magna Mater at a local Gallo-Roman sanctuary in a provincial town like Kempraten is rather surprising. Without the written evidence one would have expected, for instance, a local variant of Mercury to have been worshiped there, as is often the case at Gallo-Roman sanctuaries (Fauduet 2010, 207–14). Whilst one may presume that other deities were also worshiped at Kempraten, this cannot be verified due to the lack of written sources.

Materials, methods and data presentation

A brief overview of the geoarchaeological and bioarchaeological analyses is provided here (for a detailed outline of all methodologies see Koch in prep.). The entire finds assemblage (including the medium/large animal bones) was hand-retrieved and separated by stratigraphic units. Certain selected layers and features were sampled for geoarchaeological, archaeobotanical and small animal remains analysis. Most of the finds had come from trampled (exterior) deposits and were thus fragmented. The waterlogged conditions, on the other hand, meant that organic remains were excellently preserved in certain features (Fig. 6.1). Since hardly any Roman temple precincts with waterlogged preservation have come to light so far, the information the assemblage provides is particularly significant.

Geoarchaeology

A total of seven geoarchaeological soil samples were analysed as part of this project to provide information on the sanctuary's history of use. The samples were taken from the ambulatories and *cellae* of the temples, from a site where burnt offerings were made and its surroundings, and from the enclosure ditch of phase I (Fig. 6.1). After drying them out for a period of two months, the soil samples were vacuum-hardened in epoxy resin (Araldit DY026sp and Laromin C260) and then cut using a diamond saw to create polished sections. A total of 23 thin sections measuring 4.5 × 4.5 cm were produced from a selection of soil samples (Beckmann 1997). Polished down to a thickness of 30 microns and mounted on specimen slides, the samples were then analysed in accordance with the methodological guidelines established by Bullock *et al.* (1985), Courty *et al.* (1989) and Goldberg and Macphail (2006), using a microscope equipped with plane-polarised light, cross-polarised light and fluorescent light. The results are discussed below together with the features concerned.

Pollen analysis

Six waterlogged samples of one cubic centimetre each from the pits (Pos. 719, 79 and 1334) and from a drainage ditch (Pos. 1188) were analysed for pollen. The samples were prepared using established palynological methods by treating them with HF 40% (fluoric acid) and acetolysis and embedding them in glycerine. Depending on the sample, tree pollen varied between 48% and 78%. The high proportion of Scots pine (*Pinus sylvestris*) and elm (*Ulmus* spec.) was striking. Additionally, sparse traces of pome (Maloideae), stone fruit trees (*Prunus* spec.) and walnut (*Juglans regia*) were also present.

The palynological spectra of other plants showed a predominance of meadow plants, such as ribwort (*Plantago lanceolata*), brown-scale knapweed (*Centaurea jacea*), clover (*Trifolium*) and other Fabaceae, buttercups (*Ranunculus* sp.), and plants of the mint (Lamiaceae) and the carrot (Apiaceae) families. Ruderal species, such as stinging nettle (*Urtica dioica*) and plants of the goosefoot family (Chenopodiaceae) as well as wetland species were quite abundant. Deadnettle (*Lamium*), hedgenettle (*Stachys*), enchanter's nightshade (*Circaea*) and other forest and shrubbery species were also identified. Cerealia-type pollen was found in all samples but in relatively small amounts (3.3% max.); other cultivated species hardly occurred at all.

Plant macrofossils

Forty sediment samples with a total volume of 245.4 litres were processed using the wash-over method (Jacomet 2007; Tolar *et al.* 2010) to recover plant macrofossils and small animal remains. Sieves with aperture sizes of 2mm, 1mm and 0.35mm were used. The samples were taken from pits (Pos. 719, 79 and 1334), charcoal-rich layers in and around the altar (Pos. 1) where the burnt offerings were made, the area of the hearth (Pos. 139), several small hearths/concentrations of charcoal (Pos. 492, 332 and 997) and some charcoal-rich layers inside the temple buildings (ambulatory of temple A: Pos. 199, ambulatory of temple B: Pos. 429 and *cella* of temple B: Pos. 894). The latter, however, were almost devoid of plant remains (as well as burnt bones).

A total of 47,117 plant macrofossils and 192 taxa were identified. Accordingly, the diversity was quite significant. The vast majority (98%) of the remains were waterlogged. The fact that fruit of the Scots elm (*Ulmus glabra*) and ash (*Fraxinus excelsior*) were identified highlights the excellent preservation conditions – both are very rarely found, even in periods during which, judging by pollen analyses, they should have been abundant.

From an ecological point of view the assemblage contained the following categories in descending order: weeds 49% (29% perennial ruderals, 19% weeds of summer crops and annual ruderals, <1% weeds of winter cereals), woodland plants 24% (15% trees/shrubs, 9% undergrowth), wetland plants 22%, crops 3% (2% cereals, 1% fruit and nuts, <1% vegetables and herbs, as well as very small amounts of pulses and oil- and fibre-producing crops), grassland 2%. Figure 6.2 shows the spatial distribution of the plant remains.

Small animal remains

In total 24,670 animal remains (fragments) were extracted from 40 sediment samples. Some 60 taxa were identified. The range of species included fish (40%), reptiles (22%), birds (21%), mice and small mammals (9%), amphibians (6%), and domestic and wild small mammals (1%). Birds and fish played the most important role in the rituals. Mice, small mammals, amphibians and reptiles probably were mostly deposited in a natural way since hardly any of their bones were burnt. In support of this, the ruins of the sanctuary were populated by many slow worms (*Anguis fragilis*).

Medium/large animal bones

A total of 8,293 bone fragments from well-stratified and dated contexts throughout the entire sanctuary and its immediate surroundings were collected and examined. The bone assemblage therefore is one of the larger ones from a Roman sanctuary analysed to date (Deschler-Erb 2015). Due to the preservation conditions and the detailed recovery strategy it was only possible to identify the species of 50% of the bones. Forty-two percent of the identified fragments were from domestic cattle, 33% from domestic pig, 13% sheep/goat and 8% equid (Fig. 6.3). Domestic poultry and

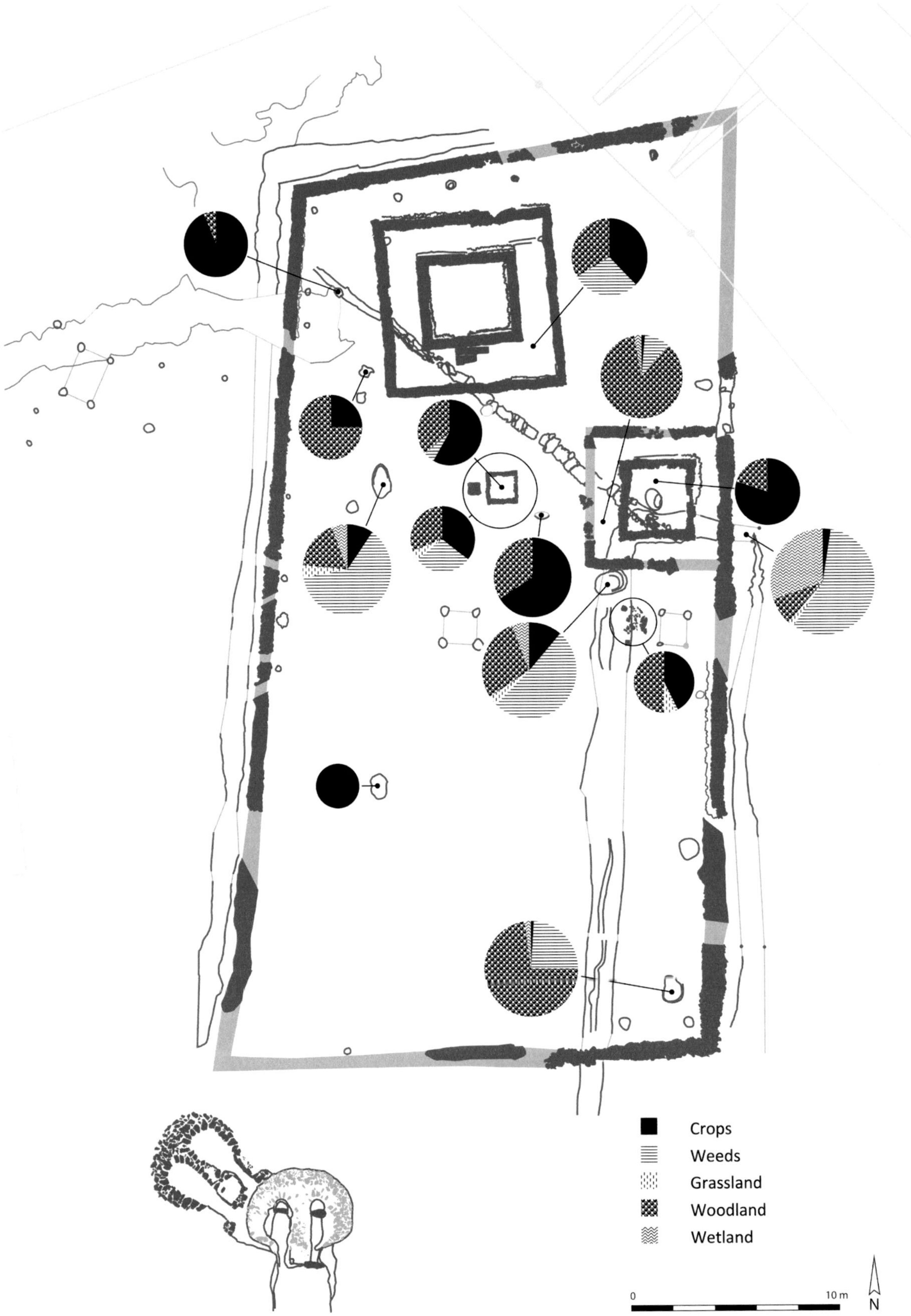

Figure 6.2 Rapperswil-Jona, Kempraten Seewiese. Spatial distribution of the plant macrofossils. The sizes of the circles indicate the macrofossil concentration on a logarithmic scale.

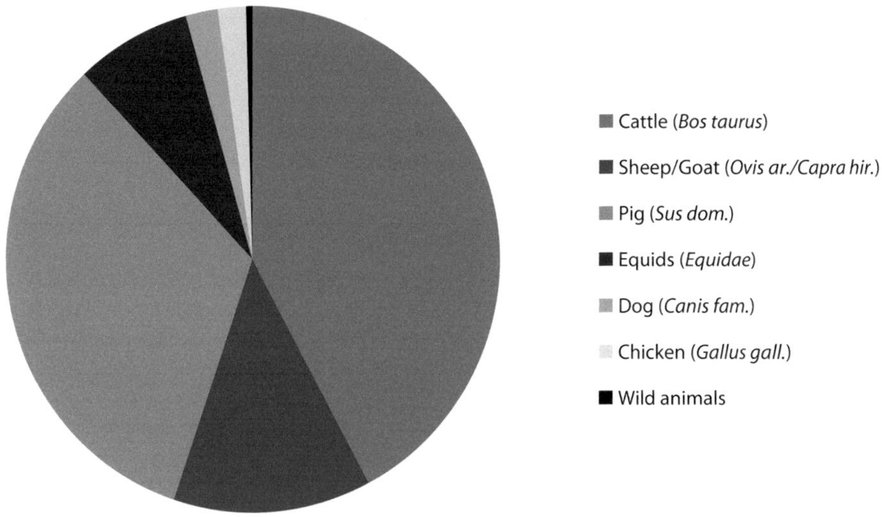

Figure 6.3 Rapperswil-Jona, Kempraten Seewiese. Pie chart showing the frequency (%) of hand-retrieved large animal bones.

dogs made up almost 2%. Wild mammals were quite sparse, constituting less than 1%. They did, however, account for a considerable range of species: bear, red deer (no antler, only postcranial bones), ibex and hare.

Reconstructing the setting of the sanctuary

The enclosure separated the temple precinct from the surrounding area and marked out the sacred space. In the earlier phase I, the sanctuary was enclosed by a ditch. At the base of this enclosure ditch an accumulation of run-off sediments with microbedded layers due to the gradation from sand to silt were present as depositional crusts (Valentin and Bresson 1992; Deák *et al.* 2017), as shown by the geoarchaeological analysis. Furthermore, the formation of slaking crusts which are typical for a bare surface rain input was also observed. Both observations clearly indicate the presence of an open trench which was exposed to the elements. Unlike the wall of phase II, the enclosure in phase I does not appear to have been visible above ground. This means that it was not important how visible the boundary between the sacred and the secular was; what counted was its existence. The geoarchaeological samples from the two temples suggest different reconstructions. The ambulatory of temple A consisted of a layer of gravel, which showed only faint signs of trampling and can therefore be viewed as a substructure, overlain by a loam floor. It was severely weathered and exhibited burnt red patches in some areas. Zones with highly degraded wood suggest a possible wooden floor, although they may also have been part of some other feature. The upper edge of the loam floor showed an indistinct level of occupation with compact areas, horizontally oriented components and faintly convolute deformations caused by the trampling of a wet sediment. The ambulatory of temple A was thus probably an open portico structure and would have been exposed to the weather during heavy rainfalls, as attested by fine, silty crusting and the translocation of fine material along pores in the sediment. The ambulatory of temple B was different in as much as those features were missing. This may have been due to the fact that a higher or even closed-off perimeter wall existed in that phase, or that this area was less exposed to the elements.

The soil samples from the ambulatory and the *cella* of temple B indicated at least three different phases of use with the usual compacted areas caused by trampling (Plate 11). The extensive use of the ambulatory of temple B during phase Ib resulted in an accretion of occupation deposits. These levels were separated by various levelled layers of gravel which could be interpreted as repairs. The topmost section of the soil sample probably showed parts of another occupation layer with evidence of a renovation of temple B in phase II.

The thin sections from inside the altar (Pos. 1) showed a sequence of two gravel layers which were older than its construction and were found throughout vast areas of the excavation. A first layer of coarse gravel (Pos. 515, phase Ib) was covered with a second layer of finer gravel (Pos. 435, phase II) in order to consolidate the ground before constructing the altar. The subsequent extensive use of this gravel layer resulted in the compaction of the sediment and the horizontal orientation of the components (Banerjea *et al.* 2015; Rentzel and Narten 2000; Rentzel *et al.* 2017), characterised by a rich accumulation mainly of burnt organic material and concentrations of ash and charcoal. A loam layer was later laid in the altar for burnt offerings. Charred remnants of porridge or bread, burnt bones and grape seeds (see below and also Plate 11) as well as reddened loam aggregates attest to its subsequent reuse as a place of sacrifice for burnt offerings. At that stage, the terrain was exposed to the elements, which was marked by two minor runoff events that resulted in fine sediments (clay, silt and fine sand) being washed in with the surface water (Courty *et al.* 1989, 125). Near the altar, a loam layer (Pos. 1070) was observed beneath the first, coarse layer of gravel, which presented an unusually high concentration of phosphate as well as a vivianite formation in its upper section (Plate 11). There were also clues to suggest that the loam had been exposed to heat. These were probably the remains of a hearth or a place of sacrifice.

Further remnants of human and/or animal faeces in the sanctuary were found in the pollen samples (particularly from pit 1334), which contained copious amounts of spores of coprophilous fungi and eggs of endoparasites, such as intestinal roundworms (*Ascaris* spec.), whipworms (*Trichuris* spec.) and lancet liver flukes (*Dicrocoelium dendriticum*), and nitrophilous ruderals. Moreover, some mammal and fish bones exhibited traces of digestion. Gnaw marks (7%) on the medium/large animal bones also showed that scavengers were present on site.

The botanical remains (macrofossils and pollen) showed that trees were planted in the temple precinct. There were more open areas in the close vicinity of the temples, as indicated by the presence of numerous ruderal plants. These plants are characteristic of locations that are highly frequented (trampled). Whilst ruderal species such as the stinging nettle (*Urtica dioica*), field mint (*Mentha arvensis*), many-seed goosefoot (*Chenopodium polyspermum*), broadleaf plantain (*Plantago major* s.str.) and chickweed (*Stellaria media*) occurred particularly often in the Kempraten samples, grassland plants were only found in very small numbers. This means that there was no meadow area, where feasting might have taken place within the sanctuary.

Trees and undergrowth were particularly dense in the southernmost part of the temple precinct. Pit 1334 contained a strikingly high proportion of Scots pine (*Pinus sylvestris*, Fig. 6.4) and elm (*Ulmus* sp.). This was surprising because without human assistance the Scots pine on the Swiss Plateau thrives mainly in extremely dry habitats or else on the edges of raised bogs (Ellenberg 1988). The same applies to elm trees, which had virtually disappeared from the forests since the Bronze Age. Both taxa must, therefore, have been deliberately planted at or near the site. While it is possible in theory that the different trees and shrubs identified in the samples came into the temple precinct in the form of branches used in ceremonies, the possibility can be excluded in the case of Kempraten, because both the macrofossils and the pollen included significant amounts of species that grow mainly under trees, for example enchanter's-nightshade (*Circaea lutetiana*), touch-me-not balsam (*Impatiens noli-tangere*) and the spotted dead-nettle (*Lamium maculatum*). These undergrowth species are typical of average forest floors and do not grow naturally under the Scots pine. This confirms that *Pinus sylvestris* could have only thrived at the sanctuary thanks to human planting.

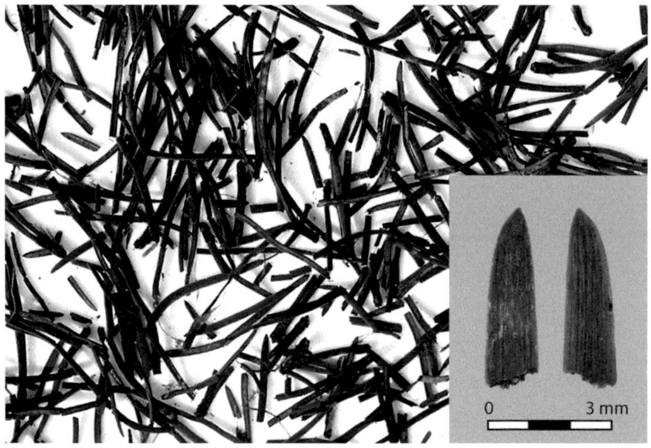

*Figure 6.4 Rapperswil-Jona, Kempraten Seewiese. Scots pine (*Pinus sylvestris*) needles from pit 1334 with detail.*

Pollen analysis indicates that the sanctuary itself was located within a highly diverse and intensively exploited landscape. Within a radius of a few kilometres there was a mosaic of grassland, fields of crops, hedges and shrubs. Semi-natural mixed deciduous forests with fir (*Abies alba*) and spruce (*Picea abies*) probably only existed at a considerable distance from the settlement. Mice and small mammals also attest to the ecology of the locality, as do reptiles and amphibians. They point to wetland areas with dry sunny places on one hand and deep-soil cropland in the immediate vicinity on the other.

Activities at the sanctuary

Burnt offerings

The charred plant remains and calcined bones can be viewed as the remnants of burnt offerings. In the sampled deposits, they were concentrated near the altar (Pos. 1) and hearth 139. In the other hearths/burnt patches they were somewhat less numerous. While only 2% of plant macrofossils were charred, these came mainly from the two areas mentioned. Sixty percent of the small animal bones retrieved from the same areas were highly calcined, while only 5% of the remains from the pits had been exposed to intense heat. These results together with the geoarchaeological analyses prove that the altar was used to make burnt sacrificial offerings.

Forty-six percent of the cultivated plants were burnt, a much higher proportion than other plant remains. This observation is testament to the importance of cultivated plants as burnt offerings. Various cereals were found as well as walnut (*Juglans regia*), grape (*Vitis vinifera*), and the remnants of stone pine cones (*Pinus pinea*), ten seeds and six scales. Besides these charred plant remains 5,326 amorphous fragments were also recovered. The larger pieces were rounded and had a diameter of approximately three centimetres. It is generally difficult to identify such objects – they can represent various foodstuffs such as bread or baked

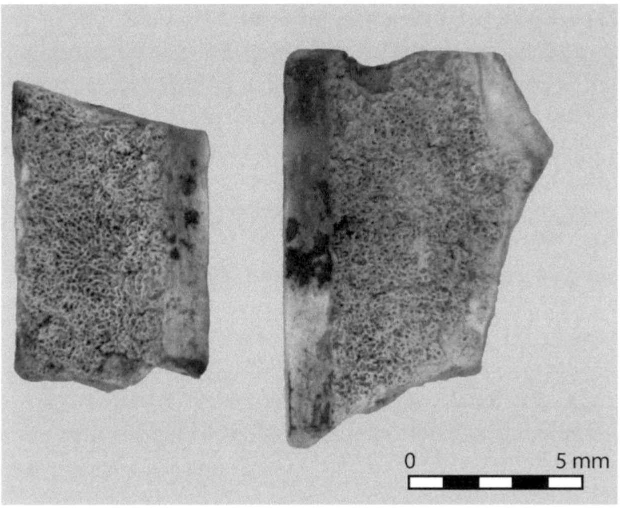

Figure 6.5 Rapperswil-Jona, Kempraten Seewiese. Long bone fragments of fowl. The medullary structure suggests that the animals were in the laying phase when they were slaughtered.

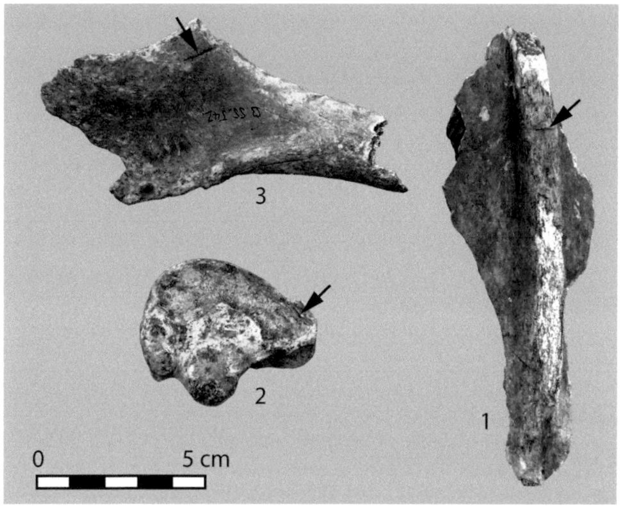

Figure 6.6 Rapperswil-Jona, Kempraten Seewiese. Equid bones with cut marks. Pelvis (1), foot bone (2), scapula (3).

goods, fruit pulp or resin. The aleuron layers of cereal grains, however, were identified in thin sections of these fragments (Plate 11). The objects were therefore the result of some type of food preparation using cereals, possibly some kind of baked dish like bread or porridge that was made as an offering or perhaps as part of a ritual meal.

The burnt and thus very often highly fragmented small animal remains were mainly from birds, fowl in particular, and to a lesser extent from domestic mammals. The medullary structure of many of the long bone fragments of fowl suggests that the animals were in the laying phase when they were slaughtered (Fig. 6.5, Peters 1997, 51). Hardly any of the chicken bones found were unburnt and they were generally rare amongst the hand-retrieved bones. As not all the sediment of the site was dry-sieved, it is possible that some chicken bones were missed from hand collection. Because of the careful excavation and in comparison to the sieved samples, however, the proportion of chicken bones would have not been significantly higher if dry-sieving had taken place. We may assume that whole birds were sacrificed by fire (*holocaust*), because all elements of *Gallus* are represented (trunk n = 21; fore n = 11; hind legs n = 16; head n = 4), which is exceptional compared to other sites. Besides chicken bones, eggs were also sacrificed by fire as indicated by a small quantity of calcined eggshells. However, if the temperature had been very high the eggshells' state of preservation would have been very poor. In a few cases fish bones with traces of severe burning were identified; these, however, were rare exceptions.

Feasting

Most of the hand-retrieved medium/large animal bones probably came from ritual meals that involved the consumption of meat from the animals that were sacrificed at the sanctuary because, in contrast to the small animal bones, they exhibited hardly any traces of burning. Whilst every major anatomical element was represented of all animal species, nearly all of them were fragmented and 7% bore traces of cutting and chopping. This shows that the animals were brought to the temple precinct whilst they were still alive and that most of their meat was consumed on site after they had been slaughtered. The meat does not actually appear to have been removed from the site and distributed outside the temple precinct, assuming it was not distributed off the bone. Fish was probably intended to enrich ritual meals. Fishing was mainly done in oxygen-rich, quick-flowing waters around the settlement but also in ponds that were moderately enriched with nutrients, and in the lake, according to the requirements of the identified taxa.

Cooking vessels with remnants of soot and food attest to the preparation of food probably within the temple precinct. The ceramic vessels were the same as those found at the *vicus*, in terms of both their shapes and sizes. This may indicate that cooking and eating was done in groups of the same size as an average household (*c.* 5–8 people) (Flutsch *et al.* 2002, 218). Mortaria were also found in significant numbers. In Roman (Mediterranean) cuisine, they were used to prepare thick sauces and cream cheese (Baatz 1977). Studies from Great Britain, however, show that a broad range of dishes were made in them (Cramp *et al.* 2011). They might also have been used to prepare particular types of dishes associated with the cult practised at the site.

Special features

Two observations were made on the bone finds from Kempraten that have not yet been fully interpreted. At 8% the amount of equid bones at Kempraten-Seewiese was comparatively high. Quite a large proportion also exhibited traces of cutting and chopping (8%), which is generally seen on cattle bones in a Roman context (Fig. 6.6). These traces of butchery suggest that their meat was also eaten.

Unusually for the Roman period, a high proportion of pike (11%, *Esox lucius*) was found amongst the fish bones identified to species level. Most of them had been caught before they had reached sexual maturity. The relatively uniform age range here suggests that they were specifically targeted. Young pike were found in samples from various locations throughout the sanctuary (Fig. 6.7). The remains were both burnt and unburnt. Therefore, young pike also played a role in the life at the sanctuary.

Discussion

Life at the sanctuary

Burnt offerings made on the altar located at the centre between the two temples probably played a crucial role in the practices and were the ritual focus of the sanctuary. Peripheral hearths/places of sacrifice attest to additional burnt offerings, indicated by the same composition of burnt remnants in these places and the altar. It is not possible, however, to exclude the use of some of these places also for cooking. For the most part the ritual acts probably took place out in the open, since no permanent facilities were found that could have accommodated a congregation. One possibility is, therefore, that the most important celebrations took place during the summer months. The age structure of the animals could provide evidence of seasonal festivities: whilst the time of these ceremonies cannot be pinpointed exactly, the evidence is indicative of the period between the end of March and September. Approximately two thirds of the cattle were less than six months old when they were slaughtered. If they were born in the spring, they must have been slaughtered during the summer months. The structures of the chicken bones retrieved at Kempraten also indicated that the chickens were most likely to have been slaughtered in the spring and early summer months. The laying period likely lasted from the spring to the early autumn and during this time the medullary structures in the bones would have consistently decreased.

Micromorphological analyses showed that the ambulatories of both temples were exposed to the elements in different ways. Whether this was due to varying architectural designs or because the sample locations were positioned differently in terms of exposure to the elements, cannot be established with certainty. According to these analyses, the ambulatory of temple B was regularly used and that of temple A also bore some traces of use. Similar observations were also made with regard to the soil samples from temple A/phase 3 and the mithraeum at Biesheim/F (Rentzel 2011). Thus, the ambulatories of the temples were actually walked on, probably as part of the ritual acts. It is generally believed that the ambulatories of Gallo-Roman temples were used to circulate around the *cella*, although this has not actually been established beyond doubt. The *cellae* themselves, on the other hand, probably would not have been open to the general public. However, due to the construction technique (mortar floors) and state of preservation of the *cellae* at Kempraten, the micromorphological samples did not yield any results in this regard.

Whole body parts amongst the animal bones and traces of faeces show that live animals were present. Phosphate traces in layers of phase Ib beneath the altar may have originated from blood spilled during sacrifices but also from urine of animals kept there before the altar was built. Whilst it is an obvious assumption to make that living animals were brought into a temple precinct and were left there for some time, this can rarely be proven. The examinations carried out at the temple precinct at Biesheim/F did show that the animals had stayed in the ambulatory of temple A/phase 3 for a certain period of time. This was suggested by remnants of faeces in the occupation surfaces (Flück 2011, 284).

The grove

The southern half of the sanctuary probably had a grove of Scots pine and elm trees, as shown by the botanical analyses. Groves are mentioned in ancient written sources (Luginbühl 2010; Scheid 1993). In Antiquity, woods and groves were considered sacred districts of the gods and were located outside the towns and cities. Pliny wrote:

> Once upon a time trees were the temples of the deities, and in conformity with primitive ritual simple country places even now dedicate a tree of exceptional height to a god; nor do we pay greater worship to images shining with gold and ivory than to the forests and to the very silences that they contain.' (Pliny, *NH*, 12:2, 3)

Particular types of trees were directly linked with certain deities; the sessile oak, for instance, was sacred to Jupiter and the poplar to Hercules (Pliny, *NH*, 12:2, 3–4). The Latin term *lucus* was generally used. Its meaning, however, is not clear and it can either signify a clearing in a forest, several trees with a religious connotation, or a sanctuary with or without a tree (Neudecker 2015, 222). Such sacred groves or gardens in Antiquity seem to all have expressed a yearning for a life in close proximity to the gods and nature (Neudecker 2015, 231).

Certain votive inscriptions from the northwestern provinces attest to the presence or donation of trees in sanctuaries. An altar from Lyon/F, for example, had an inscription to say that it was placed between two trees: *aram et signum inter duos arbores* (CIL 13,01780). Another inscription from Xanten/D, which is lost today, mentioned the establishment of a sanctuary with trees: *templum cum arborib(us) costituit/vslm* (CIL 13,08638). However, only in a very small number of cases has it been possible so far to find bioarchaeological evidence: Biesheim/F (Schlumbaum *et al.* 2011) and Empel/NL (Groenman-van Waateringe and

*Figure 6.7 Rapperswil-Jona, Kempraten Seewiese. Spatial distribution of pike (*Esox lucius*) remains.*

Pals 1994). These examples show that sacred groves also existed in or near sanctuaries in the northwestern provinces.

It is difficult, due to the state of research, to compare the results from the temple precinct at Kempraten with samples from the settlement and its surroundings. Both Scots pine and elm are unlikely to have grown in any great numbers in the area around Kempraten. This has been confirmed by the analysis of firewood carried out so far. Samples from a potter's kiln and from lime kilns, both of which, however, date from the 1st century, contained neither elm nor Scots pine (*Pinus pinea*) (Ackermann *et al.* 2015; Schärer *et al.* in press). The current state of knowledge with regard to Biesheim/F also indicates that elm trees only played a role in the temple precinct there (Schlumbaum *et al.* 2011).

We may, therefore, assume that the trees in and around the temple precinct were separate from people's everyday lives. They were used to create a different world, which set itself apart from its surroundings, also by virtue of its architecture (enclosure ditch/enclosure wall), thereby marking the extraordinary (*cf.* Rüpke 2001, 96–9). The trees were probably planted specifically to fulfil this purpose. However, we cannot exclude the possibility that they were already on site when the sanctuary was established. Future analyses on the wood should provide further insights into what types were used within the sanctuary.

Both Scots pine and elm could have been selected specifically for their symbolic value. The motive and reasoning behind the selection, however, can no longer be reconstructed. Stone pine trees (*Pinus pinea*) played an important role in the cult of Magna Mater (Vermaseren 1977, 90–1; 115). However, they do not thrive in northern climates. Perhaps the Scots pine was therefore chosen as a substitute at Kempraten because it is closely related to the stone pine tree and looks quite similar. Elm trees in a ritual context, on the other hand, are hardly mentioned in the ancient written sources. The elm, however, was also well represented in the temple precinct at Biesheim/F (Schlumbaum *et al.* 2011).

The sacrifices

The following comparison of the animal and plant remains with other sacred and secular contexts across the Gallic and Germanic provinces allows us to identify the characteristics of the cult celebrated at Kempraten based on the selection of sacrificial offerings. It must, nevertheless, be kept in mind that only a small number of studies have been carried out so far in relation to the *vicus* at Kempraten and only a few sanctuaries are known to have existed in eastern Switzerland, none of which has been studied in any great detail. The neighbouring province of *Raetia* to the east has no comparable sites or studies.

The analysis of the animal bones showed that the animals were treated in different ways after slaughtering: the mammals appear to have been cooked and eaten, since their bones exhibit hardly any traces of burning. It remains unclear whether cuts of meat were also deposited as offerings to the gods and/or whether the less desirable entrails were burnt in the fire (*cf.* Rüpke 2001, 146–7; Scheid 2003, 84). Chickens or fowl were sacrificed by fire to the gods, as shown by the high proportion of calcined bird bones.

The species and age ranges of the sacrificed animals exhibited special features compared to the *vicus* at Kempraten and to other Roman temple precincts in the Gallic and Germanic provinces. With its large proportion of cattle bones, the range of species identified is fundamentally different from that found at other sanctuaries in Roman Switzerland (*e.g.* Aventicum/CH: Lachiche and Deschler-Erb 2008, Deschler-Erb 2015; Thun-Allmendingen/CH: Rehazek and Nussbaumer 2009; Lousonna-Vidy/CH: Olive 1989) and in the province of *Germania Superior* in general (Deschler-Erb 2015). It was almost as high in the *vicus* at Kempraten/CH (Deschler-Erb 2013, 201, Abb. 230). However, there was a significant difference between the cattle from the *vicus* (Deschler-Erb 2013, 204, Abb. 238) and those from the temple precinct with regard to their ages: approximately 70% of the cattle slaughtered at the sanctuary were very young (infantile), probably less than six months old. The pigs, on the other hand, were generally at the age most suitable for slaughtering (>1 year and <3 years). This age range is also different from that seen at other sanctuaries, where the pigs were usually slaughtered at a very young age (only a few months old) and the cattle were largely adult animals by the time they were killed (*e.g.* Aventicum/CH: Lachiche and Deschler-Erb 2008). The small amount of game, however, fits in with many other sanctuaries located at the periphery of settlements. High percentages have generally only been observed at rural sanctuaries (*e.g.* Mollis-Hüttenböschen/CH: Schmid 1966; Thun-Allmendingen/CH: Rehazek and Nussbaumer 2009). Evidence to suggest that horses may have been sacrificed and eaten at the temple precinct is highly unusual, since this was more or less taboo in Roman society (André 1998, 115) and has only been found in exceptional circumstances and certain regions (*cf.* Groot 2008, 77–81). The Celts, on the other hand, did consume horse meat, but only in secular contexts (Méniel 2006, 173).

Sacrificing of fowl has been shown to have occurred at various Roman sanctuaries (Hochmuth *et al.* 2005; Hüster Plogmann 2011; Jacques *et al.* 2008; Olive 2008; Von den Driesch and Pöllath 2000). Accordingly, the proportions of poultry were higher at many sanctuaries than in secular contexts (*e.g.* Avenches/CH). Moreover, they were higher at sanctuaries of 'foreign' deities, such as Mithras, Isis and Magna Mater, also associated with mystery cults (*e.g.* ritual cave at Zillis/CH: Deschler-Erb and Stopp in prep.) compared to Gallo-Roman sites, where very high concentrations are only rarely found; the Fortuna sanctuary in Nijmegen/NL is an exception (Zeiler 1997). The sacrificial practice, however, appears to have differed from that performed at Mithraic temples, as most of the chicken bones were burnt only at Kempraten, at the Isis and Magna Mater sanctuary in Mainz/D and at the Fortuna sanctuary in Nijmegen/NL.

Pike has generally a consistent presence at Roman sites in present-day Switzerland but usually occurs in low numbers. Two samples from the residential areas of the *vicus*, for instance, only yielded three pike remains (Deschler-Erb 2013, 201; Häberle 2013). It appears, therefore, that pike was not at all popular as a fish for everyday consumption (Hüster Plogmann 2006, 192). In fact, such young animals are particularly rare elsewhere.

The charred plant remains also showed the importance of plant 'sacrifice' (cereals, grapes and nuts) and of bread/porridge. However, when compared to secular contexts, they did not stand out. An exception to this rule were stone pine cones, which are regularly found as offerings in Roman temples (Kislev 1998; Vandorpe and Jacomet 2011; Zach 2002).

These comparisons show that the selection of sacrificial offerings at Kempraten did, indeed, exhibit certain special features.

Synthesis and conclusions

Based on the analysis above we will now attempt to draw conclusions with regard to the rituals that were performed at the temple precinct. The selection of sacrificial offerings in particular gives rise to various hypotheses. The ritual acts were performed at a sanctuary, which set itself apart from its secular surroundings even by virtue of its architecture, that is its enclosure ditch or wall. The landscape design within the temple precinct (grove) created an additional space for an experience that was distinct from people's everyday lives. This area must also have been an important space for the participants in the rituals. This separation from everyday life may have been carried on into the selection of particular sacrificial animals (horses/pike).

The sacrificial meals were largely made up of beef and pork. The slaughtering of even just one of these animals would have produced up to several hundred portions of meat. This meat was probably eaten at communal feasts in the sanctuary, which would have played an important role in the local community. We do not know how the meat was prepared, whether on spits or in cooking pots. The sizes of the cooking pots found at the sanctuary suggest that parts of the sacrificial meals were prepared by smaller groups. We plan to carry out analyses on the cooking vessels, which will provide insights into the food they were used to prepare. Moreover, we cannot exclude the possibility that communal kitchens of significant size existed outside the temple precinct.

The preference for beef is surprising. Throughout the study area and in the Roman Empire in general (see *e.g.* Diocletian's Edict on Maximum Prices), the consumption of beef, particularly of old cattle, is considered to have been part of the diet of the poorer sections of the populations (Flutsch *et al.* 2002, 234–5). However, the slaughter ages of the animals in the temple precinct may serve as an argument against the notion that these were the sacrificial offerings of a lower social class. The regional subsistence strategy and/or dietary habits would have also played a role in the selection. Sacrifice and agricultural production would have been closely linked (Deschler-Erb 2015, 214; Rüpke 2001, 152–3); after all, one can only sacrifice what one is able to produce. Researchers presume that the high proportions of cattle bones in the temple precincts at Empel/NL and Elst/NL were linked with the reliance of the regional economy on cattle breeding (Lauwerier 1988; Seijnen 1994). Similar theories have been voiced in respect of the high proportions of sheep/goat bones at Zillis/CH (Deschler-Erb and Stopp in prep.- update available?). The current state of research with regard to the area around Kempraten, however, does not allow us to draw definite conclusions.

It is also possible that the selection of animals for sacrifice was in some way linked with the god that was worshiped at the sanctuary (Rüpke 2001, 150–1), although it will hardly be possible to prove this by archaeological means. The theory does exist in relation to large amounts of chicken bones found at Mithraic sanctuaries (Lentacker *et al.* 2004) or goats in the context of the worship of Mercury at Uley/GB (Woodward and Leach 1993). A connection with the cult of Magna Mater comes to mind particularly with regard to the calf bones recovered. As we know from written sources (*e.g.* Ovid, *Fasti* 4, 335), a calf (*iuvencam*) was sacrificed when the goddess was brought to Rome. However, no corresponding evidence was found at the Isis and Magna Mater sanctuary in Mainz/D.

Pikes appear to have played a role in certain feasts at Kempraten as suggested by their uniform age pattern. The consumption of young pike probably constituted a deliberate deviation from people's usual diet. We may assume that pike carried a specific (symbolic) meaning, which can no longer be determined. We can only speculate as to whether catching the fish might have been even part of the ritual act. It is a fact that young pike are difficult to catch and must be deliberately targeted.

The consumption of horse meat also stands in contrast to the everyday diet at the time, provided the interpretation of the bones recovered at Kempraten as food remains is correct. Whilst horses had a strong symbolic meaning in all cultures, according to written records, they appear to have played a rather insignificant role in Roman religious practice (the October Horse was an exception, Rüpke 2001, 108). Therefore, other reasons for dissecting horses must also be considered. Pliny, for instance, ascribed healing properties to horse blood (Pliny, *NH*, 28:41, 147). However, in order to draw blood from an animal it would not have been necessary to dissect it, as was done at Kempraten. This was probably another regional peculiarity of religious practices.

Apart from the selection of particular types of food, one must also consider particular dishes and ways of preparing food for feasts. The mortaria recovered may be a case in point. We know from written sources, for instance, that *moretum* was served at feasts in honour of Magna Mater (Ovid, *Fasti* 4, 367). The records also show that certain dishes could have particular meanings. In this case, the *moretum* is seen to represent the simple cuisine from the goddess's region of origin (Ovid, *Fasti* 4, 367).

In contrast to the food consumed during sacrificial feasts, the offerings selected for burning exhibited no regional specialities but were in line with supra-regional, Mediterranean habits. The making of burnt sacrifices was a Mediterranean custom that spread from there to the temples north of the Alps (Deschler-Erb 2015, 164–5). The fact that birds were sacrificed by fire at the Kempraten sanctuary suggests that they did not have the same function as the mammals. Large proportions of bird or chicken bones are viewed as representing the dietary habits of a well-to-do section of society (Flutsch *et al.* 2002, 234–5). However, in a religious context, chicken or bird sacrifices appear to have been less prestigious. Juvenal, for instance, mentions that in the opinion of some it is not worth sacrificing even a hen or a quail for the return to home of a poor fellow (Juvenal, *Satires* 12, 93–6). However, since birds must have played a specific role in religious practices, as shown by the fact that they were selected for sacrifice by fire, this interpretation falls short. Burnt offerings are more likely to have been 'individual' sacrificial acts because it would have been much more affordable for an individual to buy a chicken than a whole pig or cattle, despite the fact that the chicken would have been quite expensive compared to its size. The selection of laying hens points to specific rituals since they would have been probably considered to have a particular symbolic meaning. There appears to have been a rule that female animals were generally chosen to be offered as sacrifices to female deities (Scheid 2003, 79–110). The sacrifice of birds and in particular chickens can be also linked to 'foreign' cults with a strong Roman influence, such as the cult of Magna Mater, and this was also seen at the sanctuary of Fortuna, a Roman goddess, at Nijmegen/NL. In contrast to the cult of Mithras, where birds were consumed at religious feasts, they were sacrificed by fire as part of the cults of Magna Mater, Isis (Kempraten and Mainz/D as well as at Belo/E outside the study region, Lignereux *et al.* 1995), and Fortuna (Nijmegen/NL), as shown by parallels from the study region. According to written sources, the burning of whole animals, which was probably practised at Kempraten, was reserved for the gods of the underworld (Scheid 2003, 89). The same applies to the burning of sacrifices on the ground or in fire pits. Whilst Magna Mater was not a goddess of the underworld, she was linked with death (and rebirth), as were all 'foreign' deities.

Pine nuts were used in Mediterranean cuisine and offering pine cones derived from Mediterranean customs. North of the Alps they are frequently found in Roman religious contexts too (Kislev 1988; Lodwick 2015). When pine cones are burnt, they produce a pleasant smell and emit sparks. This could have, therefore, contributed significantly to the overall experience of witnessing a burnt sacrifice. Furthermore, pine cones seem to have been symbolically linked with death, as shown by the fact that they were often depicted on funerary monuments. Pine trees also played an important role in the cult of Magna Mater; their cones were often depicted on altars and clay pine cones were found as votive offerings at Magna Mater sanctuaries at Ostia/I and Rome/I (Rieger 2004, 133; Rieger 2007, 103).

In summary, the religious community at Kempraten used a specific selection and combination of votive offerings, rituals, and related activities, which were probably influenced by a variety of sources. Whilst some of the offerings (horses/pike) appear to have been a local speciality, future research will hopefully determine whether the preference for (young) cattle at communal feasts was actually an expression of regional dietary habits or whether it was influenced by customs linked with the worship of Magna Mater. Burnt offerings, on the other hand, fit well into the supra-regional system of symbolic meaning, which was strongly influenced by customs of Mediterranean origin. These sacrifices may represent individual acts that would have had different function to that of communal feasts. There may have also been links to the cult of Magna Mater. Whilst the goddess would have been worshiped throughout the Roman Empire, there were different regional characteristics (Rieger 2007). We must not forget, however, that other, unknown deities may have been also worshipped in the temple precinct. Overall, the analyses carried out at the Kempraten temple precinct have provided new, fascinating insights into the considerable plurality that existed in the polytheistic religion of the Roman Empire.

Acknowledgements

Archaeology Department of Canton St. Gallen and Lottery Fund St. Gallen, Philippe Rentzel, Thomas Beckmann. Translation: Sandy Hämmerle.

References

Ackermann, R. (2013) *Der römische Vicus von Kempraten, Rapperswil-Jona: Neubetrachtung anhand der Ausgrabungen Fluhstrasse 6–10 (2005–2006)*. St Gallen, Kantonsarchäologie St Gallen.

Ackermann, R., Koch, P., Gubler, T., Mändli, T. and Schlumbaum, A. (2015) Römische Kalkbrennereien im Vicus von Kempraten (Rapperswil-Jona SG). *Minaria Helvetica* 36/2015, 55–89.

Alvar, J. (2008) *Romanising Oriental Gods: Myth, Salvation and Ethics in the Cults of Cybele, Isis and Mithras*. Religions in the Graeco-Roman World 165. Leiden, Brill.

André, J. (1998) *Essen und Trinken im alten Rom*. Stuttgart, Reclam.

Arweck, E. and Keenan, W. (eds.) (2006) *Materializing Religion. Expression, Performance and Ritual*. Aldershot, Ashgate Publisher.

Baatz, D. (1977) Reibschüssel und Romanisierung. *Acta Rei Cretariae Romanae Fautorum* 17/18, 147–58.

Banerjea, R., Bell, M., Matthews, W. and Brown, A. (2015) Applications of micromorphology to understanding activity areas and site formation processes in experimental hut floors. *Archaeological and Anthropological Sciences* 7, 89–112.

Beckmann, T. (1997) Präparation bodenkundlicher Dünnschliffe für mikromorphologische Untersuchungen. In K. Stahr (ed.) *Mikromorphologische Methoden in der Bodenkunde. Ergebnisse eines Workshops der DBG, Kommission VII, 9–11 Okt. 1995 an der Universität Hohenheim*, 89–104. Hohenheim, Universität Hohenheim.

Blänsdorf, J. (2012) *Jürgen Blänsdorf: Die Defixionum Tabellae des Mainzer Isis- und Mater Magna-Heiligtums*. Mainz, Generaldirektion Kulturelles Erbe, Direktion Landesarchäologie.

Bonnet, C., Rüpke, J. and Scarpi, P. (2006) *Religions orientales – culti misterici: Neue Perspektiven – nouvelles perspectives – prospettive nuove*. Stuttgart, Steiner.

Bullock, P., Fedoroff, N., Jongerius, A., Stoops, G., Tursina, T. and Babel, U. (1985) *Handbook for Soil Thin Section Description*. Wolverhampton, Waine Research.

CIL Corpus Inscriptionum Latinarum.

Courty, M.-A., Goldberg, P. and Macphail, R. (1989) *Soils and Micromorphology in Archaeology*. Cambridge, Cambridge University Press.

Cramp, L. J. E., Evershed, R. P. and Eckardt, H. (2011) What was a mortarium used for? Organic residues and cultural change in Iron Age and Roman Britain. *Antiquity* 85, 1339–52.

Deák, J., Gebhardt, A., Lewis, H., Usai, M. R. and Heejin, L. (2017) Soils disturbed by vegetation clearance and tillage. In C. Nicosia and G. Stoops (eds.) *Archaeological Soil and Sediment Micromorphology*, 233–64. Chichester, Wiley-Blackwell.

Deschler-Erb, S. (2013) Untersuchungen zu den römerzeitlichen Tierknochen. In R. Ackermann (ed.) *Der römische Vicus von Kempraten, Rapperswil-Jona: Neubetrachtung anhand der Ausgrabungen Fluhstrasse 6–10 (2005–2006)*, 197–210. St Gallen, Kantonsarchäologie St Gallen.

Deschler-Erb, S. (2015) *Tier und Kult: Spezielle Tierknochendeponierungen der Spätlatène-und Römerzeit aus Aventicum/Avenches (CH) im nordalpinen Vergleich*. Documents du Musée Romain d'Avenches 27. Avenches, Association Pro Aventico.

Dietler, M. (1996) Feasts and commensal politics in the political economy. In P. Wiessner and W. Schiefenhövel (eds.) *Food and the Status Quest: An Interdisciplinary Perspective*, 87–125. Providence RI, Berghahn Books.

Douglas, M. (1975) Deciphering a meal. In M. Douglas *Implicit Meanings: Essays in Anthropology*, 249–75. London, Routledge and Kegan Paul.

Ellenberg, H. (1988) *Vegetation Ecology of Central Europe*. Cambridge, Cambridge University Press.

Fauduet, I. (2010) *Les temples de tradition celtique en Gaule romaine: Nouvelle édition revue et augmentée*. Paris, Editions Errance.

Flück, M. (2011) Der gallorömische Tempelbezirk im Spiegel der Mikromorphologischen Untersuchungen. In M. Reddé (ed.) *Oedenburg II. Fouilles Françaises, Allemandes et Suisses à Biesheim et Kunheim, Haut-Rhin, France. L'Agglomeration civile*, 243–50. Mainz, Römisch-Germanisches Zentralmuseum.

Flutsch, L., Niffeler, U. and Rossi, F. (eds.) (2002) *Bäder – Reben – Legionen. Die Römerzeit in der Schweiz*. Basel, Schweizerische Gesellschaft für Ur- und Frühgeschichte.

Frei-Stolba, R., Koch, P. and Lieb, H. (2015) Eine neue Fluchtafel aus Kempraten (Kt. St. Gallen/CH). In M. Scholz (ed.) *Lesen und Schreiben in den römischen Provinzen. Schriftliche Kommunikation im Alltagsleben. 2. Internationales Kolloquium DUCTUS im RGZM, 15.–17. Juni 2011*. 119–29. Mainz, Römisch-Germanisches Zentralmuseum.

Goldberg, P. and Macphail, R. I. (2006) *Practical and Theoretical Geoarchaeology*. Oxford, Wiley.

Groenman-van Waateringe, W. and Pals, J.-P. (1994) De vegetatie op en rondom het heiligdom. In N. Roymans and T. Derks (eds.) *De tempel van Empel: Een Hercules-heiligdom in het woongebied van de Bataven*, 72–81. 's-Hertogenbosch, Stichting Brabantse Regionale Geschiedbeoefening.

Groot, M. (2008) *Animals in Ritual and conomy in a Roman frontier community: Excavation in Tiel-Passewaaij*. Amsterdam Archeological Studies 12. Amsterdam, Amsterdam University Press.

Häberle, S. (2013) Auswertung der archäozoologischen Schlämmreste aus der Grube HB6.II_GR5. In R. Ackermann (ed.) *Der römische Vicus von Kempraten, Rapperswil-Jona: Neubetrachtung anhand der Ausgrabungen Fluhstrasse 6–10 (2005–2006)*, 190–6. St Gallen, Kantonsarchäologie St Gallen.

Hayden, B. (1996) Feasting in prehistoric and traditional societies. In P. Wiessner and W. Schiefenhövel (eds.) *Food and the Status Quest: An Interdisciplinary Perspective*, 127–46. Providence RI, Berghahn Books.

Hochmuth, M., Benecke, N. and Witteyer, M. (2005) Cocks and song birds for Isis Panthea and Mater Magna: The bird remains from a sanctuary in Mogontiacum/Mainz. In G. Grupe and J. Peters (eds.) *Feathers, Grit and Symbolism: Birds and Humans in the Ancient, Old and New Worlds*, 319–29. Rahden, Leidorf.

Hüster Plogmann, H. (2006) *Fisch und Fischer aus zwei Jahrtausenden: Eine fischereiwirtschaftliche Zeitreise durch die Nordschweiz*. Forschungen in Augst 39. Augst, Römerstadt Augusta Raurica.

Hüster Plogmann, H. (2011) Die Mikrofauna. In M. Reddé (ed.) *Oedenburg II. Fouilles Françaises, Allemandes et Suisses à Biesheim et Kunheim, Haut-Rhin, France. L'Agglomeration civile*, 176–94. Mainz, Römisch-Germanisches Zentralmuseum.

Jacomet, S. (2007) Plant macrofossil methods and studies: Use in environmental archaeology. In S. A. Elias (ed.) *Encyclopedia of Quaternary Science*, 2384–412. Amsterdam, Elsevier.

Jacques, A., Lepetz, S., Van Andringa, W., Matterne, V. and Tuffreau-Libre, M. (2008) Vestiges de repas et identification d'un siège de collège à Arras-Nemetacum. In S. Lepetz and W. Van Andringa (eds.) *Archéologie du sacrifice animal en*

Gaule romaine: Rituels et pratiques alimentaires, 237–55. Montagnac, éditions monique mergoil.

Juvenal (nd) *Satires* [In *Juvenal and Persius*]. Ed. and transl. S. Morton Braund [2004, Loeb Classical Library]. Cambridge MA, Harvard University Press.

Kislev, M. E. (1988) Pinus pinea in agriculture, culture and cult. In H. Küster (ed.) *Der prähistorische Mensch und seine Umwelt: Festschrift für Udelgard Körber-Grohne zum 65. Geburtstag*, 73–9. Stuttgart, Konrad Theiss.

Lauwerier, R. (1988) *Animals in Roman Times in the Dutch River Area*. Nederlandse Oudheiden 12. Amersfoort, ROB.

Lachiche, C. and Deschler-Erb, S. (2008) De la viande pour les hommes et pour les dieux: Sa gestion dans deux villes de la Suisse romaine. In W. Van Andringa (ed.) *Sacrifices, marché de la viande et pratiques alimentaires dans les cités du monde romain*, 107–34. Turnhout, Brepols.

Lentacker, A., Ervynck, A. and Van Neer, W. (2004) Gastronomy or religion? The animal remains from the mithraeum at Tienen (Belgium). In S. O'Day, W. Van Neer and A. Ervynck (eds.) *Behaviour Behind Bones: The Zooarcheology of Ritual, Religion, Status and Identity*, 77–94. Oxford, Oxbow Books.

Lepetz, S. and Van Andringa, W. (2008), *Archéologie du sacrifice animal en Gaule romaine: Rituels et pratiques alimentaires*. Montagnac, éditions monique mergoil.

Lignereux, Y., Peters, J., Bubień-Waluszewska, A. and Sillières, P. (1995) Sacrifice d'oiseaux en l'honneur d'Isis au Ier siècle après Jésus-Christ à Bélo en Bétique (Bolonia, Cadix, Andalousie). *Revue de Médecine Vétérinaire* 146, 575–82.

Livarda, A. (2013) Date, rituals and socio-cultural identity in north-western Roman provinces. *Oxford Journal of Archeology* 32(1), 101–17.

Lodwick, L. (2015) Identifying ritual deposition of plant remains: A case study of stone pine cones in Roman Britan. In T. Brindle, M. Allen, E. Durham and A. Smith (eds.) *TRAC 2014: Proceedings of the Twenty-Fourth Annual Theoretical Roman Archeology Conference*, 54–69. Oxford, Oxbow Books.

Luginbühl, T. (2010) Arbres et bois 'sacrés' dans le monde celtique et gallo-romain: Approche taxinomique. In F. Bertholet and K. Reber (eds.) *Jardins antiques: Grèce – Gaule – Rome*, 49–86. Gollion, Infolio.

Méniel, P. (2006) Religion et sacrifices d'animaux. In Ch. Goudineau (ed.) *Religion et société en Gaule*, 165–76. Paris, Editions Errance.

Neudecker, R. (2015) Gardens. In R. Raja and J. Rüpke (eds.) *A Companion to the Archeology of Religion in the Ancient World*, 220–34. Oxford, Wiley.

Olive, C. (1989) L'habitat et le lieu cultuel: Étude comparative des faunes. In D. Paunier, J. Bernal, D. Castella, L. Flutsch, J.-B. Gardiol and F. Rossi, *Le vicus gallo-romain de Lousonna-Vidy: Le quartier occidental, le sanctuaire indigène: Rapport préliminaire sur la campagne des fouilles 1985*, 165–80. Lausanne, Bibliothèque historique vaudoise.

Olive, C. (2008) Honorer Mithra en mangeant: La cuisine du mithraeum de Septeuil. In S. Lepetz and W. Van Andringa (eds.) *Archéologie du sacrifice animal en Gaule romaine: Rituels et pratiques alimentaires*, 267–73. Montagnac, éditions monique mergoil.

Ovid (nd) *Fasti* [Festkalender]. Ed. and transl. N. Holzberg [1995]. Zürich, Artemis+Winkler.

Peters, J. (1997) Zum Stand der Hühnerhaltung in der Antike. In *Beiträge zur Archäozoologie und Prähistorischen Anthropologie* I, 42–58.

Pliny. *Natural History*. Transl. H. Rackham, W. H. S. Jones, and D. E. Eichholz [1949–54, Loeb Classical Library No. 352]. Cambridge MA, Harvard University Press.

Reddé, M. (ed.) (2011) *Oedenburg II. Fouilles Françaises, Allemandes et Suisses à Biesheim et Kunheim, Haut-Rhin, France. L'Agglomeration civile*. Mainz, Römisch-Germanisches Zentralmuseum.

Rehazek, A. and Nussbaumer, M. (2009) Tierknochen. In S. Martin-Kilcher and R. Schatzmann (eds.) *Das römische Heiligtum von Thun-Allmendingen, die Regio Lindensis und die Alpen*, 174–8. Bern, Bernisches Historisches Museum.

Rentzel, P. and Narten, G.-B. (2000) Zur Entstehung von Gehniveaus in sandig-lehmigen Ablagerungen: Experimente und archäologische Befunde. *Jahresbericht der Archäologischen Bodenforschung des Kantons Basel-Stadt* 1994, 31–52.

Rentzel, P. (2011) Spuren der Nutzung im Mithraeum von Biesheim: Mikromorphologische Untersuchungen. In M. Reddé (ed.) *Oedenburg II. Fouilles Françaises, Allemandes et Suisses à Biesheim et Kunheim, Haut-Rhin, France. L'Agglomeration civile*, 248–55. Mainz, Römisch-Germanisches Zentralmuseum.

Rentzel, P., Nicosia, C., Gebhardt, A., Brönnimann, D., Pümpin, C. and Ismail-Meyer, K. (2017) Trampling, poaching and the effect of traffic. In C. Nicosia and G. Stoops (eds.) *Archaeological Soil and Sediment Micromorphology*, 281–98. Chichester, Wiley-Blackwell.

Rieger, A.-K. (2004) *Heiligtümer in Ostia: Architektur, Ausstattung und Stellung öffentlicher Heiligtümer in einer römischen Stadt*. Studien zur antiken Stadt. München, Pfeil.

Rieger, A.-K. (2007) Lokale Tradition versus überregionale Einheit: der Kult der Magna Mater. In C. Bonnet (ed.) *Religioni in contatto nel mediterraneo antico*. Mediterranea IV, 89–120. Pisa, Fabrizio Serra.

Rüpke, J. (2001) *Die Religion der Römer: Eine Einführung*. Munich, C. H. Beck.

Schärer, L., Helfert, M., Ackermann, R. and Koch, P. (in press) *Keramikproduktion in Kempraten, Rapperswil-Jona. Die römischen Töpfereien des 1. und 2./3. Jahrhunderts*. St Gallen, Kantonsarchäologie St Gallen.

Scheid, J. (1993) Lucus, nemus: Qu'est-ce qu'un bois sacré. In O. De Cazanove and J. Scheid (eds.) *Les bois sacrés: Actes du colloque international, Naples, 23–25 novembre 1989*, 13–20. Naples, Centre Jean Bérard.

Scheid, J., transl. J. Lloyd (2003) *An Introduction to Roman Religion*. Edinburgh, Edinburgh University Press.

Schlumbaum, A., Wick, L., Schucany, C., Schwarz, P.-A. and Bockius, R. (2011) Die Ulmen und andere besondere Gehölze aus dem Römischen Tempelbezirk und der Zivilsiedlung. In M. Reddé (ed.) *Oedenburg II. Fouilles Françaises, Allemandes et Suisses à Biesheim et Kunheim, Haut-Rhin, France. L'Agglomeration civile*, 73–102. Mainz, Römisch-Germanisches Zentralmuseum.

Schmid, E. (1966) Ergebnisse der Ausgrabung 1964 beim gallorömischen Tempel auf Hüttenböschen (Mollis). *Jahrbuch des Historischen Vereins des Kantons Glarus* 61, 34–41.

Schucany, C. (2010) Oppida, camps romains et agglomérations secondaires sur le Plateau Suisse. In P. Ouzoulias and L. Tranoy (eds.) *Comment les Gaules devinrent romaines*, 97–108. Paris, Editions La Découverte.

Seijnen, M. (1994) Dierebotten en rituele maltijden. In N. Roymans and T. Derks (eds.) *De tempel van Empel: Een Hercules-heiligdom in het woongebied van de Bataven*, 162–74. 's-Hertogenbosch, Stichting Brabantse Regionale Geschiedbeoefening.

Tolar, T., Jacomet, S., Velušček, A. and Čufar, K. (2010) Recovery techniques for waterlogged archaeological sediments: A comparison of different treatment methods for samples from Neolithic lake shore settlements. *Vegetation History and Archaeobotany* 19, 53–67.

Valentin, C. and Bresson, L.-M. (1992) Morphology, genesis and classification of surface crusts in loamy and sandy soils. *Geoderma* 55, 225–45.

Vandorpe, P. and Jacomet, S. (2011) Remains of burnt vegetable offerings in the temple area of Roman Oedenburg (Biesheim-Kunheim, Haut-Rhin, Alsace): First results. In J. Wiethold (ed.) *Carpologia: Articles réunis à la mémoire de Karen Lundström-Baudais*, 87–100. Glux-en-Glenne, Bibracte.

Vermaseren, M. J. (1977) *Cybele and Attis, the Myth and the Cult*. London, Thames and Hudson.

Von den Driesch, A. and Pöllath, N. (2000) Tierknochen aus dem Mithrastempel von Künzing. Lkw. Deggendorf. In K. Schmotz (ed.) *Vorträge des 18. Niederbayerischen Archäologentages*, 145–62. Rhaden, Leidorf.

Witteyer, M. (2013) Rituelle Niederlegungen im Heiligtum für Isis und Magna Mater in Mainz. In A. Schäfer and M. Witteyer (eds.) *Rituelle Deponierungen in Heiligtümern der hellenistisch-römischen Welt: Internationale Tagung Mainz, 28.–30. April 2008*, 317–52. Mainz, Generaldirektion Kulturelles Erbe, Direktion Landesarchäologie.

Woodward, A. and Leach, P. (1993) *The Uley Shrines. Excavation of a Ritual Complex on West Hill, Uley, Gloucesershire. 1977–1979*. London, English Heritage.

Zach, B. (2002) Vegetable offerings on the Roman sacrificial site in Mainz, Germany: Short report on the first results. *Vegetation History and Archaeobotany* 11, 101–6.

Zeiler, J. T. (1997) Offers en slachtoffers: Faunaresten uit de Fortunatempel te Nijmegen (2e eeuw n.Chr.). *Paleo-aktueel* 8, 105–7.

Chapter 7

Ritual Meals and Votive Offerings: Shells and Animal Bones at the Archaic Sanctuary of Apollo at Ancient Zone, Thrace, Greece

Rena Veropoulidou and Daphne Nikolaidou

Introduction

The prevalence of functional explanations in the study of religious and ritual practices have traditionally dominated the field of archaeology, leading to the downplaying of the material dimensions of belief and of the contextualised dimensions of religion in several study areas (Raja and Rüpke 2015a, 2–3), including ancient Greece. During the last three decades, however, the situation is gradually changing and issues related to the experience of ritual, its impact on social relations and its role in promoting or transforming social orders and dominant ideologies have started being addressed (*e.g.* Fogelin 2007, 55–8).

Regarding ancient Greek religion, particular emphasis has been placed on official and collective religious practices organised and controlled by higher social and political institutions, such as the Greek *polis* (city-state), while the informal, personal modes of belief and ritual performance have been largely neglected (Kindt 2015). In this context, the focus has been traditionally on written sources and iconographic evidence, which are, nevertheless, subjective and highly selective. The last 15 years, however, have seen a proliferation of studies seeking to gain insights into the fluid nature of rituals and the ways in which these were performed and experienced (*e.g.* Fogelin 2007, 61; Stavrianopoulou 2008, 3).

In this framework, bioarchaeological analyses in several ritual contexts have addressed old research questions in novel ways (see *e.g.* Ekroth and Wallensten 2013), including the ritual use of animals (cycle of sacrifice) and its ultimate expression, the *thysia* (the actual killing) of the animal. Drawing upon current archaeological theories and integrating textual descriptions, visual representations and archaeological evidence, recent studies have exposed the varied nature of cultic actions (*e.g.* Larson 2007, 1–2) that indicate the existence of differing ways of perceiving gods and performing rituals. Despite these advances, there are still several theoretical and material aspects of ritual practices, such as fish bones, marine shells and charred plant remains, which are rarely discussed. In this sense, the heterogeneity of the ancient Greek worship still remains concealed, the personal and local parameters of ritual and religion are not thoroughly discussed, while the primacy of ritual practice in constructing, creating or modifying religious beliefs and social relations is little explored.

This paper endeavours to add new information on the use of terrestrial and marine faunas in the materialisation of cultic practices in a Greek colony in the northern Aegean with the aim of highlighting the heterogeneity and social aspects of religious actions and rituals. The discussion is based on the contextualised analysis of the animal bones and shells recovered from the Sanctuary of Apollo at the coastal city of Zone, Thrace, which dates to the Archaic period (6th century BC). Considering the colonial context of the sanctuary, the analysis aims to investigate the possible uses of animals, whether the composition and characteristics of the assemblages could be linked to specific rituals known from written and iconographical evidence, and to explore whether the mixture of local dwellers, colonists and visitors led to the creation of site-specific cultic practices. The ultimate aim is to allow insights into the role of these animals in the belief system of this colony and in the performance of rituals, and thus to point out the similarities and discrepancies between archaeological, literary and iconographical records, and eventually to discuss the dynamic processes in which historical factors, local cultural conditions and ritual participants were entangled.

Thysia, theoxenia and trapezomata

According to textual and epigraphic sources and vase paintings the most common ancient Greek rituals that involved foods were *thysia, theoxenia* and *trapezomata*. The various meanings and interpretations of these practices may be regarded as controversial while recent work suggests that their nature was fluid and they had multiple economic, social and political effects on the participants (Stavrianopoulou 2008, 1–4). A general overview of their structure is provided here.

Thysia, the principal act of the sacrificial cycle, involved the ceremonial slaughter of an animal, almost exclusively a domestic animal, such as sheep, goat, cattle and pig (*e.g.* Ekroth 2007, 256; Van Straten 1995, 171–5). It is firstly mentioned by Homer (*Iliad* I, 458–67, II, 421–9; *Odyssey* III, 447–63), while the burning of (wrapped in fat) bones on the altar is described by Hesiod (*Theogony*, 535–57). More details are given by Sophocles (*Antigone*, 1005–11), according to whom the *meria* (thigh bones: femora) were chosen for burning. Aeschylus (*Prometheus bound*, 505–11) also writes that god's share consisted of legs wrapped in fat and *osphys* (tail: caudal vertebrae and sacrum). Drawing upon literary evidence and vase paintings, Burkert (1983, 1–12) provided a summary of the *thysia*: after slitting the animal's throat, its internal organs were inspected and part of its blood was sprinkled on the altar; the thigh bones, the tail and the sacrum were cut out and burnt on the altar, thus letting gods inhale the smoke (*knissa*) from the burning of bones. The complete burning of the god's share was followed by ritual meals, during which the meat of the sacrificed animal was consumed by the priests and the participants of the ceremony. Hence, mortals feasted in honour of the gods, but gods did not participate in this meal. The general scholarly view is that animal sacrifice stood at the very centre of ancient Greek religious thought, as it was considered the main means by which mortals honoured their gods, expressed their gratitude, and sought to acquire favours or avert their anger (Bremmer 2007, 139).

In relation to *thysia*, texts and myths further inform on other food offerings for the deities, such as raw or cooked meat, and bloodless foods, including cakes, breads, pots of cooked grain and fruits (*e.g.* Bruit 1990, 173). Some of these foods could be offered as part of the *aparchai* custom, that is the first-offerings that include 'a portion of the proceeds from a wide range of human activities' (Jim 2011, 40). For instance, from the Homeric period and in some cases up to the 6th century AD hunters offered a share of the animal and fishermen offered the first catch to god(s) to ensure good hunting and catches for the rest of the year (Rouse 1902, 40–58). Either *aparchai* or not, these foods were offered both in simple and structured ways, such as the *theoxenia* and *trapezomata* rituals. During *theoxenia*, a couch or a throne was set next to a table on which people placed the offerings, foods and drink, for the god or his/her lesser associates to dine. Therefore, the god(s) were taking the role of guests and were sharing a meal with the mortals that comprised foods usually consumed by the latter with the aim to bring them closer together. *Theoxenia* was most probably performed in or near the temple, usually after the *thysia*, thus food offerings placed honour to the deity and enhanced the meagre god's burnt portion, who was simultaneously the recipient of the sacrifice and a guest (Bremmer 2007, 139; Gill 1974, 136). The food offerings were cooked, and it is assumed that after the end of the ceremony they would be available to the priests and the cult personnel (*e.g.* Ekroth 2011, 26, 27; Jameson 2014, 145–76). It must be noted, however, that *theoxenia* is mainly mentioned in relation to particular festivals and gods, such as Apollo at Delphi and Pellene, and the Dioscuri at Acragas and Paros (Jameson 2014, 146).

The practice of *trapezomata,* on the other hand, was related to gifts of raw meat, which were probably placed on a table at the altar, but also in front of the cult statues (Gill 1974, 123–33; Jameson 1994, 56). The epigraphic record, the basic source of information on this practice, is not clear on the timing of *trapezomata* and on how it was related to the *thysia*, but it is assumed that the meat was displayed on a table during the burning of the sacrificed animal. After the end of the ritual activity, the meat was probably given to cult personnel. Raw meat, according to Segal (1974), is a metaphor for cruelty and retains associations with the beast-world, but Ekroth (2008, 101) suggests that it

> was intended as a marker of awesome power inherent in the divine, since what was given could not be eaten by humans without grave consequences.

The above descriptions ascribe certain norms, qualities and attributes to the performance of rituals, suggested mainly by literary sources, but they do not take into account the complex sequence of rites inherent in ritual practices nor the nature of and the possibilities offered by the latter in structuring the social arena.

The archaeology of Zone

The colony of Zone was established on the beachfront of Thrace in mainland Greece (Fig. 7.1) in the late 7th century BC by Greek people of Samothrace, and was one of the several colonies of the Samothracian *peraia*, the colonial state of the island. The reasons put forward for the colonial expansion of people of Samothrace on the mainland during the Archaic period (800–480 BC) include access to fertile land and other resources, such as certain minerals and oak forests, expansion of stockbreeding activities and the establishment of networks of communication and trade between the Greek world of the Aegean and the Thracian hinterland (*e.g.* Tiverios 2008).

The excavated urban centre (6th–4th centuries BC) comprises a fortified enclosure, surrounding large residential

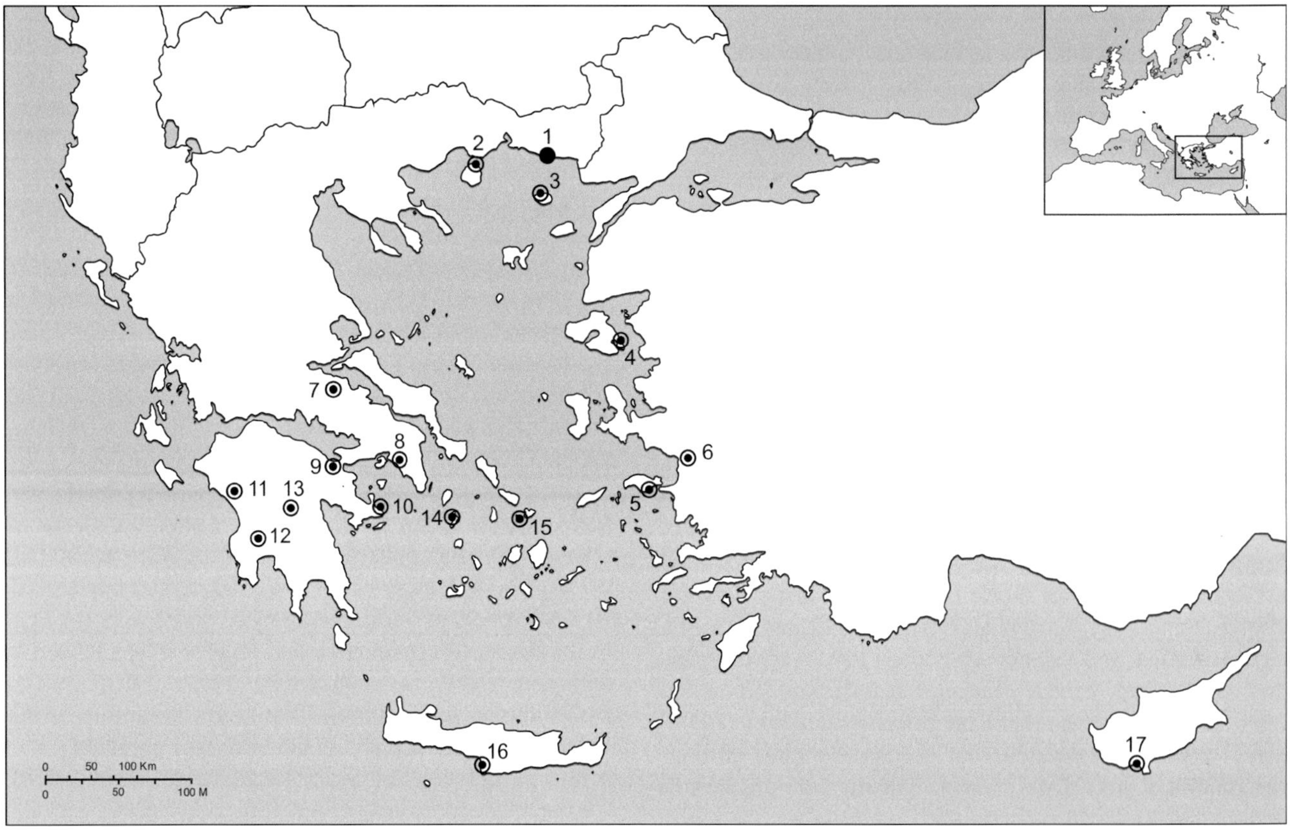

Figure 7.1 Map of Greece with the location of ancient Zone (1) and other sites mentioned in the text: 2) Sanctuary of Hercules, Thassos (Gardeisen 1996); 3) Sanctuary of Great Gods, Samothrace (Lehmann 1998); 4) Sanctuary of Demeter and Kore, Mytilene (Ruscillo 1993); 5) Heraion, Samos (Boessneck & von den Driesch 1988); 6) Artemision, Ephessos (Bammer et al. 1978, 107-157); 7) Sanctuary of Artemis and Apollo, Kalapodi (Stanzel 1991); 8) Altar of Aphrodite Ourania, Athens (Reese 1989); 9) Sanctuary of Demeter and Kore, Korinthos (Bookides et al. *1999); 10) Temple of Poseidon, Kalavreia, Poros (Mylona 2013); 11) Artemision, Olympia (Benecke 2006); 12) Sanctuary of Demeter and Kore (Nobis 1997) and other sanctuaries, Messene (Nobis 1994); 13) Athena Alea, Tegea (Vila 2000); 14) Archaic sanctuary, Kythnos (Theodoropoulou 2013); 15) Artemision, Delos (Bevan 1985, 134); 16) Sanctuary of Artemis, Hermes and Poseidon, Kommos (Reese 2000); 17) Sanctuary of Apollo Hylates, Kourio (Davis 1996).*

building blocks organised in accordance with the Hippodamic system, industrial areas and two sanctuaries, one of Demeter and another of Apollo, while the cult of Aphrodite is also corroborated by epigraphic evidence. Extensive cemeteries mainly dated to the 5th and 4th centuries BC lay outside the western part of the enclosure. The coastal position of the city along well-established routes and its richness in natural resources allowed the development of several trading activities. It also encouraged the peaceful co-existence and the mutual cultural influence between colonists and local Thracian populations, as suggested by imported and locally made pottery, Thracian inscriptions in Greek letters and persistence of local technologies (*e.g.* iron-working) (Blakely 2013, 158–61; Tsatsopoulou 1996, 922). The large number of exceptional finds reveals the prosperity of the city and its development into an important trade centre during the 5th and the 4th centuries BC. Zone is mentioned in the tribute lists of the First Athenian League in the 5th century BC, being assessed at two talents, an indication that it was considered the richest if not the largest city of the Samothracian *peraia* (Loukopoulou 2004, 881–2). However, it never became a city of great power; the late 4th–early 3rd century BC saw its economic and social decline, followed by its conquest by the Macedonian king Phillip II (Tsatsopoulou 2005).

The most important civic building of the Archaic period (6th century BC) was the Sanctuary of Apollo (Plate 12), the poliad deity of the city, as evidenced by the 4th-century BC coinage. It is a rectangular building, probably a *prostyle* temple, and belongs to a large building complex (35 × 45m) with a central slab-paved courtyard surrounded by an L-shaped *stoa* (*galleria*). The excavation of the sanctuary revealed a large number of local and imported finds highlighting its importance. These include parts of marble sculptures (*kouroi*), which are not frequently found in northern Greece, iron votives, clay figurines, and large quantities of ceramics, some imported from Attica and some that preserve dedicatory inscriptions to Apollo in Greek or in the Thracian dialect (Tsatsopoulou 1996, 2005;

Tsatsopoulou *et al.* 2015). On the basis of the architecture and the findings, it is hypothesised that the sanctuary was the focus of religious and socio-political life, but also a meeting point for the local Thracians, seafarers and colonists (Blakely 2013, 166–7).

Materials and methods

During the excavation of the Sanctuary of Apollo in the early 1980s animal bones and shells were recovered. No dry-sieving was practiced and no soil samples were processed with a flotation system, thus the recovery rates of small mammals, birds, reptiles, young animals, small molluscs and less robust anatomical elements are expected to be low, while data on charred plant remains are absent. An integrated bioarchaeological approach is thus impossible, but given the rarity of such assemblages of this period, this zooarchaeological study opens new interpretational avenues and highlights the importance of careful recovery of all lines of evidence.

The study of the faunal material followed standard laboratory procedures (Boessneck 1969; Schmidt 1972; Veropoulidou 2011, 25–118; Von den Driesch 1976). All animal bones and shells that bore sufficient diagnostic features were identified to taxon/species level and their analysis followed the quantification methods of NISP (Number of Identified Specimens) and MNI (Minimum Number of Individuals). The main argument against the use of these methods in animal bone analysis is that they often fail to take into account the different fragmentation rates of bones of different species and, in the case of MNI, can lead to an overrepresentation of animals represented by very few specimens (Halstead 2011, 744). However, the NISP and MNI were deliberately chosen for this assemblage, given that some anatomical elements, especially vertebral fragments, could not be discounted from analysis, as according to literary and iconographical sources they constituted a vital part of the sacrificial ritual (see below). Thus, ribs and vertebrae were recorded along with the selected for analysis anatomical parts (Halstead 2011, 744). The calculation of the MNI was conducted simply by separating the most abundant anatomical element of every taxon into side (left/right), using the greatest number as the unit of calculation (White 1953, 397). As the elements of the trunk (ribs, vertebrae) pose serious problems of quantification, they were not included in the calculation of the MNI and the taxonomic representation. They were only examined in terms of taphonomy.

The age at death of the animals was calculated on the basis of dental wear and epiphyseal fusion (Deniz and Grant 1982; Payne 1973; 1982; Silver 1969). The specimens with diagnostic features for sex identification were extremely few. No pathologies were encountered. For the shells, analysis involved identification of body parts and side (left/right valves and umbones), and recording of any natural alterations (abrasion, corrosion, bio-erosion and encrustation) to allow categorisation of specimens as fresh or beach-worn. In both classes of material, the level and mode of fragmentation and any other modifications (gnawing, trampling, burning, butchery marks, incisions, perforations) were recorded (for animal bones see Binford 1981; for shells see Veropoulidou 2011, 77–95) with the aim to understand the cultural and natural taphonomic processes.

The zooarchaeological assemblage

The studied assemblage includes 2,241 animal bones, more than half of which (1,451) could be identified to family or species level (Table 7.1), and 131 marine molluscan remains identified to species level (Table 7.1, Plate 13a) (Nikolaidou 2015; Veropoulidou 2015). The animal bone assemblage includes eight identifiable species: cattle (*Bos taurus*), pig (*Sus scrofa domesticus*), sheep (*Ovis aries*), goat (*Capra hircus*), fox (*Vulpes vulpes*), dog (*Canis familiaris*), red deer (*Cervus elaphus*) and fallow deer (*Dama dama*). The remains of equids (*Equus* sp.) and some of the deer (*Cervus* sp.) bones did not bear sufficient diagnostic features to be identified to species level. Domestic species (NISP: 98.8%, MNI: 136) clearly outnumber wild; the most frequently represented are sheep and goats, followed by cattle. The shell assemblage exhibits average taxonomic variation with 15 different marine species, among which scallops (*Flexopecten glaber*) clearly outnumber all others (NISP: 65.6%, MNI: 47), followed by purple-shells (*Bolinus brandaris*), cockles (*Cerastoderma glaucum*) and thorny-oysters (*Spondylus gaederopus*). All other species, including oysters (*Ostrea edulis*), other purple-shells (*Hexaplex trunculus, Stramonita haemostoma*), mussels (*Mytilus galloprovincialis*) and rayed trough shells (*Mactra stultorum*), are represented by few specimens.

Almost half of the animal remains (NISP: 50.3%, MNI: 71) and one quarter of shells (NISP: 22%, MNI: 13) come from two areas: the *sekos* (*cella*) and the area around the temple. The area north of the temple also had a substantial amount of animal bones (NISP: 15.9%, MNI: 25) along with the largest concentration of shells (NISP: 63%, MNI: 43) (Table 7.1). The remaining were found dispersed and in low percentages throughout the ritual complex. The good state of preservation of bones and shells and the low number of fragments suggest minimal post-depositional disturbance and little, if any, trampling, thus allowing insights into discrete episodes of consumption and deposition in areas where human traffic must have been rather limited.

Animal meat: sacrifices and ritual meals

The taxonomic representation of animals in the Sanctuary of Apollo (Table 7.1) is characterised by the dominance of

Table 7.1 Spatial distribution of animal and molluscan remains (Number of Identified Specimens: NISP, and percentage of NISP, Minimum Number of Individuals: MNI) at the ritual complex of Apollo. Numbers next to molluscan species refer to Plate 13.

	Sekos (Cella)			*North of temple*			*Dispersed*			*Total*			*Common name*
Species	*NISP*	*NISP%*	*MNI*	*NISP*	*NISP%*	*MNI*	*NISP*	*NISP%*	*MNI*	*NISP*	*NISP%*	*MNI*	
Equus sp.	4	0.5	2	1	0.4	1	28	5.7	4	33	2.3	7	Equids
Bos taurus	259	35.5	13	121	52.2	6	190	38.7	12	570	39.3	31	Cattle
Sus scrofa domesticus	5	0.7	1	4	1.7	1	15	3.1	5	24	1.7	7	Pig
Ovis/Capra	300	41.1	24	50	21.6	5	185	37.7	18	535	36.9	47	Sheep/Goat
Ovis aries	122	16.7	25	37	15.9	4	52	10.6	5	211	14.5	34	Sheep
Capra hircus	36	4.9	4	9	3.9	1	15	3.1	4	60	4.1	9	Goat
Canis familiaris	1	0.1	1		0.0			0.0		1	0.1	1	Dog
Vulpes vulpes		0.0		2	0.9	1		0.0		2	0.1	1	Fox
Cervus elaphus		0.0		2	0.9	1	2	0.4	1	4	0.3	2	Red deer
Dama dama		0.0		3	1.3	2	1	0.2	1	4	0.3	3	Fallow deer
Cervus sp.	3	0.4	1	3	1.3	1	3	0.6	1	7	0.5	5	Cervids
Total (animal bones)	730	100	71	232	100	23	491	100	51	1451	100	147	
Flexopecten glaber (1)	19	65.5	9	59	71.1	33	6	31.6	5	84	64.1	47	Scallop
Bolinus brandaris (2)	1	3.4	1	8	9.6	6				9	6.9	7	Purple shell
Cerastoderma glaucum (3)				1	1.2	1	7	36.8	3	8	6.1	4	Cockle
Spondylus gaederopus (4)	1	3.4	1	3	3.6		2	10.5		6	4.6	1	Thorny-oyster
Hexaplex trunculus	2	6.9	1	3	3.6	1				5	3.8	2	Purple shell
Arca noae				2	2.4		2	10.5		4	3.1		Noah's arc
Glycymeris glycymeris (7)				3	3.6					3	2.3		Dog-cockle
Pecten jacobaeus (8)				2	2.4	2				2	1.5	2	St James scallop
Mytilus galloprovincialis (9)	2	6.9	1							2	1.5	1	Mussel
Tonna galea (10)	2	6.9								2	1.5		Tun shells
Patella ulyssiponensis (11)	1	3.4		1	1.2					2	1.5		Limpet
Mactra stultorum							1	5.3	1	1	0.8	1	Rayed trough shell
Stramonita haemostoma							1	5.3	1	1	0.8	1	Purple shell
Cassidaria echinophora				1	1.2					1	0.8		Helmet shell
Ostrea edulis (15)	1	3.4								1	0.8		Oyster
Total (shells)	29	100	13	83	100	43	19	100	10	131	100	66	
Grand total	759		84	315		66	510		61	1582		213	

sheep and goats, the former outnumbering the latter by a ratio of 4:1, a high percentage of cattle and few bones of pigs. This representation accords with textual evidence, as for example the Homeric hymn to Hermes (22, 71) that connects Apollo to pastoral life, being *inter alia* the protector of flocks and cattle, and thus mostly sheep/goats and cattle were sacrificed in his honour (Hatzinikolaou 2007, 116).

Regarding the low amount of pig bones, the zooarchaeological evidence from ancient Greek sanctuaries reveals that these are more numerous in consumption debris than in altar deposits (Ekroth 2007, 261). For example, the altar deposit from the sanctuary of Apollo Hylates at Kourion contained low numbers of pig bones, while they were abundant in the food consumption refuse from the sanctuaries of Demeter and Kore at Corinth, Hercules at Thassos, Apollo and Artemis at Kalapodi, Athena Alea at Tegea and at the temples of Kommos (Fig. 7.1). Meanwhile, low numbers of pigs were encountered in the consumption debris in the Archaic Artemision at Olympia and the Heraion of Samos (Fig. 7.1). It has been suggested that the sacrifice of pigs may have been sometimes performed in contexts other than those of public sacrifice and/or with a different ritual, often resulting in low frequencies of pig bones in altar deposits (*e.g.* Jameson 2014, 215). Homer's *Odyssey* (XIV, 419–38) provides an idea of what those rituals might have involved, like the description of Eumaios' sacrifice of a pig that refers to the burning of hair and meat and to the offering of cooked portions of meat, but does not mention the cutting and burning of thigh bones and tails, as was usually the case for the common sacrificial animals. On the other hand, textual and zooarchaeological evidence often associates the sacrifice of pigs with the cult of Demeter (Forstenpointner 2001, 50). The sanctuary of Demeter at Knossos offers such an example, as in the deposits of the 5th to 2nd centuries BC pigs made up more than 90% of the zooarchaeological material (Moody 2012, 238, Table 14.3). The zooarchaeological assemblage from the sanctuary of Demeter at Zone could shed more light on the issue of the pig representation when its study is completed. Notably, however, a tendency for the slaughtering of young pigs has been attested in ritual deposits (Moody 2012, 239), and therefore, there remains a possibility that the few pig remains in the sanctuary of Apollo are partly due to the lack of recovery of less robust anatomical elements.

The Zone assemblage also yielded a range of a few unusual for ritual contexts animals according to textual and iconographical evidence (Table 7.1). The scarcity of references to wild animals in the ancient sources and the distinction between wild and domestic animals in ancient Greek thought (*e.g.* Segal 1974) has led researchers to conclude that they were never chosen for the *thysia* ritual (Detienne 1989, 8). This is corroborated by zooarchaeological data, as with the exception of the sanctuary of Apollo and Artemis at Kalapodi (Fig. 7.1), there are no indications that wild animals were actually sacrificed. However, recent zooarchaeological analyses have provided evidence for the consumption of wild and other unusual animals' meat, probably after the *thysia*, thus implying that they were brought to the sanctuary possibly to increase the amount of meat eaten. These indications come from several ritual contexts, such as the sanctuaries of Poseidon at Kalaureia, Apollo and Artemis at Kalapodi, Demeter and Kore at Messene, at the Artemision of Ephesos and at the Heraion of Samos (Fig. 7.1). On this basis, it has been suggested that the categorisation of meat as more or less sacred would be influenced by the origin of the animal, the species, the way of killing and cooking, and the context of consumption (Ekroth 2007, 263–7). This pattern fits for instance with Xenophon's writings (*Anabasis*, 5.3.37), mentioning the hunt of wild boars and deer in order to be consumed in sacrificial meals at the sanctuary of Artemis at Skillous, although there is an established association between this particular goddess and wild fauna.

Regarding wild and unusual species representation, equids, red deer, fallow deer, fox and a single dog mandible were found at the sanctuary of Apollo. Of these, red deer has been associated with Apollo (Burkert 2011, 225). An interesting example comes from the sanctuary of Apollo Hylates at Kourion (Fig. 7.1), where iconographic representations and votive offerings underline this close association, the god being the protector of deer which act as symbol of masculine fertility and regeneration (Vernet 2011, 260). Therefore, their presence in the sanctuary, perhaps offered as *aparchai*, may be seen as reflecting particular traits of the god.

In the context of the *thysia* the literary sources describing the burning of god's share on the altar refer mainly to the femora and less to the *osphys* (Ekroth 2009, 130). The pattern differs on vase paintings, which usually depict the burning of the *osphys* on the altar, possibly due to its higher potential of being recognised (Van Straten 1995, 130). The ritual burning of femora and *osphys* is expected to create a zooarchaeological assemblage bearing a series of diagnostic features. Indeed, examples of recently published zooarchaeological assemblages from altar deposits include highly fragmented, carbonised or calcined bones of both anatomical units of sheep, goats, cattle and/or pigs, though there is a clear preference for the burning of femora (Ekroth 2009, 136–9; Forstenpointner 2003, 204; Reese 1989, 68).

In the sanctuary of Apollo at Zone, however, the overall condition and skeletal representation of the zooarchaeological material is rather different. In particular, the condition of animal bones is good, though a small amount bears traces suggestive of gnawing (NISP = 11.2%, NISP = 9.6% from the *sekos*) (Plate 14a). The fragmentation patterns reveal a focus on meat consumption and marrow extraction, as the bones were mainly recovered in the form of shaft splinters and shaft and end fragments (Table 7.2, Plate 14b). Traces of

Table 7.2 Fragmentation patterns (based on NISP) in the sanctuary of Apollo.

Anatomical elements	*Equids*		*Cattle*		*Pig*		*Sheep/Goat*		*Cervids*		*Fox*		*Fallow deer*		*Total*
	NISP	*NISP%*	*NISP*	*NISP%*	*NISP*	*NISP%*	*NISP*	*NISP%*	*NISP*	*NISP%*	*NISP*	*NISP%*	*NISP*	*NISP%*	
Whole	2	13.3	1	0.4			7	1.7							10
Some shaft missing			1	0.4			4	1.0							5
Epiphysis & shaft	2	13.3	24	10.5			82	20.3	3	60.0	2	100	1	25.0	114
Epiphysis splinter			5	2.2			6	1.5							11
Shaft splinter	6	40.0	99	43.2	1	25.0	84	20.8							190
Cylinder	1	6.7	11	4.8			75	18.6							87
Epiphysis	1	6.7	1	0.4			2	0.5							4
Epiphysis & shaft splinter	1	6.7	36	15.7			10	2.5							47
New break	2	13.3	51	22.3	3	75.0	134	33.2	2	40.0			3	75.0	195
Total	15	100.0	229	100.0	4	100.0	404	100.0	5	100.0	2	100	4	100.0	663

Following Binford's (1981) observations on carnivore attrition and marrow extraction, only the long bones were considered relevant to the study of fragmentation. Therefore, species (dog, red deer) only represented by other anatomical parts are not included in the table.

burning were detected on a moderate percentage of specimens (NISP = 19.6%, NISP = 23.3% from the *sekos*), but they bore a greyish colour suggestive of indirect contact with fire, while no carbonised/calcined bones were recovered. With the exception of a chopped deer mandible and a skinned proximal metatarsal of fallow deer, butchery marks were only identified on bones of sheep, goats and cattle (NISP = 7.2%, NISP = 5.6% from the *sekos*). Those included chopping, dismembering, filleting and skinning marks and they were not limited to selected anatomical elements, thus implying that all the carcass processing took place within the area of the sanctuary. Regarding the skeletal representation (Table 7.3), all major parts of the skeleton of cattle, sheep and goats were recovered from the *sekos* of the temple, including femora, but only one cattle caudal vertebra.

The variety of species and body parts, the fragmentation patterns, the absence of carbonised/calcined bones, and the presence of various butchery marks identified on the animal bones appear to be incompatible with an altar context of use. Animal bones rather seem to be the remains of food consumption events that took place within the ritual sphere of the sanctuary, probably after the *thysia*, taking also into account that the largest amounts were found in the *sekos* and the area north of the temple.

These suggestions combined with the skeletal representation of the animals can further provide some interesting insights into the sanctuary rituals. If we accept that the consumption debris is indirect evidence of the actions taking place near the altar, the missing sacra and caudal vertebrae may suggest either that they were completely burnt on the altar or that the actual altar deposit has not been discovered. The first hypothesis accords partially with epigraphic evidence about rituals that involved the complete destruction by fire of the whole animal (*holocausts*), usually associated with hero cults, while the possibility of the total burning of only a part of the animal has also been suggested (Ekroth 1998, 129). Therefore, the missing *osphys* could be an indirect indication of their total burning as the god's portion on the altar. This interpretation was also put forward for the missing *osphys* at the sanctuary of Apollo and Artemis at Kalapodi and at the Archaic Heraion of Samos (Fig. 7.1). Interestingly, the burning of sacra and caudal vertebrae is usually attested in deposits dated to the Classical period and later (Ekroth 2009, 144), such as at the sanctuary of Aphrodite Ourania at Athens (Fig. 7.1), where tail bones made up a major part of the altar deposit, thus being in accordance with iconographic evidence suggesting that curving tails on the altars satisfied deities (Van Straten 1995, 122). Therefore, the skeletal representation in the consumption debris of the sanctuary of Apollo could potentially indicate the sacrificial burning of the *osphys* from as early as the Archaic period in Thrace, reflecting either a local tradition or rituals related to visitors and traders from different parts of the Aegean and the north. The validity of this hypothesis, however, could only be tested if the actual altar deposit were available.

The high frequency of thighbones in the Sanctuary of Apollo needs to be considered in the context of ritual

Table 7.3 Skeletal representation of the main anatomical elements (NISP) at the ritual complex of Apollo.

Anatomical elements	*Equids*	*Cattle*	*Pig*	*Sheep/Goat*	*Sheep*	*Goat*	*Cervids*	*Dog*	*Fox*	*Red deer*	*Fallow deer*
Mandible		32	6	41	59	12	2	1			
Scapula		7		22	9	4	1			1	
Humerus	3	28		32	17		1		1		
Radius	6	33		44	8		2				
Ulna		4	1	1							
Metacarpus	1	38		39	24	9					1
Pelvis	1	10	2	9	6						
Femur	1	35	2	52	1						
Tibia	1	45	1	82	15	5			1		
Metatarsus	3	35	1	52	13	7	2				3
Metapodium		15		4							
Calcaneus		10			3	1				3	
Astragalus	1	15		3	5	2					
Phalanx 1		24		8			1				
Phalanx 2		16									
Phalanx 3		7	1								
Cervical vertebra		9		8							
Thoracic vertebra		6									
Lumbar vertebra		7									

practices other than those performed on the altar. Thighs and leg joints, some of the fleshier body parts of the sacrificial victims, may have been treated as priestly prerogatives, or placed on a table near the altar as food offerings during the performance of the *theoxenia* ritual, and at a later stage probably taken by the priests to be consumed *in situ*. The latter is implied by the remarkably high frequency of traces of contact with fire detected on femora (NISP = 35.8%, NISP = 41.8% from the *sekos*), suggesting that the meat of domestic animals was cooked. The very low presence of wild and unusual species does not allow any safe conclusive remarks regarding their possible ritual treatment. The taphonomic attributes of their bones, and especially the presence of burnt deer bones, could indicate that their meat was also prepared to be consumed *in situ,* although the evidence is too little to confirm this. Additionally, it could be suggested that apart from purely sacrificial meat that was undoubtedly available, meat of non-sacrificial origin might have also been present in order either to increase the amount of meat eaten in the sanctuary and/or to be offered to the god, perhaps as *aparchai*, along with the cooked meat of domestic animals during secondary rituals, like *theoxenia*. Although it is impossible to identify whether such a ritual was performed in honour of Apollo himself or some associate of his, it is interesting that Apollo is one of the gods that have been often linked with *theoxenia* (Jameson 2014, 146, 171).

Molluscan flesh and shells: ritual meals and votive offerings

The taxonomic representation of the shell assemblage in the Sanctuary of Apollo (Table 7.1, Plate 13.a) is characterised by the predominance of scallops (NISP = 65.6%, MNI = 47), a small amount of three different species of purple-shells (NISP = 11.5%, MNI = 10) and some cockle shells (NISP = 6%, MNI = 4). The remaining species (oysters, mussels, thorny-oysters, rayed trough shells) are represented by only a few specimens. The majority of the identified taxa belongs to species that prefer sandy substrates in shallow water zones, such as those that dominate the modern coast of the site. They are native in north Aegean coasts (Zenetos 1996) and they have been encountered at other archaeological sites in Thrace (*e.g.* Boutsidis 2004) and at the various sub-assemblages of Zone (Veropoulidou in prep.). Therefore, it may be suggested that people gathered molluscs from easily accessible areas, possibly not far from the settlement. The identification of similar taxa in other parts of Zone, though in different frequencies, may be taken as an indication of local people's preferences in species selection and as indirect evidence of specific values attributed to these (see below). Nevertheless, the possibility of a non-local provenance cannot be ruled out taking into account their widespread distribution in north Aegean coasts and the assumed visitors at Zone.

A large number of shells (NISP: 83%, n = 109) had been gathered fresh from the sea, thus being suitable for food consumption. The majority of them are preserved intact and in pristine or moderate condition. The dominant species, the scallop, is represented by large specimens (mean: 48.3mm), almost half of which (38 out of 81 valves) bear a dark grey or black mark on their outer concave surface. These marks must have resulted from the direct contact of each individual scallop with fire (Plate 13b), most probably from roasting on coals. Most of these remains were found concentrated in the *sekos* (NISP = 19, MNI = 9) and at the north area of the temple (NISP = 59, MNI = 30). Interestingly, each deposit includes a similar amount of right and left valves (*e.g.* *sekos*-deposit 1, right: 6, left: 9; *sekos*- deposit 2, right: 1; north area of the temple, right: 30, left: 29), thus suggesting that each individual, after being roasted and opened, was deposited *in situ*, while also indicating at least one episode of deposition at *sekos* and one at the north area of the temple.

The total absence of heavily burnt shells indicates that molluscs have not been burnt on the altar, a suggestion that fits with written sources that considered undomesticated and uncultivated foods out of the limits of social order (*e.g.* Seagal 1974), and thus, not appropriate to be sacrificed to gods. On the other hand, roasting of scallops accords with Athenaeus (*Deipnosophists* 86b–92d), who states that scallops were cooked inside their shells to become more nutritious and digestible. Therefore, scallops may represent remains of food consumption events in the *sekos* and the north area of the temple that were left there possibly as reminders of these events. In both contexts there is also a variety of a few other edible molluscs (*sekos* NISP = 11, MNI = 5, north of temple NISP = 15, MNI = 4), including cockles, mussels, thorny-oysters and purple-shells, most of which had been gathered fresh. They do not bear any traces of scorching, and thus it is impossible to determine if they had been cooked, for example by boiling, an appropriate cooking method for a ritual context (Ekroth 2008, 99–100), but one that does not leave any traces on shells (Veropoulidou 2011, 62–8).

The presence of cooked scallops and other edible molluscs inside the *sekos* and within the boundaries of the sanctuary, though puzzling on the basis of textual and iconographic evidence, accords with other archaeological finds from Greek sanctuaries, as for instance those of Artemis, Hermes and Poseidon at Kommos, Demeter at Korinth, Demeter and Kore at Mytilene, and an unknown deity at Kythnos (Fig. 7.1). In these ritual contexts, seafood may have been a bloodless foodstuff offered to the divinities and then consumed by participants (see also Gill 1974, 117–9). Whether mollusc offerings were part of an official or individual initiative of the worshipers cannot be answered, but it is well known that people offered various foods, according to their beliefs and their economic and social status (Antonaccio 2005; Gill 1974).

The cooked scallops, therefore, might have been offered, perhaps as *aparchai*, in *theoxenia* rituals to honour the god(s) with all foods people consumed, while the possible raw molluscs could fall in the case of *trapezomata* rituals. Offerings that were freely available and easy to access are considered inexpensive compared to animal sacrifice but they seem to have been a common part of ritual practice according to textual evidence, usually related to votaries that could not make rich offerings (Antonaccio 2005, 100; Gill 1974). At coastal Zone, the fruits of the sea may have been considered significant, taking into account the maritime setting of the city, while the consumption of molluscs as food in most domestic contexts (Veropoulidou in prep.) indicates that they were a common foodstuff and as such they could have been offered to the god(s) and then consumed by the participants alongside other foods.

The rest of the shells (NISP: 17%, n = 22) bear traces of bio-erosion, encrustation and wear, which means that they had been gathered during beach-combing and without the mollusc (beach-worn). These include one scallop, some thorny-oysters and purple-shells, as well as dog-cockles (*Glycymeris glycimeris*), Noah's arcs (*Arca noae*), limpets (*Patella ullysiponensis*) and oysters. There are also two specimens that have been artificially pierced. The wear in the periphery of the perforation suggests that these shells have been strung and, thus, possibly used as pendants. How and why these specimens ended up in the sanctuary is difficult to establish. However, their large size and good condition implies that they may represent votive offerings dedicated to the divinity, perhaps to ask favours or avert anger. Taking into account the several parts of marble sculptures found in the sanctuary, it would be tempting to suggest that the shells were placed in front of these. Inside sanctuaries there is usually a range of expensive votive offerings, while shells as votive offerings are not featured in literary or epigraphic sources. However, they are still found in several coastal ritual settings, such as at the sanctuaries at Kythnos, of Demeter and Kore at Messene, of Artemis and Apollo at Kalapodi, and at the Artemision at Delos (Fig. 7.1).

The recovery of marine molluscs inside the Sanctuary of Apollo should not come as a surprise considering the marine nature of the god, who was hailed by fishermen, seafarers and colonists (Miller 1939), but also the maritime character of the city and the variable range of worshippers that visited the sanctuary. The finding of marine shells at the sanctuary of Great Gods at Samothrace, the metropolis of Zone, has been also directly associated with the importance of the sea for the islanders (Lehmann 1998, 40). Therefore, an explanation would be that marine molluscs and shells represented offerings and dedications of the local community (both Thracians and colonists), of fishermen for abundant catches or perhaps of travellers/sailors to ensure safe journeys.

Regarding species selection, apart from ecological factors that played a significant role, written sources, ethnographic and archaeological evidence indicates that certain molluscs had specific uses (*e.g.* industrial, medicinal) and/or symbolic attributes that should be taken into account in their interpretation. A prominent example is the clear preference for scallops in the Sanctuary of Apollo, considering that a preference for cockles and only a moderate number of scallops have been documented in the domestic contexts at Zone (Veropoulidou in prep.). It has been suggested that scallop represents the 'womb' (Cook 1925, 132–3 n. 2, 302 n. 2), thus, being interpreted as the ultimate feminine symbol of fertility, usually associated with Demeter and Aphrodite (Avramova 2007, 28). According to one version of the myth, Aphrodite, a goddess originating from the east, rose out of a scallop shell (Flory 1988), an iconographic motif known from the Classical (*e.g.* Kefalidou 2009) and Hellenistic periods (Thompson 1952, 148–9) that persisted until the Renaissance (Wheeler 1957). This connection of Aphrodite with scallops, however, cannot be extended to the Archaic period on current evidence. In contrast, the connection of scallops with Demeter and Kore is dated to as far back as the Archaic period on the basis of scallop shells or imitations of scallops in clay and metal that were found in Archaic and Classical Greek sanctuaries dedicated to them. Examples include two beach-worn scallops at Messene, imitations of scallops in metal at Cyrene (Warden 1990) and Korinth (Stroud 1968, 326–7), and a concentration of cooked scallops and oysters at the acropolis at Mytilene (Fig. 7.1). Furthermore, an association of scallops with the mysteries of Demeter celebrated at Eleusis has been suggested, as for example the *Aloa* festival, where women dedicated the first fruits of land and sea and then they marched holding scallops and phallic symbols to honour Demeter (Johnson 1994, 230), thus allowing insights into a link between the symbolism of human sexual vigour and fertility with that of agriculture (Jameson 2014, 75). In addition, scallops and their imitations in clay and metal were used as *pyxides* (part of women's toilette paraphernalia, probably for keeping jewellery and cosmetic substances) and have been found in female graves dated to the late Classical and Hellenistic periods (Nankov 2011, 8–13). This has led researchers to suggest that scallops symbolised immortality and the expectation for happy life after death (Thompson 1958, 148). Interestingly, a clay vessel that imitates a scallop shell has been also found in a burial tentatively dated to the Hellenistic period at the cemetery of Zone (Tsatsopoulou pers. comm.).

In ancient Zone, no scallops were found in the Sanctuary of Demeter (Veropoulidou in prep.), and hence an association with Demeter cannot be established. On the other hand, epigraphic evidence in Zone indicates the cult of Aphrodite. Considering that the phenomenon of visiting gods seems to have been a casual practice of Greeks (Antonaccio 2005, 108), a tempting explanation could be that in the sanctuary of Apollo people performed also rituals in honour of Aphrodite that involved scallops as early as the Archaic period. This practice could reflect a Thracian local tradition or imply visitors from the eastern Mediterranean.

Cultic practices in the sanctuary: local aspects and social implications

The detailed and contextual analysis of terrestrial and marine faunal remains indicates both similarities and differences between the archaeological data and the practices testified in ancient texts and iconography. The main correspondence is the preference for domestic animals, especially sheep and goats, species that have been associated with Apollo.

However, the discrepancies are many more. The variety of species identified archaeologically that can be interpreted as remains of ritual meals is notable, although it is unclear if these animals played an active role in the ritual performance. So, for instance, the offering of such a large variety of foods to the god(s), some maybe as *aparchai*, during *theoxenia* and/or *trapezomata* rituals could be the result of an anticipation to create closer bonds with the god(s), of the worship of different divinities or it was simply a mechanism to increase the amount of meat eaten. Furthermore, there is evidence for the consumption of particular parts of the animal skeleton (thighs) that could have been offered to the priests as honorary portions and to the worshippers along with meat from wild animals and molluscs. Remarkably, the total burning of the *osphys* during the *thysia* may be potentially extended to as early as the Archaic period. Moreover, the observed preference for the consumption of scallops could be related to the maritime nature of Apollo, but possibly also to Aphrodite. Finally, the several beach-worn marine shells that could represent votive offerings highlight the variety of products and artefacts dedicated to the god(s). Therefore, the zooarchaeological data point towards ritual meals with various ingredients and different kinds of votive offerings that may be taken as indicative of local traditions, folk beliefs or group interests, which, though hard to reconstruct, shed light on the complexity and the polymorphism of ancient Greek cultic practices, while also highlighting the particularities that emerged from mixing local and colonial rituals.

In the sanctuary of Apollo at Zone, animal sacrifices, possible secondary rituals, such as *theoxenia*, and ritual meals and offerings would have created a terrain of social interactions, but also they could have helped promote the interconnectedness of people from various backgrounds, contributing to the exchange of traditions and ideas from across the Aegean. Furthermore, the range of votive items, including prestigious artefacts and 'humble' marine shells, allows insights into the performance of the cult through their accumulation and display. The diversity of the offerings, a form of both sacrifice and a gift that was not so rule-bound as the sacrifice, may indicate the active construction of cult and the shaping or challenging of social categories, thus providing a framework of action (Antonaccio 2005, 100–2). The variety in ritual practices and performance, similarly also for example to the persistence of local traditions in iron-working and the Thracian inscriptions in Greek letters

(Blakely 2013, 160–2), could be taken as indications that local cultural conditions and colonial ideologies were entangled.

To conclude, this zooarchaeological assemblage provided a rare opportunity to discuss the creation and performance of ritual activity in a Greek colony during the Archaic period. It demonstrated that the materialisation of ritual activities was expressed in multiple ways that depended on the period, place and context. The variety of ritual acts, foodstuffs and offerings documented at the sanctuary of Apollo at Zone accords with the overall results of other zooarchaeological studies in ritual contexts across the Aegean but underlines the existence of various ways of honouring and perceiving gods. This study has highlighted the complexity of Greek ritual and religion and emphasised the potential of bioarchaeological analysis to shed new light on such concepts and practices.

Acknowledgements

We would like to thank the Honorary Director of the Ephorate of Antiquities at Evros Mr. Polyxeni Tsatsopoulou for the invitation to study and publish the zooarchaeological material. We are most grateful to Prof. Stelios Andreou for providing the facilities of the Toumba Thessaloniki Excavation Laboratory for the study. Our warmest thanks also to Prof. Paul Halstead, Anastasia Vasileiadou and Stefania Michalopoulou for their help with references, and to Nikos Valasiadis for drawing Figure 7.1 and processing the images. Finally, we would like to thank the two anonymous reviewers for their constructive comments. All errors remain our own.

References

Aeschylus (nd) *Prometheus Bound*. Transl. T. Roussos [1991]. Athena, Kaktos.

Athenaeus (nd) *Deipnosophistae*. Ed. and transl. C. B. Culick [1927–41, Loeb Classical Library]. Cambridge MA, Harvard University Press

Antonaccio, C. M. (2005) Dedications and the character of cult. In R. Hägg and B. Alroth (eds.) *Greek Sacrificial Ritual, Olympian and Chthonian. Proceedings of the Sixth International Seminar on Ancient Greek cult, 25–27 April 1997, organized by the Department of Classical Archaeology and Ancient History, (Acta Ath-8°, 18)*, 99–112. Stockholm, Göteborg University.

Avramova, M. (2007) The symbolism of shells. In A. Iakovidou (ed.) *Thrace in the Graeco-Roman World. Proceedings of the 10th International Congress of Thracology, Komotini – Alexandroupolis, 18–23 October 2005*, 24–35. Athens, National Hellenic Research Foundation.

Bammer, A., Brein, F. and Wolff, P. (1978) Das Tieropfer am Artemisaltar von Ephesos. In S. Sahin, E. Schwertheim and J. Wagner (eds.) *Studien zur Religion und Kultur Kleinasiens, vol.1, Festschrift für Friedrich Karl Dörner*, 107–57. Leiden, E. J. Brill.

Benecke, N. (2006) Animal sacrifice at the late archaic Artemision of Olympia: the archaeozoological evidence. In U. Tecchiati and B. Sala (eds.) *Archaeozoological Studies in Honour of Alfredo Riedel*, 153–60. Province of Bolzano, Ripartizione Beni Culturali.

Bevan, E. (1985) Representations of Animals in Sanctuaries of Artemis and other Olympian Deities. PhD thesis. University of Edinburgh.

Binford, L. R. (1981) *Bones: Ancient Men and Modern Myths*. New York: Academic Press.

Blakely, S. (2013) *Daimones* in the Thracian Sea: Mysteries, Iron, and Metaphors, *Archiv für Religionsgeschichte* 14(1), 155–82.

Boessneck, J. (1969) Osteological differences between sheep (*Ovis aries* Linné) and goat (*Capra hircus* Linné). In D. Brothwell and E. S. Higgs (eds.) *Science in Archaeology: a Survey of Progress and Research*, 2nd edn, 331–58. London, Thames and Hudson.

Boessneck, J. and von den Driesch, A. (1988) *Knochenabfall von Opfermahlen und Weihgaben aus dem Heraion von Samos (7. Jh. v. Chr.)*. Munich, Institut für Palaeoanatomie, Domestikationsforschung und Geschichte der Tiermedizin der Universität München.

Bookides, N., Hansen, J., Snyder, L. and Goldberg, P. (1999) Dining in the Sanctuary of Demeter and Kore at Corinth. *Hesperia* 68, 1–54.

Boutsidis, C. (2004) *The Malacological Material from the Neolithic Makri, Evros*. Master diss., Aristotle University of Thessaloniki.

Bremmer, J. N. (2007) Greek normative animal sacrifice. In D. Ogden (ed.) *A Companion to Greek Religion*, 132–44. Oxford, Blackwell.

Bruit, L. (1990) The meal at the Hyakinthia: ritual consumption and offering. In O. Murray (ed.) *Sympotica. A Symposium on the Symposion*, 162–74. Oxford, Oxford University Press.

Burkert, W. (1983) *Homo Necans. The Anthropology of Ancient Greek Sacrificial Ritual and Myth*. Berkeley CA, University of California Press.

Burkert, W. (1985) *Greek Religion. Archaic and Classical*. Transl. J. Raffan. Oxford, Blackwell.

Burkert, W. (2011) [1977] *Griechische Religion der Archaischen und Klassischen Epoche, Zweite Überarbeitete und Erweiterte Auflage*. Stuttgart, Kohlhammer.

Davis, S. (1996) Animal sacrifices. In D. Buitron-Oliver (ed.) The Sanctuary of Apollo Hylates at Kourion: excavations in the archaic precinct. *Studies in Mediterranean Archaeology* CIX, 181–2.

Deniz, E. and Payne, S. (1982) Eruption and wear in the mandibular dentition as a guide to ageing Turkish angora goats. In B. Wilson, C. Grigson and S. Payne (eds.) *Ageing and Sexing Animal Bones from Archaeological Sites*, BAR British Series 109, 155–205. Oxford, Archaeopress.

Detienne, M. (1989) Culinary practices and the spirit of sacrifice. In M. Detienne and J.-P. Vernant (eds.) *The Cuisine of Sacrifice Among the Greeks*, 1–20. Chicago IL, University of Chicago Press.

Ekroth, G. (1998) Altars in Greek Hero-Cults: A Review of the Archaeological Evidence. In R. Hägg (ed.) *Ancient Greek Cult Practice from the Archaeological Evidence, Proceedings of the Fourth International Seminar on Ancient Greek Cult, organized by the Swedish Institute at Athens October 1993*, 22–24. Stockholm, Åströms Förlag.

Ekroth, G. (2002) *The Sacrificial Rituals of Greek Hero-Cults in the Archaic to the Early Hellenistic Periods*. Kernos Supplement 12. Liège, Centre International d'Étude de la Religion Grecque Antique.

Ekroth, G. (2007) Meat in ancient Greece: sacrificial, sacred or secular? *Food and History* 5(1), 249–22.

Ekroth, G. (2008) Burnt, cooked or raw? Divine and human culinary desires at Greek animal sacrifice. In E. Stavrianopoulou, A. Michaels and Cl. Ambos (eds.) *Transformations in Sacrificial Practices. From Antiquity to Modern Times*, 87–111. Berlin, Lit Verlag.

Ekroth, G. (2009) Thighs or tails? The osteological evidence as a source for Greek ritual norms. In P. Brulé (ed.) *La Norme en Matière Religieuse en Grèce Ancienne, Actes du XIe Colloque du CIERGA (Rennes, Septembre 2007)*, 125–51. Kernos Supplément 21. Liège, Centre International d'Étude de la Religion Grecque Antique.

Ekroth, G. (2011) Meat for the gods. In V. Pirenne-Delforge and F. Prescendi (eds.) *'Nourrir les dieux?' Sacrifice et représentation du divine*, 15–41. Kernos Supplément 26. Liège, Centre International d'Étude de la Religion Grecque Antique.

Ekroth, G. and Wallensten, J. (2013) *Bones, Behavior and Belief. The Zooarchaeological Evidence as a Source for Ritual Practice in Ancient Greece and Beyond*. Acta Instituti Atheniensis Regni Sueciae, Series in 4/55. Stockholm, Swedish Institute at Athens.

Felsch, R. C. S. (2001) Opferhandlungen des Alltagslebens im Heiligtum der Artemis Elaphebolos von Hyampolis in den Phasen SH IIIC – Spätgeometrisch. In R. Laffineur and R. Hägg (eds.) *Potnia. Deities and Religion in the Aegean Bronze Age, Proceedings of the 8th International Aegean Conference Göteborg, 12–15 April 2000, Aegaeum 22*, 193–8. Liège–Austin TX, Université de Liège.

Flory, M.B. (1988) Pearls for Venus. *Historia* 37, 498–504.

Fogelin, L. (2007) The archaeology of religious ritual. *Annual Review of Anthropology* 36, 55–71.

Forstenpointner, G. (2001) Demeter im Artemision? – Archäozoologische Überlegungen zu den Schweineknochenfunden aus dem Artemision. In U. Muss (ed.) *Der Kosmos der Artemis von Ephesos*, 49–70. Sonderschriften Band 37. Vienna, Österreichisches Archäologisches Institut.

Forstenpointner, G. (2003) Promethean legacy: investigations into the ritual procedure of 'Olympian' sacrifice. In E. Kotjabopoulou, Y. Hamilakis, P. Halstead, C. Gamble and P. Elefanti (eds.) *Zooarchaeology in Greece: Recent Advances*, 203–13. London, British School at Athens.

Gill, D. (1974) *Trapezomata*: a neglected aspect of Greek sacrifice. *The Harvard Theological Review* 67(2), 117–37.

Grant, A. (1982) The Use of Tooth Wear as a Guide to the Age of Domestic Ungulates. In B. Wilson, C. Grigson and S. Payne (eds.) *Ageing and Sexing Animal Bones from Archaeological Sites*, BAR British Series 109, 94–108. Oxford, Archaeopress.

Halstead, P. (2011) The faunal remains. In D. J. Pullen (ed.) *The Early Bronze Age Village of Tsoungiza Hill (vol. 1)*, 741–803. Princeton NJ, The American School of Classical Studies at Athens.

Hatzinikolaou, K. G. (2007) Oi latreies ton Theon kai ton Eroon stin Ano Makedonia kata tin arhaiotita (Elimeia, Eordaia, Orestida, Ligkistida). PhD thesis, Aristotle University of Thessaloniki.

Hesiod (nd) *Theogony*. Transl. P. G. Lekatsas [1938]. Athens, Papyros.

Homeri Opera (nd) *Hymni Cyclvs Fragmenta Margites Batr. Vitae, V* [1965]. Oxford Classical Texts. Oxford University Press

Homeri Opera (nd) *Iliadis I–XII* [1966]. Oxford Classical Texts. Oxford University Press.

Homeri Opera (nd) *Odysseae I–XII* [1965]. Oxford Classical Texts. Oxford University Press.

Jameson, M. H. (2014) *Cults and Rites in Ancient Greece. Essays on Religion and Society*. Cambridge, Cambridge University Press.

Jameson, M. H. (1994) Theoxenia. In R. Hägg (ed.) *Ancient Greek Cult Practice from the Epigraphical Evidence. Proceedings of the Second International Seminar on Ancient Greek Cult, Organized by the Swedish Institute at Athens, 22–24 November 1991*, 35–57. Stockholm, Acta Instituti Atheniensis Regni Sueciae 8–13.

Jim, T. S. F. (2011) The vocabulary of ἀπάρχεσθαι, ἀπαρχή and related terms in Archaic and Classical Greece. *Kernos* 24, 39–58.

Johnson, B. (1994) *Lady of the Beasts: the Goddess and her Sacred Animals*. Rochester VT, Inner Traditions International.

Kefalidou, E. (2009) O Eros sti gennisi tis Aphroditis. In N. C. Stampolidis and G. Tasoulas (eds.) *EROS. Apo tin Kosmogonia tou Hesiodou stin Estere Archeotita*, 79–80. Athens, Museum of Cycladic Art.

Kindt, J. (2015) Personal religion: a productive category for the study of ancient Greek religion? *Journal of Hellenic Studies* 135, 35–50.

Larson, J. (2007) *Ancient Greek Cults. A Guide*. New York, Routledge.

Lehmann, K. (1998) *Samothrake. Odigos ton anaskafon kai tou Mouseiou*. Transl. I. M. Akamatis. Thessaloniki, Institouto Kalon Technon tou Panepistimiou tis Neas Yorkis.

Loukopoulou, L. (2004) Thrace from Nestos to Hebros. In M. H. Hansen and T. H. Nielsen (eds.) *An Inventory of Archaic and Classical Poleis*, 870–83. New York, Oxford University Press.

Miller, R.D. (1939) The Origin and Original Nature of Apollo. PhD thesis, University of Pennsylvania.

Moody, J. (2012) Hinterlands and hinterseas: resources and production zones in Bronze Age and Iron Age Crete. In G. Cadogan, M. Iacovou, K. Kopaka and J. Whitley (eds.) *Parallel Lives: Ancient Island Societies in Crete and Cyrpus*, 233–71. London, British School at Athens.

Mylona, D. (2013) Dealing with the unexpected: unusual animals in an Early Roman cistern fill in the Sanctuary of Poseidon at Kalaureia, Poros. In G. Ekroth and J. Wallensten (eds.) *Bones, Behaviour and Belief: The Zooarchaeological Evidence as a Source for Ritual Practice in Ancient Greece and Beyond*, 149–66. Acta Instituti Atheniensis Regni Sueciae, Series in 4/55. Stockholm, Swedish Institute at Athens.

Nankov, E. (2011) Berenike bids farewell to Seuthes III: the silver-gilt scallop shell pyxis from the Golyama Kosmatka Tumulus. *Archaeologia Bulgarica* XV(3), 1–22.

Nikolaidou, D. (2015) Thysies kai prosfores: ta zooarcheologika dedomena. In P. Tsatsopoulou-Kaloudi, C. Brixhe, C. Pardalidou, S. Iliopoulou, K. Kaloudis, M. Galani-Krikou, A. Zournatzi, P.

Tselekas, R. Veropoulidou and D. Nikolaidou (eds.) *Archaia Zone I – To Iero tou Apollona*, 944–63. Komotini, Ypourgeio Politismou kai Athlitismou – Eforeia Archaiotiton Evrou.

Nobis, G. (1994) Die Tierreste aus dem antiken Messene-Grabung 1990/91. In M. Kokabiand J. Wahl (eds.) *Beiträge zur Archäozoologie und Prähistorischen Anthropologie8. Arbeitstreffen der Osteologen*, 297–313. Stuttgart, Theiss.

Nobis, G. (1997) Tieropfer aus einem Heroen- und Demeterheiligtum des antiken Messene (SW Peloponnes, Griechenland). Grabungen 1992 bis 1996. *Tier und Museum* 5(4), 97–111.

Pariente, A., des Courtils, J. and Gardeisen, A. (1996) Sacrifices d' animaux à l' Hérakleion de Thasos. *Bulletin de Correspondence Hellénique* 120(2), 799–820.

Payne, S. (1973) Kill-off patterns in sheep and goats: the mandibles from Aşvan Kale. *Anatolian Studies* 23, 281–303.

Raja, R. and Rüpke, J. (2015a) Archaeology of religion, material religion, and the ancient world. In R. Raja and J. Rüpke (eds.) *A Companion to the Archaeology of Religion in the Ancient World*, 1–26. West Sussex, Wiley-Blackwell.

Reese, D. S. (1984) Faunal remains from the Kommos temples. *American Journal of Archaeology* 88(2), 257.

Reese, D. S. (1989) Faunal remains from the altar of Aphrodite Ourania, Athens. *Hesperia* 58(1), 63–72.

Reese, D. S. (2000) Marine invertebrates. In J. Shaw and M. Shaw (eds.) *The Greek Sanctuary. Kommos: An Excavation on the South Coast of Crete, Vol. IV*, 571–642. Princeton NJ, Princeton University Press.

Rouse, W. H. D. (1902) *Greek Votive Offerings: An Essay in the History of Greek Religion*. Cambridge, Cambridge University Press.

Ruscillo, D. (1993) Faunal remains from the Acropolis site, Mytilene. *Echos du Monde Classique* XXXVII (12), 201–10.

Schmid, E. (1972) *Atlas of Animal Bones*. Amsterdam, Elsevier.

Segal, C. (1974) The raw and the cooked in Greek literature: structure, values, metaphor. *The Classical Journal* 69(4), 289–308.

Silver, A. (1969) The ageing of domestic animals. In D. Brothwell and E. S. Higgs (eds.) *Science in Archaeology: A Survey of Progress and Research*, 283–302. London, Thames and Hudson.

Sophocles (nd) *Antigone*. Transl. K. Georgousopoulos [1994]. Athena, Kaktos.

Stanzel, M. (1991) Die Tierreste aus dem Artemis-/Apollon-Heiligtum bei Kalapodi in Böotien/Griechenland. PhD thesis, Ludwig-Maximilians-Universität.

Stavrianopoulou, E. (2008) Dynamic sacrifices? Some introductory remarks. In E. Stavrianopoulou, A. Michaels and C. Ambos (eds.) *Transformations in Sacrificial Practices from Antiquity to Modern Times. Proceedings of an International Colloquium, Heidelberg, 12–14 July 2006*, 1–16. Berlin, Lit Verlag.

Stroud, R. S. (1968) The sanctuary of Demeter and Kore on Acrocorinth: preliminary report II: 1964–1965. *Hesperia* 37, 299–330.

Theodoropoulou, T. (2013) The sea in the temple? Shells, fish and corals from the sanctuary of the ancient town of Kythnos and other marine stories of cult. In G. Ekroth and J. Wallensten (eds.) *Bones, Behaviour and Belief: The Zooarchaeological Evidence as a Source for Ritual Practice in Ancient Greece and Beyond*, 197–222. Acta Instituti Atheniensis Regni Sueciae, Series in 4/55. Stockholm, Swedish Institute at Athens.

Tiverios, M. (2008) Greek colonisation of the Northern Aegean. In G. R. Tsetskhladze (ed.) *Greek Colonisation. An Account of Greek Colonies and other Settlements Overseas. Vol. 2*, 1–154. Leiden–Boston MA, Brill.

Thompson, D. B. (1952) Three centuries of Hellenistic terracottas. *Hesperia* 21(2), 116–64.

Tsatsopoulou, P. (1996) Mesimvria-Zone, 1987–1997. *Archaeological Work in Macedonia and Thrace* 10(A), 917–26.

Tsatsopoulou, P. (2005) Zoni. In L.D. Loukopoulou, A. Zournatzi, M.G. Parisaki and S. Psoma (eds.) *Epigraphes tis Thrakis tou Aigaiou Metaksi ton Potamon Nestou and Evrou (nomoi Ksanthis, Rodopis and Evrou)*, 505–30. Athens, National Institute of Research and Ministry of Culture.

Tsatsopoulou-Kaloudi, P., Brixhe, C., Pardalidou, C., Iliopoulou, S., Kaloudis, K., Galani-Krikou, M., Zournatzi, A., Tselekas, P., Veropoulidou, R. and Nikolaidou, D. (eds.) (2015) *Archaia Zone I – To Iero tou Apollona*. Komotini, Ypourgeio Politismou kai Athlitismou – Eforeia Archaiotiton Evrou.

Van Straten, F. (1995). *Hierà Kalá. Images of Animal Sacrifice in Archaic and Classical Greece*. Leiden, Brill Academic Publishers.

Vernet, Y. (2011). L'Apollon chypriote, de la nature et des animaux, *Cahier du Centre d' Études Chypriotes* 41, 251–64.

Veropoulidou, R. (2011) Shells from the settlement of the Thermaic Gulf. Reconstructing molluscan exploitation during the Neolithic and the Bronze Age. PhD thesis, Aristotle University of Thessaloniki.

Veropoulidou, R. (2015) Trofes kai anathimata: i symvoli tis ostreoarheologikis meletis. In P. Tsatsopoulou-Kaloudi, C. Brixhe, C. Pardalidou, S. Iliopoulou, K. Kaloudis, M. Galani-Krikou, A. Zournatzi, P. Tselekas, R. Veropoulidou and D. Nikolaidou (eds.) *Archaia Zone I – To Iero tou Apollona*, 965–63. Komotini, Ypourgeio Politismou kai Athlitismou – Eforeia Archaiotiton Evrou.

Vila, E. (2000) Bones remains from sacrificial places: the temples of Athena Alea at Tegea and of Asea on Agios Elias (the Peloponnese, Greece). In M. Mashkour, A. M. Choyke, H. Buitenhuis and F. Poplin (eds.) *Archaeozoology of the Near East IVB. Proceedings of the Fourth International Symposium on the Archaeozoology of Southwestern Asia and Adjacent Areas*, 197–205. ARC-Publicatie. Groningen, Center for Archaeological Research and Consultancy.

Von den Driesch, A. (1976) *A Guide to the Measurement of Animal Bones from Archaeological Sites*. Cambridge, Peabody Museum of Archaeology and Ethnology.

Warden, P. G. (1990) The small finds. In D. White (ed.) *The Extramural Sanctuary of Demeter and Persephone at Cyrene, Libya. Final Reports* IV, 36–86. Philadelphia, University of Pennsylvania Museum of Archaeology and Anthropology.

Wheeler, M. (1957) A symbol in ancient times. In I. Cox (ed.) *The Scallop: Studies of a Shell and its Influences on Humankind*, 33–48. London, Shell Transport and Trading Company.

White, T. E. (1953) A method of calculating the dietary percentages of various food animals utilized by aboriginal peoples. *American Antiquity* 18, 396–8.

Xenophon (nd) *Anabasis*. Transl. C. L. Brownson [1998, Loeb Classical Library]. Harvard University Press.

Zenetos A. (1996) The marine Bivalvia (Mollusca) of Greece. In *Fauna Greciae VII. Athens*, Hellenic Zoological Society and NCMR.

Chapter 8

Animals and Rituals in Iron Age Iberian Settlements in the Region of Valencia, Spain

Maria Pilar Iborra Eres

Introduction

In the last few decades considerable advances have been made in the study of Iron Age faunal assemblages in the Iberian Peninsula, especially in regards to economic strategies and animal husbandry (*e.g.* Albizuri and Nadal 1999; Iborra 2004; Valenzuela 2008), but also, to some extent, to ritual practices involving animals in domestic spaces (Albizuri 2011; Casellas 1995; Iborra 2004; Miro and Molist 1990; Nieto 2013; Valenzuela 2008). This study adds much needed new data to the existing corpus of ritual deposits by presenting six newly excavated assemblages of faunal remains, interpreted as part of Iron Age domestic and community rituals in the modern-day region of Valencia, Spain. In particular, the focus of the paper is on rituals performed in houses and defensive structures in three Iberian *oppida:* El Molón (Lorrio *et al.* 2014), La Bastida de les Alcusses (Vives-Ferrándiz *et al.* 2015), and El Puig d'Alcoi (Grau and Segura 2013), all dated to between the 5th and the 2nd centuries BC and located in two different Iberian Iron Age territories. The common features of the deposits under study are the inclusion of slaughtered and/or consumed domestic animals and their formal deposition. Such rituals are not necessarily related to religious beliefs (Bradley 2003; Brück 1999), but they can be related to daily life. In this framework, ritual practices are understood here as symbolic actions, which may have very diverse functions, often repeated in a regular pattern (Bradley 2003; 2005; Grant 1989; Merrifield 1987).

Iberian culture: Iron Age society, animals and economy in the region of Valencia

Before tackling any ritual practices it is necessary to sketch the context of Iron Age society in the study area. The term Iberian culture, which is essentially an archaeological designation, has been employed to define the group of people who inhabited the eastern strip of the Iberian Peninsula roughly between 600 and 100 BC. The Iberian society was highly urbanised and socially stratified (*e.g.* Ruiz 2008). Archaeological studies on funerary spaces, settlement patterns and households have expanded knowledge of the Iberian Iron Age society and have helped establish the presence of economic elites, landowners, and peasantry (*e.g.* Grau 2002; Vives Ferrándiz 2013). From a political point of view, this world was structured into a variety of territories characterised by a complex pattern of hierarchical settlements. The study of landscape and settlement structures in combination with archaeobotanical and zooarchaeological data allowed the recognition of a variety of settlement patterns. During the first period of the Iberian culture (550–400 BC) centralised political units emerged, which were fortified sites, *oppida*, and were located on hilltops, controlling their surrounding valley. These *oppida* were associated with a series of lesser rural sites (Bonet and Vives Ferrándiz 2011, 246; Grau 2002, 250). Between the second half of the 5th century BC and the Roman conquest, a new organisational structure related to the exploitation of resources and the protection of the territory developed. According to current theories the sites in this period showed a hierarchical structure and the functional specialisation of the settlements was geared towards the protection and exploitation of resources and the storage and distribution of surplus (Iborra and Pérez Jordà 2013; Perez Jordà and Peña-Chocarro 2013). The main population nucleus was the city, a large site of around 10–2 hectares and the capital of the political area. Other settlements in the same territory were small, and, depending on their size, have been categorised as small towns, hamlets, farmsteads, hillforts or military establishments. Ritual sites, such as sanctuaries and cemeteries, have also been found.

The overall area had been incorporated into the Mediterranean trading routes by the Iron Age and the inhabitants used some of the local agricultural produce, such as cereals, grapes and olives (Pérez Jordá *et al.* 2000, 158) as well as other goods, like metallurgical products, in order to participate in these commercial networks (Sanmarti 2009). However, the society was still predominantly agrarian, with agriculture and livestock husbandry constituting the subsistence basis (Bonet and Vives Ferrándiz 2011; Pérez Jordà *et al.* 2000).

The extensive study of animal bones from Iron Age settlements located in Iberian territories in the present-day region of Valencia, from the mouth of the Segura River to the Palancia River, has revealed a considerable diversity in terms of the animals raised (Iborra 2004; Pérez Jordà *et al.* 2000). The differences in the species represented, the anatomical units preserved, and the age at which animals were killed have indicated different systems for managing animals. Changes observed in species proportions and in the mortality profile showed the potential for production of animals or their by-products for sale or barter, which could reflect surplus production. The main species present were domestic sheep (*Ovis aries*), goat (*Capra hircus*), pig (*Sus domesticus*), cattle (*Bos taurus*), horse (*Equus caballus*), donkey (*Equus asinus*, and the hybrid forms), dog (*Canis familiaris*) and chicken (*Gallus gallus domesticus*). There was, however, a clear emphasis on sheep/goat livestock, with a predominance of sheep. The presence of cattle and pigs varied according to territory, and the decision to breed cattle and equids depended on the availability of water resources (Iborra 2004; Iborra and Pérez Jordà 2013).

Wild resources were always present although their importance varied. Hunting may have been a leisure activity and/or could have been carried out to protect crops or supplement the diet, as the age and gender profiles of carcases indicate (Iborra 2004, 367). The main species hunted were red deer (*Cervus elaphus*) and rabbit (*Oryctolagus cuniculus*). Other wild species were also found but in low numbers, including roe deer (*Capreolus capreolus*), wild boar (*Sus scrofa*), bear (*Ursus arctos*), badger (*Meles meles*), red fox (*Vulpes vulpes*), hare (*Lepus granatensis*), and lynx (*Lynx pardinus*). The Iron Age animal bone assemblage of the area contains also birds, such as partridge (*Alectoris rufa*), golden eagle (*Aquila chrysaetos*), griffon vulture (*Gyps fulvus*), mallard (*Anas platyrhynchos*), little bustard (*Tetrax tetrax*), pigeon (*Columba livia/oenas*), gull (*Larus sp.*) and Cory's shearwater (*Calonectris diomedea*) (Iborra 2004; Iborra and Pérez Jordà 2013).

The *oppida* under study

Deposits from three *oppida* have been selected for detailed analysis for the purposes of this study. The first one, El Molón, is located in the northwestern corner of the modern-day province of Valencia, near the regional border with Cuenca. The site occupied a privileged position in an area lying between the territories of the Iberians to the south and east (the *Edetania*) and the Celts to the north (Fig. 8.1). Throughout the first millennium BC, from the 7th to the 1st century, a long sequence of occupation has been recorded. During the fourth century, the site was fortified with powerful defences in the form of a dry-stone perimeter wall at the most accessible areas. The eastern approach to the site was protected by a rectangular tower, preceded by forecourts and a ditch with three levels (Lorrio *et al.* 2014).

The La Bastida de les Alcusses *oppidum* was located on a hilltop at the western part of the Iberian territory of *Contestania* (Fig. 8.1). The settlement covered four hectares and was surrounded by a wide wall with a second wall located at the west part – dated to a later phase – bounding an area of 1.5 hectares. It had four gates and three towers and its plan shows a main road, smaller streets, and squares with houses of varying sizes. It was occupied during the 4th century BC and was abandoned as a result of violent conflict with other Iberian groups in the area. The excavation of the site revealed the presence of weapons scattered throughout the site, walled gates and houses destroyed by fire with household items *in situ* (Bonet and Vives Ferrándiz 2011).

The last *oppidum* studied here is Puig d'Alcoi (Alcoi, Alicante), which was located in the central area of ancient *Contestania*, in the northern inland area of the present-day province of Alicante, on a high hill 888m.a.s.l. (Fig. 8.1). The site covered approximately 1.5 hectares; it was occupied for a long period of time from the 7th to the 4th centuries BC and controlled the Iberian territory in the surrounding valley (*e.g.* Grau 2002). In this oppidum house 200 stood out, as it differed from the rest of the buildings in its construction methods, because it followed the Punic architectural tradition, similar to that documented in the Phoenician *colonia* of La Fonteta de Guardamar del Segura (Alicante) located 90km away on the coastline (Gonzalez Prats 2014; Rouillard 2010).

Materials and methods

All six deposits selected for the purpose of this study are from closed, stratigraphically well-defined, contexts. Each of these can be considered as 'special animal deposit' (Grant 1984; Hill 1995) and at least some of them as 'associated bone groups' (ABGs) following Morris (2011, 13). Their study was carried out by means of standard zooarchaeological analysis. The species identification was conducted using the reference collection of the IVC+R Laboratory of Archaeology and Palaeontology in Valencia. The taxon, age, gender, and anatomical unit were recorded, following zooarchaeological criteria (Boessneck 1969; Desbiez and Keuroghlian 2011; Grant 1982; Halstead *et al.* 2002; Helmer 2000; Levine 1982; Payne 1973; Prummel 1987; Ulrey *et al.* 1965).

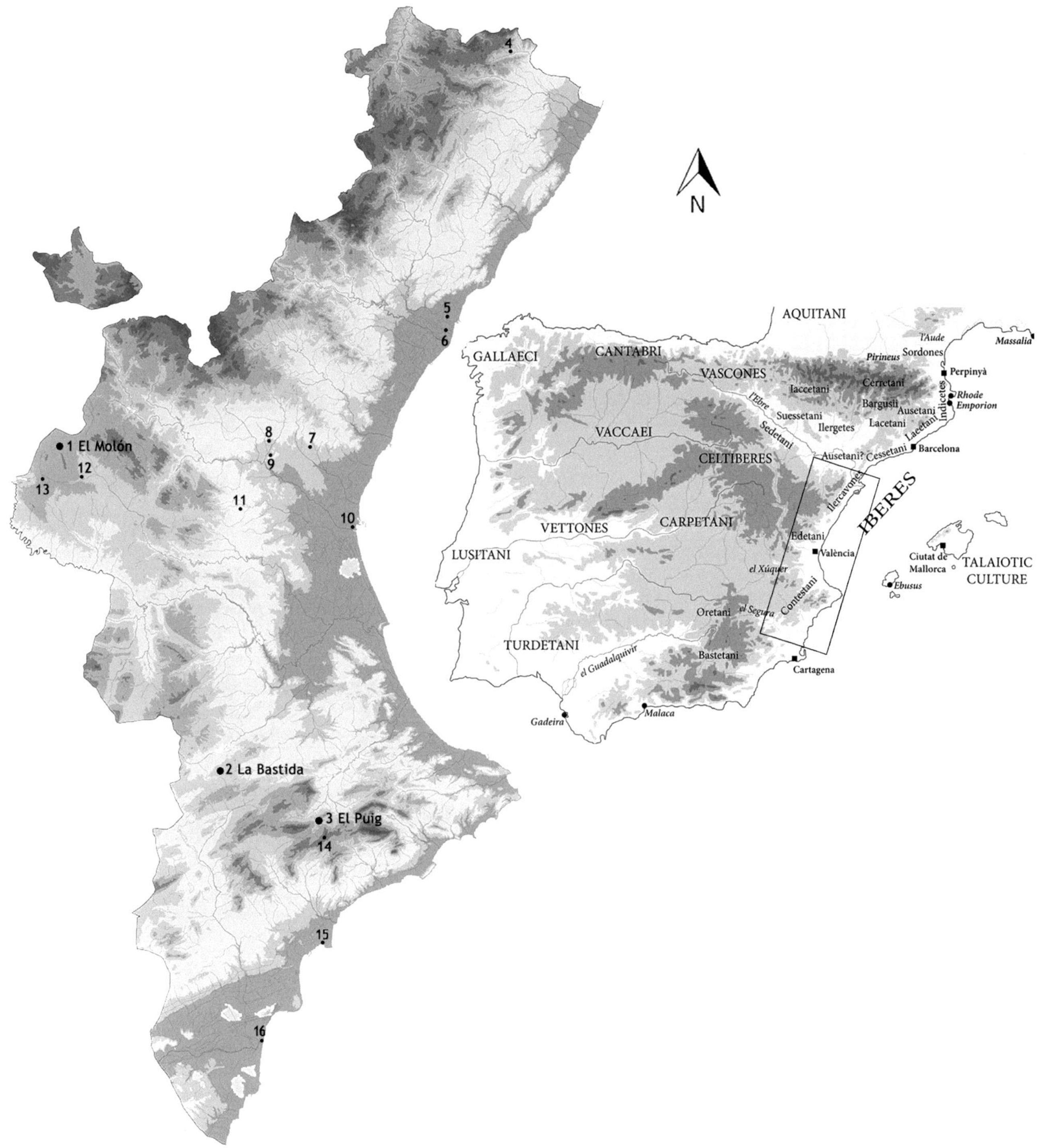

Figure 8.1 Location of the sites mentioned in the text. 1) El Molón; 2) La Bastida de les Alcusses; 3) El Puig d'Alcoi; 4) La Morranda; 5) El Torrelló del Boverot; 6) La Regenta; 7) El Puntal dels Llops; 8) Castellet de Bernabé; 9) La Seña; 10) Ruaya; 11) Cueva del sapo; 12) Los Villares; 13) Puntal del Horno Ciego; 14) La Serreta; 15) Tossal de les Basses; 16) La Fonteta.

All bone remains were studied using a stereomicroscope in order to record and describe taphonomic marks of anthropogenic origin, such as butchery marks (Lyman 1994), potential human bites (*e.g.* Fernández Jalvo and Andrews 2011; Martinez 2009; Pickering *et al.* 2013; Saladie 2009), marks of carnivore origin (Binford 1981), and diagenetic marks (Lyman 1994). The study of anthropogenic marks included details of their location, typology, depth, orientation and development in order to investigate animal processing, such as slaughter, evisceration, skinning, dismembering and meat exploitation.

Butchery marks made by metal knife-edges were identified by the presence of straight lines, their marked depth and prevailing parallel edges. Marks indicating slaughter were identified by their location on inner areas of the spinal canal, both on the occipital condyles and on the atlas. Evisceration marks were difficult to identify; they were recorded on the inner parts of the ribs, although such marks may have also been caused by consumption. Skinning marks were identified by their depth (usually deep) and their location on the distal ends of legs, on skulls and on jaws. Dismembering marks were also identified by their depth (deep) and their position on the joint areas. Finally, shallower marks, occasionally found in series and located mostly along the bone diaphysis were identified as evidence of meat extraction.

Bones that have been bitten and chewed by humans exhibit grooves produced by tooth scraping, teeth imprints, perforations and edges caused by crenelated fractures (Andrews and Fernández Jalvo 1997). Such grooves exhibit an open 'U' section, the depth and width of which may vary according to the strength and direction of the biting action, and can include inner microgrooves. These are located on the joint areas of bones and along the diaphysis of long bones, and may be associated to teeth imprints with crescent, semi-circular or sub-quadrangular shape. Perforations are usually circular and oval, they have bone flakes adhered to the edges and their dimensions range from 1 to 2.5mm. Human bite marks are usually associated to butchery incisions. The respective outline and shape of grooves made by humans are different to those made by fungi and bacteria. Biochemical action produces grooves of irregular outline that turn into dendrites, with the inner surface presenting flakes and without microgrooves (Domínguez Rodrigo *et al.* 2006). Differences between human and small carnivore bites on small preys are difficult to establish (Landt 2007). Carnivores make perforations, long scrapes and splinters on the diaphysis, and often, due to the small size of the prey, consume all of it. The pattern of disruption is different since gastric juices corrode bones and bone spicules (Andrews and Evans 1983; Payne and Munson 1985).

Results

The deposits studied were found in either private spaces, such as beneath floors of houses and work areas, or beneath public and communal spaces, such as walls, gates and towers. The animal remains appear either in isolation or alongside other archaeological material. Their taphonomic study suggests that these archaeological contexts were ritual deposits rather than the result of everyday food waste (see below). The features that defined animal deposits as parts of a ritualised event in the domestic sphere were the combination of species selection, sex, age at death, anatomical units represented (complete or partial), treatment of the carcass, the manner in which the deposit was buried or created, and the associations with other artefacts and structures.

El Molón (Camporrobles, Valencia)

The deposits studied from this site are dated to between the 3rd and 2nd centuries BC and they were associated with two different contexts: a domestic area and the fortification of the *oppidum*.

Domestic deposits (A5/504 and C16/16008)

In the domestic area two deposits, A5/504 and C16/16008, were found inside shallow pits beneath the floors of houses that were located close to a perimetral wall and next to a hearth respectively (Fig. 8.2). Deposit A5/504 was located in the department A5.1, inside a 25cm deep pit (UE 504). The remains of two piglets aged less than one month old were deposited there after consumption (see below). The assemblage included 96 bones and bone fragments grouped together with the jawbones placed on top. The completeness of the skeletons and their good state of preservation allowed their detailed butchery analysis. One individual displayed evidence of charring on the ischial tuberosity of the right pelvis, the occipital condyles, and the tympanic bullae. No signs of fire action were identified on the bones of the other individual. This limited evidence of fire on the bones may indicate that at least one animal was roasted but hardly came into contact with the flames: perhaps it was laid on embers or cooked in an earth oven. There is no evidence of carnivore marks on the surface of the bones. Processing marks are abundant and connected to the slaughter, dismembering and consumption of the animals, as described below (Fig. 8.3). Possible human bite marks were identified on some bones of both individuals (Plate 15).

Deposit C16/16008 was located next to a house (C10/C12), which according to its location and size was interpreted as the living quarters of a powerful individual within the *oppidum* (Lorrio *et al.* 2014). The animal bone deposit was found within an irregularly-shaped shallow pit measuring 20 × 30cm (UE 16006), which was located to the southwest of another pit (UE 16015a) with a child's burial (Fig. 8.2). It included 78 bones and bone fragments from two suckling piglets. The bones were in a poorer state of preservation than the set from deposit A5/504. The pig bones had been gathered together to form distinct deposits in the assemblage, and there were no signs of carnivore gnaw marks: butchery marks and possible human bite marks similar to those in pit A5/504 have been identified.

Both deposits (A5/504 and C16/16008) share exactly the same context type, number of individuals, species type and their anatomical representation and age at death. The processing of the animals' carcass was similar in both assemblages. Slaughter marks were located around and in the inner surface of the foramen magnum and of the atlas, and evisceration marks could be seen on the inside of the ribs. Marks associated with dismembering were abundant and located on the scapula, ulna, femur, radius and humerus joints (see examples in Plate 15). Finally, meat extraction marks could be seen on the internal and external rib surfaces,

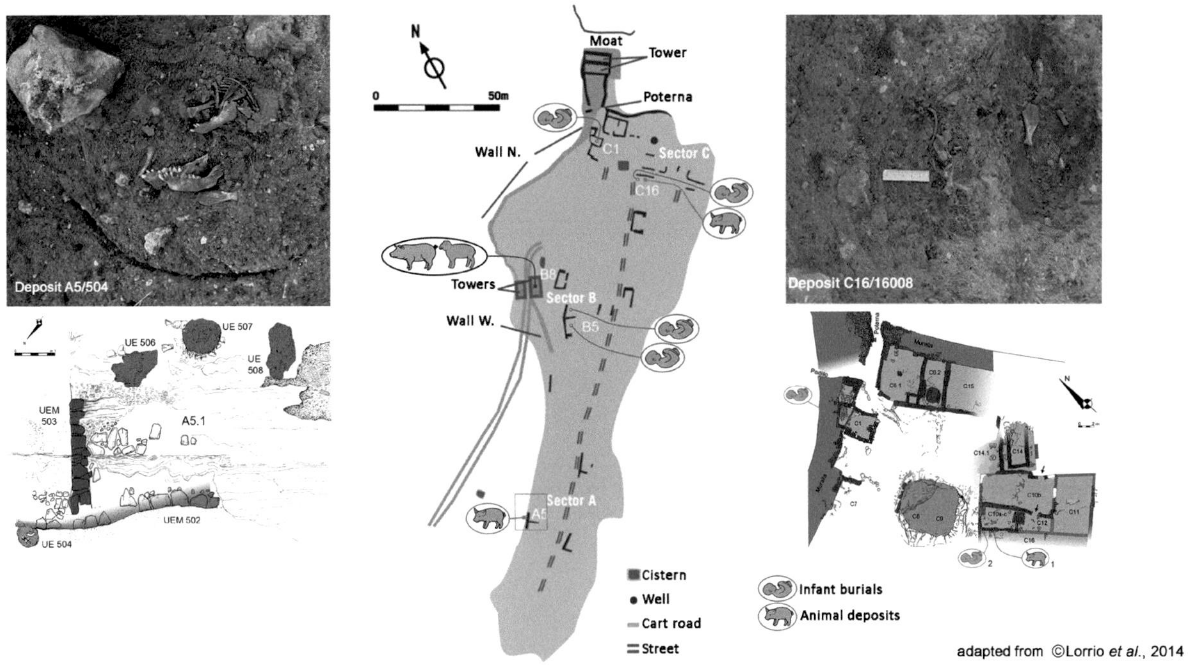

Figure 8.2 El Molón (4th–1st c. BC). Location of infant burials and animal deposits. Plan of deposits A5/504 and C16/16008.

on the transverse and perpendicular processes identified on the vertebrae, on the long diaphyses, and even on lateral surfaces of the metapodia.

Impressions of possible human bite marks on the bones have been identified on at least three of the pigs of the two deposits. These marks were located on the proximal end of a humerus, on a tibia diaphysis (where they were associated with the fracture line produced by the possible bite) on a rib and on a lumbar vertebra, below the transverse process, and on femurs. The modifications included semi-circular section marks that were quite shallow and were often associated with fresh fractures. Crenulated fracture edges and U section scrapes were also identified (see examples in Plate 15 and Fig. 8.4).

Although the animals were small and only a few weeks old, they were processed with metal implements and consumed. Their bones were carefully deposited in small shallow pits deliberately excavated, which might be interpreted as a '*bothros*', a pit designed to accommodate offerings. In one of the cases, the jaws were gathered together and covered with soil in such a way as to be protected from other agents and to seal or safeguard the act done. All these characteristics indicate a special treatment of the bone remains, which can be classified as a 'special animal deposit'. In addition, as both houses were associated with a deposit exhibiting the same features, it may be suggested that these activities were the result of a special event of unique character that was practiced only once during the lifetime of these domestic structures.

The Tower Deposit (B8.1a/8032)

El Molón's defence system was built in the late 4th century BC, and was modified at the end of the 2nd century and the beginning of the 1st, when the tower next to the main gate was built (Lorrio *et al.* 2014). Two rather unusual faunal specimens were identified in a pit associated with this defensive structure. Other archaeological material found in the same pit included a chalice-shaped vase, the remains of a flat base from an oxidation-fired ceramic, probably part of a jug, and a fragment of a perforated iron plaque (Fig. 8.5).

The sole bone contents of this deposit were two mandibles, one of a sheep and one of a pig. Both bones were placed facing down between the stones of the wall in a pit. This pit was a closed context associated with the remodelling of the defensive system to protect the main entrance, by the construction of two powerful towers that flanked the gate (Lorrio *et al.* 2014). The sheep mandible is right-sided and belongs to a four-year-old individual, based on dental attrition. Some butchery marks were identified on the lingual surface and a long, thin cut, oblique to the teeth, was also present. The pig mandible is also a right-sided bone, of a male individual. Dental wear indicates that the animal was also about four years old. The butchery marks were similar to those found on the sheep, located on the lingual and labial surfaces. There were also a long, thin cut, perpendicular to the teeth on the lingual surface, and a short, thin cut on the labial surface near the mental foramen, indicating butchery processes to extract meat and the tongue. The slaughter profile of these

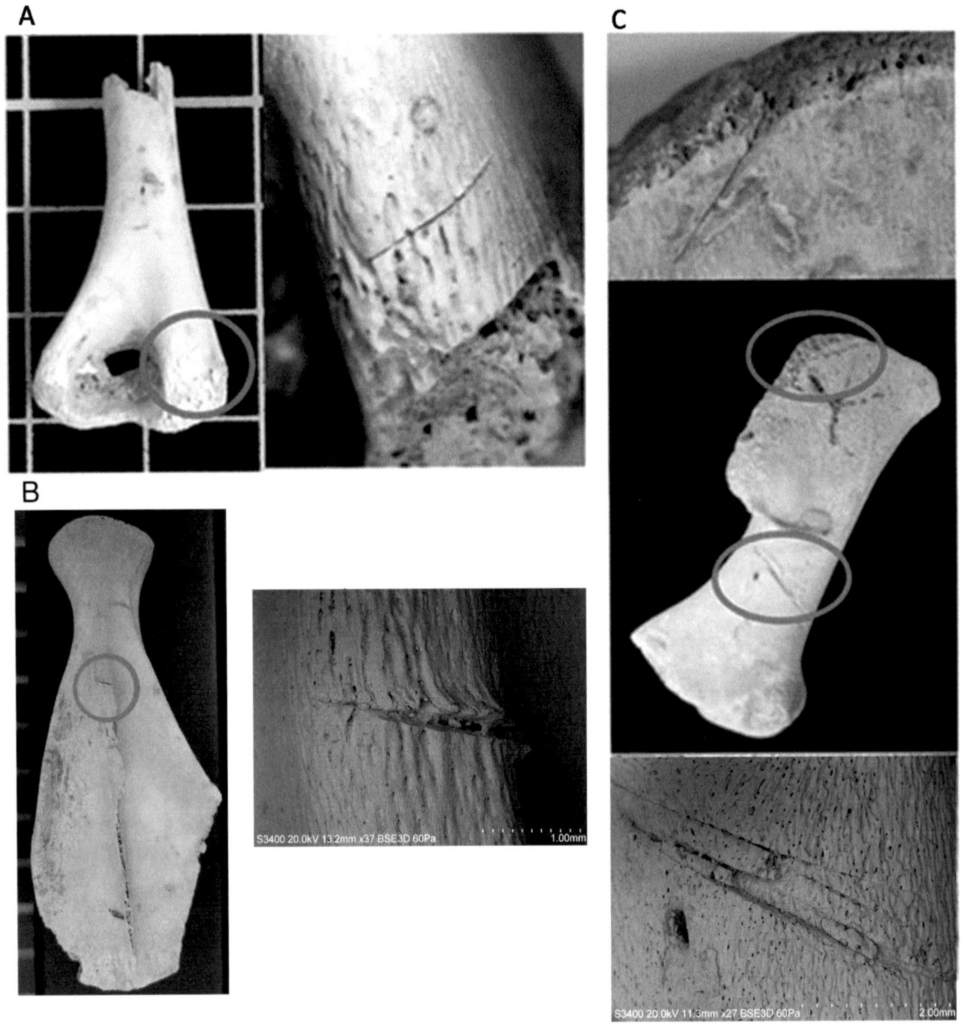

Figure 8.3 EL Molón. Cuts marks on pig bones. A) distal left humerus; B) cuts on left scapula spine; and C) cuts on the left pelvis.

two animals differs from that of other animals recorded in the settlement's domestic waste, which were all juveniles.

This deposit, made at the time of the construction of the tower, can be interpreted as a foundation offering related to the defences of the town (Lorrio *et al.* 2014).

La Bastida de les Alcusses (Moixent, Valencia)

Under the west gate of this *oppidum*, in a space of 12 sq. m, an assemblage of artefacts and bioarchaeological remains was found. The context was dated to a short period between 375 and 350 BC (Vives Ferrándiz *et al.* 2015) and was associated with the restoration of the main gate, ramparts and towers. The material was variable and included iron weapons, burnt gate remains (cloves and ironworks door fittings), animal bones, seeds and fruits, and ceramic vessels, some of which imported, that had been destroyed by various means, such as fragmentation or fire. Five panoplies stood out in this deposit, consisting of a sickle-shaped sword, shield, spear and *soliferreum* (heavy throwing spear, entirely forged in iron). All these objects were placed on the floor of the main entrance and covered (Fig. 8.6).

The faunal remains of the deposit included 102 bones and bone fragments that were separated into two clusters according to their spatial distribution and taphonomic features. The first bone cluster consisted of 91 remains dispersed across the deposit, mostly of caprine bones from at least one young individual (cranial, axial and appendicular bones). Other fragments were identified as the appendicular bones from an unidentified larger animal (*Bos/Cervus/Equus*). The fragments showed butchery marks (chopping and meat extraction) and the brownish colour of three of them indicated the use of fire. This assemblage could be interpreted as the waste from a meal.

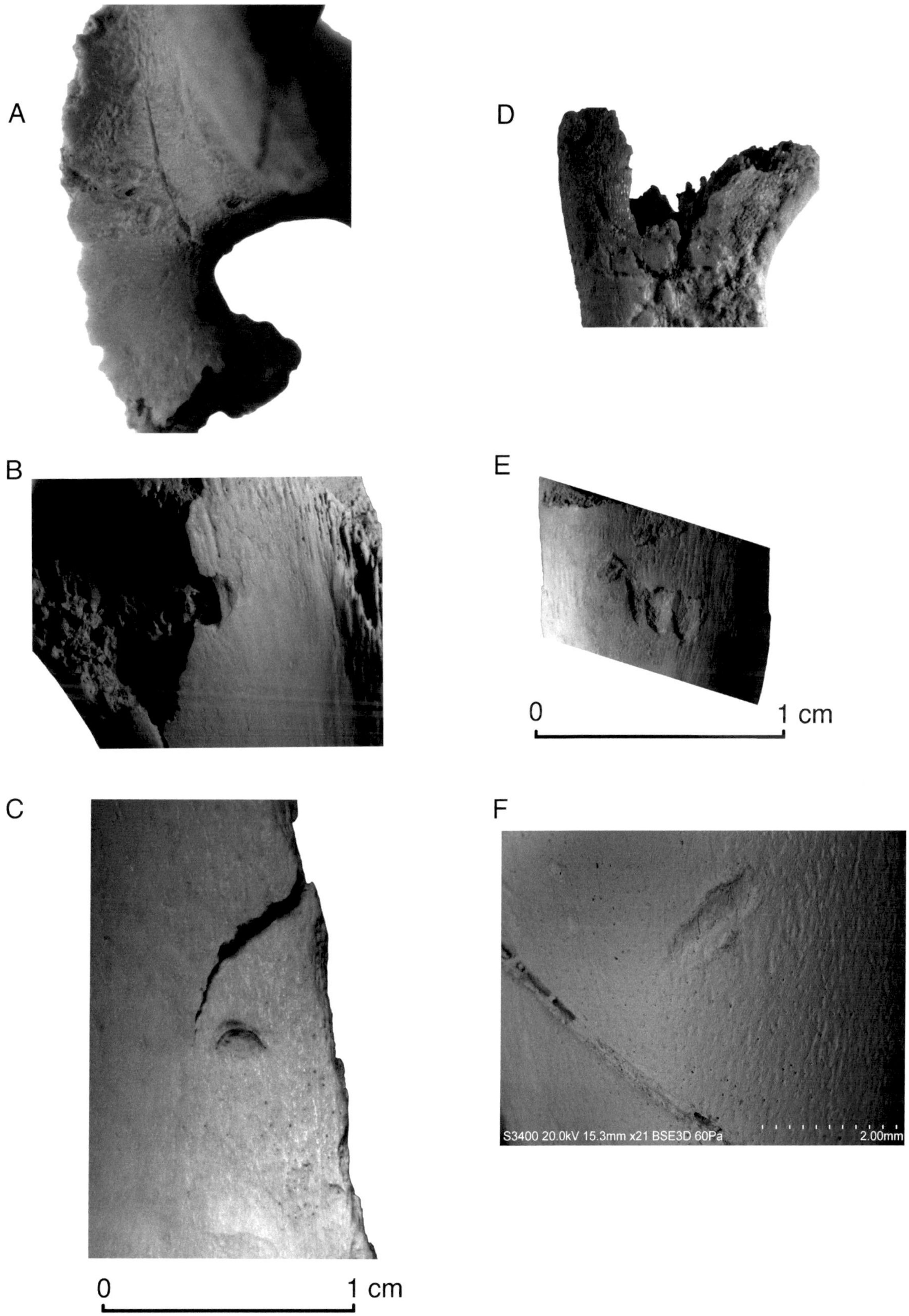

Figure 8.4 EL Molón. Possible human bite marks on pig bones. A) Lumbar vertebra with crenelated shape; B) right proximal tibia with possible human bite perforation; C) diaphysis of right tibia with tooth imprint and bite fracture; D) proximal right humerus with bite alterations; E) proximal left femur with teeth imprints; F) left femur with grooves produced by tooth scraping and cut.

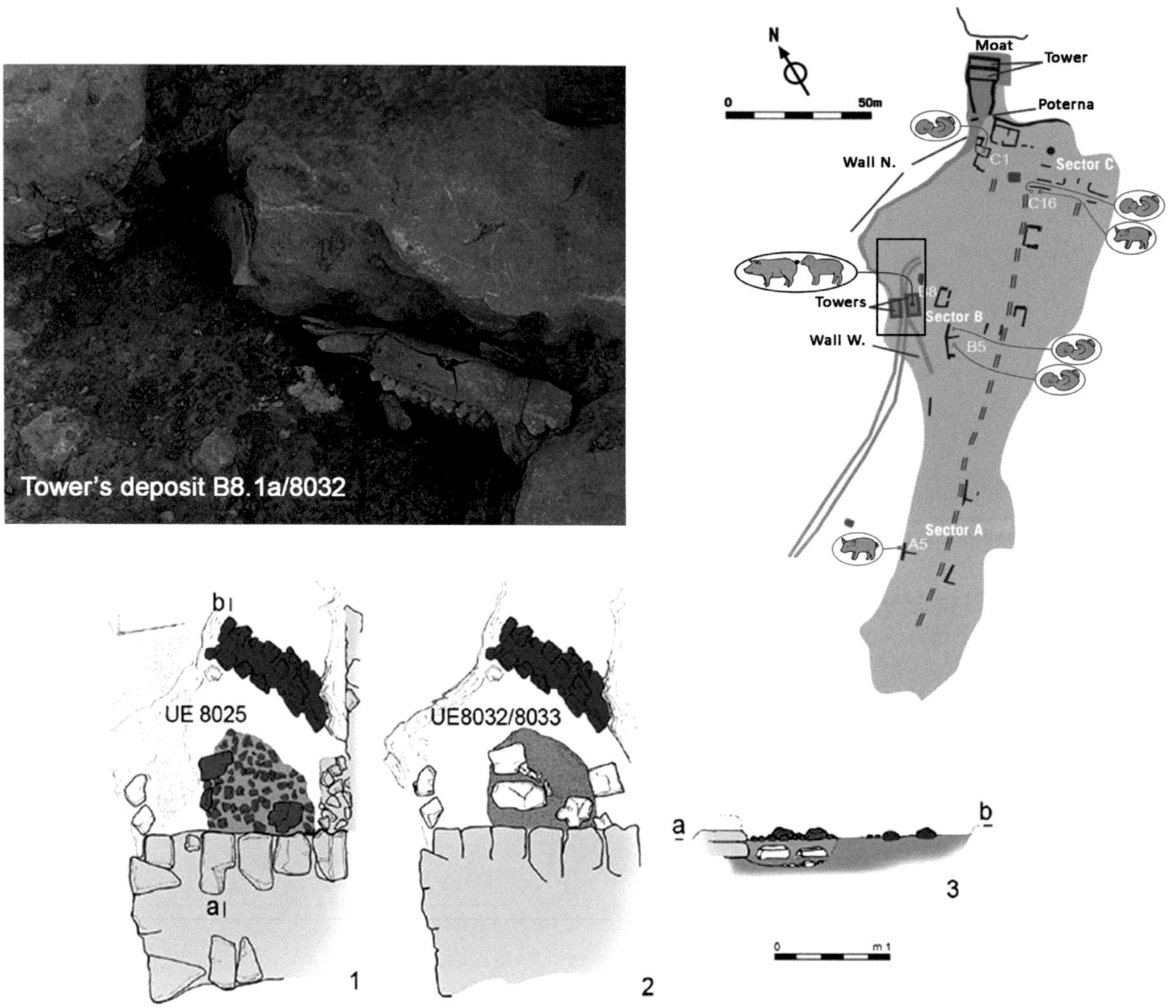

adapted from ©Lorrio *et al.*, 2014

Figure 8.5 El Molón. Location and plan of the tower deposit B8 1a/8032.

The second assemblage included a left metatarsal of a red deer with its distal joint missing and a right astragalus of a caprine that was burned to a high temperature (and turned white), both of which seemed to be placed next to the two panoplies. No butchery marks were found on these two animal bones. These bones differ from the bone cluster analysed above in the absence of butchery marks and the degree of combustion of the astragalus, which was calcined.

In this particular case, the treatment of the buried materials that were destroyed by various means, the place selected (main gate of the *oppidum*) and the single animal remains deposited as offerings (caprine astragalus and red deer metatarsal) along with the implements used, and the food consumed, possibly in a feast, indicate that a special event took place in association to the reinforcement of the main gate, and thus, possibly to the safety and security of the settlement.

Puig d'Alcoi (Alcoi, Alicante)

Two separate special deposits were found at this site beneath houses dated to the 5th and 4th centuries BC. The first deposit was dated to the second half of the 5th century BC and was found beneath the floor on the threshold of house 200 (Fig. 8.7). The assemblage contained ceramic vessels, a hearth-burnt clay plaque with irregular profile, and two clusters of animal remains. The first cluster was a set of caprine remains, and the second one included a complete equid skull with the two mandibles articulated (Fig. 8.7). All the materials were buried and protected from subsequent alterations when the area was levelled off to construct the wall of house 200 (Grau *et al.* 2015). The set of caprine remains included cranial and postcranial bones from at least one sheep and one goat with butchery and consumption marks (butchery marks on humerus and metatarsal diaphysis

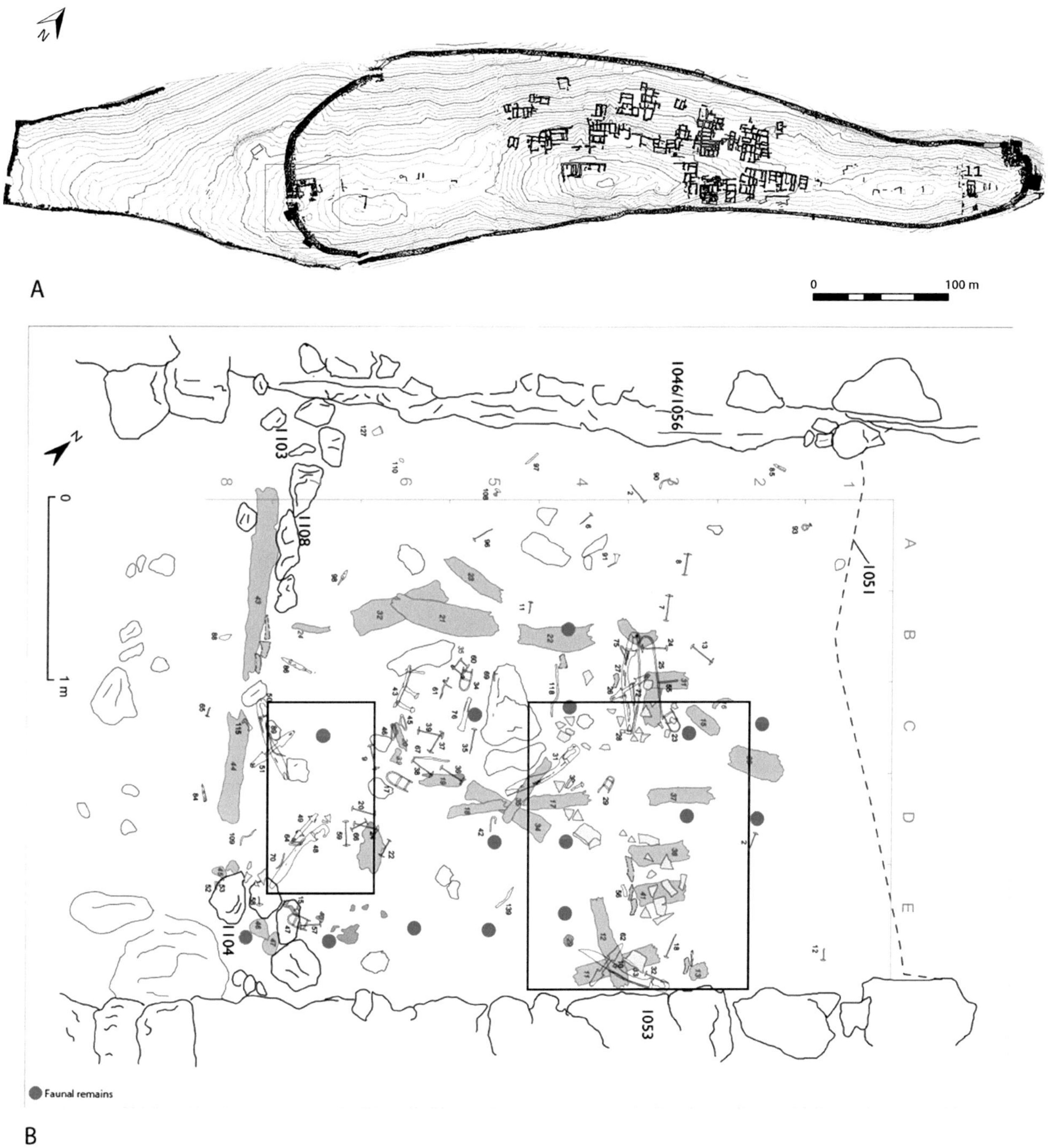

*Figure 8.6 La Bastida. A) Location and plan of the west gate deposit; B) Distribution of the materials; iron weapons (*e.g. *25, 27, 31, 33, 48, 49, 50, 77, 148); cloves and iron works door fittings (*e.g. *7, 39, 43, 34, 12, 123); ceramics (*e.g. *1, 3, 2, 4, 9, 13, 20); animal bones (grey circles); charcoal (numbers in italics); seeds and fruits (rectangle indicates areas with higher density) adapted from © Vives-Ferrándiz* et al., *2015.*

and incisions over the vertical apophysis of thoracic vertebra), which suggest that they may represent food waste. The equid was from an individual around 12 years old, according to dental attrition. This assemblage from under the threshold can be interpreted as the vestige of a special event that took place during the building of the house that would have involved the deposition of the remains and leftovers of a possible feast and the offering of an equid's head.

The second special deposit was located in a pit beneath the floor of a workshop (area 7000) inside a large dwelling, dated to the 4th century BC. The dwelling had four spaces, with a hearth and basic household equipment, which shared

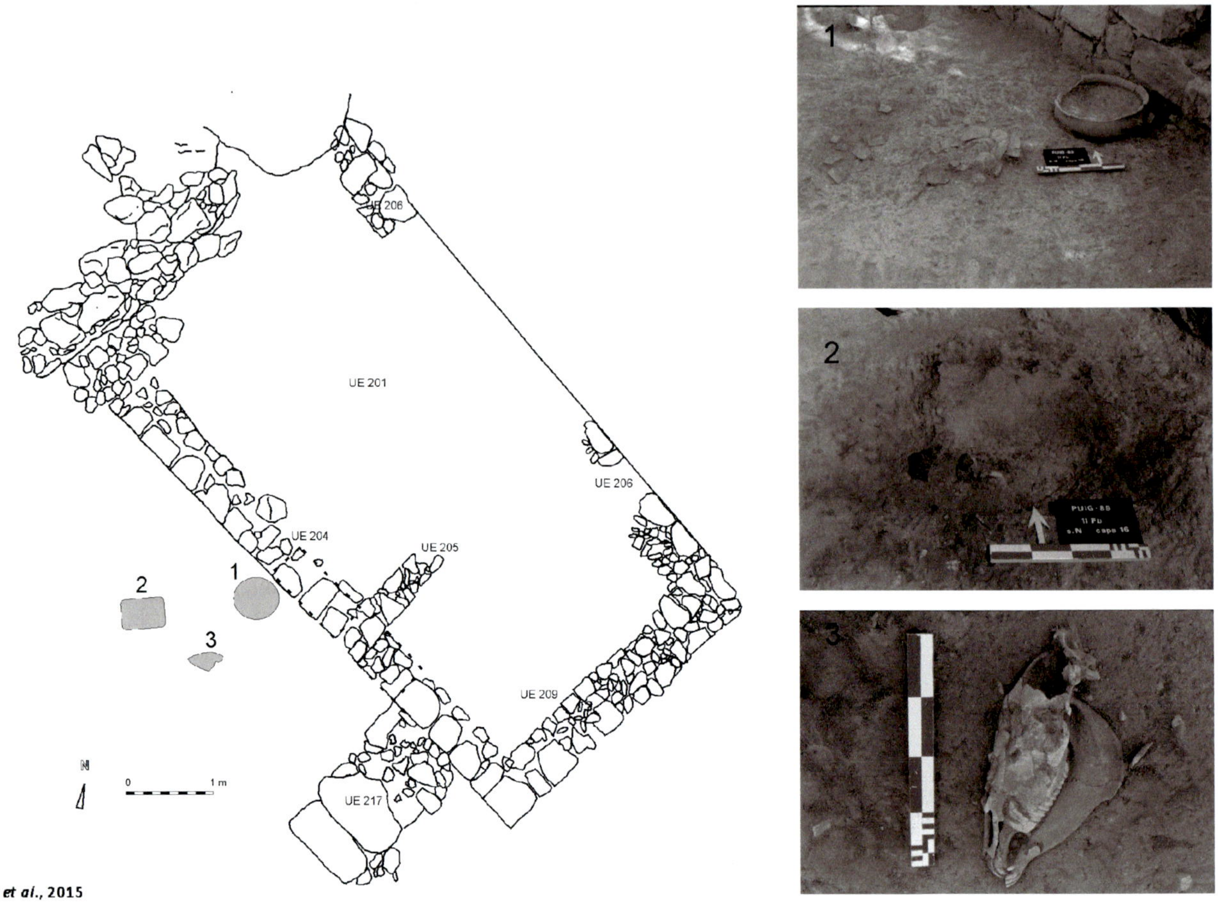

© Grau *et al.*, 2015

Figure 8.7 El Puig d'Alcoi. House 200. 1) Lebes; 2) Hearth with the remains of a feast; 3) Equid skull.

common workshops and food storage rooms. Its layout suggests that there was space for four nuclear families (Grau *et al.* 2015). The pit contained the skeletons of three female adult sheep and one sheep foetus together with perinatal human bones from a premature foetus, two terracotta loom weights and one sickle-shaped knife (Fig. 8.8). The three slaughtered sheep were aged four, six and seven years old and their skeletons were almost complete (only one left pelvis and some vertebrae and rib fragments were missing, although all the anatomical units were present), grouped together but not articulated.

Butchery marks were identified on the inner part of the foramen magnum and on the internal surface of the atlas, in both cases affecting the spinal canal. Marks related to skinning were also present and located on the distal ends of metapodials and the proximal ends of first phalanges, while dismembering marks appeared on proximal radii and distal humeri (Fig. 8.8a). Meat extraction marks were identified on a transverse process of a cervical vertebra, the internal surfaces of ribs, along the diaphyses of long bones (Fig. 8.8b) and also on scapulae and pelvis. Evidence of thermal alterations was only recorded on some carpals and tarsals (Fig. 8.8c). These bones have scarce muscular cover and were, therefore, more sensitive to the effects of fire during the roasting of the animal. All the evidence indicates that the animals had been consumed. The height at the withers of the individuals ranged from 53 to 57cm. This height, considering the related morphotype characteristics of the Merina breed with a weight of 35–40kg, would have provided 50% of consumable meat (Sanchez and Sanchez 1986). That means that the three caprines could have provided in total around 50–60kg of meat.

The remains of the animals were purposefully deposited along with the loom weights and the knife in a pit associated with an extension of the house. The combination of these elements may be thus interpreted as evidence of a ritual event related to the enlargement of the house.

Discussion

In the animal bone deposits presented here two domestic taxa, caprine and pig, predominate and represent the evidence of unusual events performed and preserved in three *oppida*. The contexts where the deposits were found were very

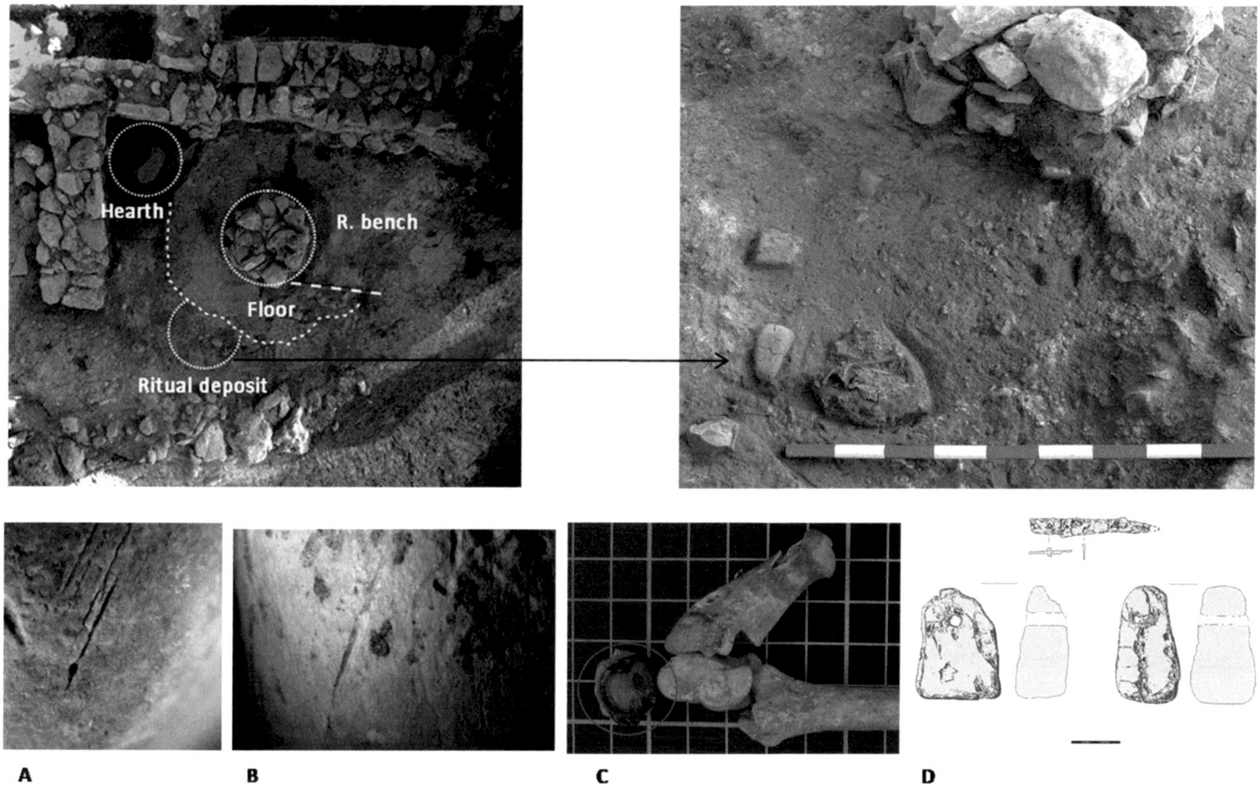

Figure 8.8 Area 7000. A) Cuts on distal humerus; B) sheep cut marks on the shaft of the humerus; C) sheep burned bones; and D) two terracotta loom weights and a sickle-shaped knife adapted from © Grau et al., *2015.*

diverse: small pits under the floors of houses and in working areas, usually in wall intersections or next to hearths, under benches and thresholds, and within defensive structures. The finds of complete animals or of noteworthy animal parts in such deposits are clear illustrations of some of the public and private ritual practices of the Iberian people.

Other case studies indicate that similar deposits are in fact widely documented throughout the Iberian territory, from Hérault in France to Andalusia in southern Spain, from the 6th to the 2nd century BC. The various deposits have been interpreted as offerings to celebrate the foundation or remodelling of houses, or public and private rites related to festivals in honour of deities, birth and fertility (Albizuri 2011; Casellas 1995; Dumezil 1977; Durand 1987; Iborra 2004; Miró and Molist 1990; Nieto 2013; Valenzuela 2008). Such rituals may have also played a role in creating alliances for social cohesion, for instance, under the control of the elites, through festivals and commensality events (Belarte and Valenzuela 2013; Dietler 1996, 200; López-Bertran and Vives-Ferrándiz 2009).

Across the whole of the modern-day region of Valencia the main species represented in such deposits are also sheep, goats and pigs. The only wild species identified in the assemblages studied here is red deer, which is present only in the deposit at La Bastida. In unusual deposits at other settlements of the period, other species, such as dog and domestic fowl, have been also documented but only rarely (Iborra 2004, 363–4).

Zooarchaeological analysis on species selection, age, sex, anatomical parts, and specific treatment for such assemblages interpreted as ritual, is crucial to identify patterns in depositional practice. Part of the ritual or ceremonial event was the preservation of the remains as evidence of the act itself. In some of the analysed cases, the animal carcass appeared complete with the bones collected and grouped. The skeleton exhibited evidence of processing and consumption marks and was deposited alone or with other artefacts. In other cases, isolated bones were deposited in what may constitute an example of *pars pro toto* along with other implements that had been damaged or fragmented, some of them related to commensality.

The deposits of El Molón, with its suckling piglets, and El Puig d'Alcoi, with sheep, show signs of consumption events. Complete pig and caprine skeletons or articulated parts of animals that range in age from neonates to adults are frequently found in domestic spaces within Iberian settlements, and often in the same area as neonatal or perinatal human remains. Deposits with unborn and neonatal human skeletons have often been interpreted as propitiatory sacrifices (Gusi *et al.* 2008), while the finding of neonatal animals has been seen as a reflection of a substitutive practice to avoid the use of human newborns in the foundation ritual (*e.g.* Belarte and Sanmartí 1997, 25–6). The importance of pigs in Mediterranean Iron Age cultures is well established; they were used in many propitiatory rituals connected to the building of houses, the advent of good harvests and even births (Collins 2006). However, the fact that they share space with the burial of neonatal or perinatal humans does not prove that they were part of the same act; their presence in a domestic space may just be a coincidence since, in many cases, these locations had a long chronological development. Deaths during gestation and after birth, and the custom of burying foetuses beneath houses, have been documented throughout the Iberian Peninsula for the Urnfield and later the Iberian Culture (Lorrio *et al.* 2010). This practice is also mirrored in the Greek (Coldstream 1977) and Italic worlds (Gierow 1966).

Evidence of the rituals involving pigs, slaughtered newborn or very young and deposited in pits beneath houses, was found at the site of El Molón and also in several other settlements located in the Valencian region, such as El Torrelló del Boverot (Almassora, Castellón), La Seña (Villar del Arzobispo, Valencia), El Puntal del Llops (Olocau, Valencia), Los Villares (Caudete de las Fuentes, Valencia) and La Morranda site (El Ballestar, Castellón), between the 7th and 2nd centuries BC (Iborra 2004, 348) (Fig. 8.1). At Molón in spite of their young age and small size, and thus their limited meat contribution, these animals had been processed and consumed. It is significant that the practice was repeated in different household units across settlements, and that the animals were slaughtered at a young age. According to Columella's agricultural calendar (*Re Rusticae* VII, 9) pig births took place during spring and therefore the age at death (under one month) indicates that the rituals took place probably in May and June, which may mean that the animal slaughter was associated with specific rituals rather than to the natural death of unborn and newborn humans. The killing of pigs in particular cults and festivities has been documented in other contexts in Mediterranean cultures (Bookides and Stroud 1997; Burkert 2013; Collins 2006; Di Stefano 2008; Hamilakis and Konsolaki 2004). In some Greek sanctuaries, the presence of pig remains with butchery marks and evidence of exposure to fire has been interpreted as part of puberty passage rites (Allegro *et al.* 2008, 119). However, in the domestic domain, when interpreting the El Molón deposits associated to two houses, there is a lack of firm elements to interpret their meaning. Their similarities, however, could suggest that these deposits were part of a common pattern that took place only once within each household's lifetime.

The evidence at El Puig d'Alcoi with three female sheep and an equid head beneath the floor on the threshold of a new house, buried deliberately together with other artefacts, suggests particularly complex rituals in the domestic sphere. The differences noted in the two examples of rituals performed in the domestic space of El Puig may be the result of differences in the nature of the rituals themselves. Alternatively, they may be due to the differences in the social status of the houses' inhabitants. In area 7000, the number of sheep deposited and their age at the time of their slaughter suggests that a large quantity of meat would have been consumed and that a substantial number of people would have been able to participate in this ritual. This deposit was associated to the extension of a large dwelling with new workspaces. Therefore, the event performed could have been related to family solidarity (Grau *et al.* 2015). More deposits with caprine bones have been identified in other Iron Age sites of the study area, such as the city of Los Villares, the village of la Seña, the fortified farm of Castellet de Bernabé, the hillfort of el Puntal dels Llops and the trade site of el Torrelló del Boverot, all in contexts dated to between the 7th and 2nd centuries BC (Fig. 8.1).

In regards to the equid head buried along with other artefacts and the leftovers of a meal in the threshold of house 200, the evidence suggests that it could be part of a feast, celebrating a house foundation event. This house differs architecturally from all the rest across the site, including Punic architectural elements, and this could provide a framework for the interpretation of the equid head offering. Within the study area, there is only one other documented parallel: the post-cranial horse carcass found in a pit covered with stone slabs at the city of Tossal de Manisses (Alicante), a site with a strong Punic character in regards to architectural constructions (Olcina *et al.* 2010). Equid remains have been recorded also at non-domestic contexts of other sites, in votive pits located at Tossal de les Basses (Alicante) and Ruaya (Valencia), where partial skeletons of equids were found together with Punic, Italic and Iberian ceramic vessels. Another important finding is the buried horse at La Regenta (Burriana, Castellón), which appeared isolated and having a bronze ring in the mouth dated to the 3rd century BC (Sarrion 2003). Additionally, complete or partial skeleton deposits of equids have been found in other territories of the Iberian Culture, such as at domestic spaces of the settlements of Els Vilars (Arbeca, Lleida: Nieto 2013) and Moleta del Remei (Alcanar, Tarragona: Belarte and Sanmartí 1997, 12); open areas like the silo field of Serrat dels Espinyers d'Isona (Lleida: Belmonte *et al.* 2013); necropolis like La Pedrera (Vallfogona de Balaguer, Lleida: Graells 2008); and the Tartessian and Celtic sanctuaries of Cancho Roano and Capote (Badajoz: Celestino and Cabrera 2008; Morales and Liseau 1994).

The deposits documented at La Bastida and El Molón with animal remains placed in public spaces associated with defensive structures have parallels in Catalonia (Belarte and Sanmarti 1997), Andalusia (Ruiz *et al.* 2015), and Celtiberian areas (Alfayé 2007). At the La Bastida *oppidum* the recorded ritual involved the fragmentation and bending of iron weapons of five panoplies. Panoplies were symbols of prestige for

Iberian warriors, and thus the examples found here may have represented five distinct power factions participating in the ritual (Vives *et al.* 2015). The other elements, especially the leg of the red deer (*pars pro toto*), probably had some symbolic value. Red deer has been often found in funerary contexts and shrines, such as at La Serreta necropolis (Alcoy, Alicante), El Molar necropolis (Alicante) and the sanctuaries of La Cueva del Sapo and Puntal del Horno Ciego (Requena, Valencia: Iborra 2004, 367; Machause *et al.* 2015; Monraval and Lopez 1984). Decorations of deer hunting scenes are also common on fine-ware vessels and seem to have played an important part in the symbolic representation of certain social classes (Aranegui *et al.* 1997).

In reference to the calcined caprine astragalus found at La Bastida, it should be stressed that these elements are commonly associated with special contexts. Groups of caprine astragali, worked and non-modified, burned and unburned, have been found in funerary contexts (tombs and votive pits), sacred areas and domestic contexts of several Iberian Iron Age settlements. Examples include the necropolis of the sites of Serreta (Alcoi, Alicante), Coimbra del Barranco Ancho (Jumilla, Murcia) and El Cigarralejo (Mula, Murcia: Iborra 2004, 336); the votive pit of Libisosa (Lezuza, Albacete: Iborra 2012); and domestic areas of several sites like Los Villares (Caudete de las Fuentes) and Puntal dels LLops (Olocau, Valencia: Iborra 2004, 303). The use of the astragalus as a symbolic element that served as an offering or gaming piece has been suggested for the Early Bronze Age onwards (*e.g.* Bartosiewicz 1997–8; De Grossi Mazzorin *et al.* 2013; Gilmour 1997; Hesse 1995; Reese 1985). The treatment and final deposition of the material under Bastida's west gate might have been related to a ritual involving a series of objects of great social importance at the time (Vives Ferrándiz *et al.* 2015).

The deposit recovered at the South Tower of the main gate of El Molón, comprising of two right jawbones of an adult sheep and pig, has a more recent chronology (2nd century BC) and was made at the time of the construction of the tower. Therefore, it can be interpreted as a *pars pro toto* foundation offering, related to the defences of the town that would have a public character and possibly a protective purpose (Lorrio *et al.* 2014; von Nicolai 2009, 85–6). Offerings performed in gates and defensive systems have also been recorded at other Iron Age settlements (Buchsenschutz and Ralston 2007; von Nicolai 2009). These might have been related to the delimitation and construction of walled perimeters or urban boundaries, which according to Alfayé (2007, 9) would thus acquire 'added value as protector, purifier while becoming sacred in character'.

Conclusions

In this chapter six deposits of faunal remains dated to between the 5th and the 2nd centuries BC have been analysed, and related to rituals of both the private and the public domain. Some of these types of deposits have been recorded for the first time in Iron Age sites in the studied territory. All assemblages appeared to be ritual in character according to contextual information, taphonomic data and detailed bioarchaeological analyses, which allowed the partial reconstruction of the possible story behind the various animal burials. These burials have been preserved as reminders of specific actions in selected important contexts, but their exact nature is difficult to discern.

The ritual deposits in the domestic domain during the studied period were variable, including complete or partial animal skeletons, some slaughtered and consumed and others not. Similar deposits have been found in all types of Iron Age sites, including *oppida*, towns, farms, hillforts and villages, and no temporal changes can be detected on current data. The rituals performed in the private domains of the *oppida* of El Molón and El Puig d'Alcoi included commensality activities. These involved mostly caprine and pigs, which were the basis of Iron Age livestock systems in the area, but at least in the case of pigs, they were processed differently to the standard practices observed in Iberian *oppida*, as they were slaughtered at a young age. In some cases, these deposits were found together with human perinatal or infant burials. A direct relationship, however, between these and the slaughter, consumption and deposition of animals could not be established. At El Puig d'Alcoi the two domestic deposits dated to the 5th and the 4th centuries seemed to be related with the social status of the houses.

The rituals at the gates, walls and other defensive structures of the studied settlements included both offerings in the form of *pars pro toto* and evidence of commensality events. They have been interpreted as foundational offerings related to the renovation of the defences of these *oppida*, and in certain cases, such as at La Bastida, have provided glimpses into the power factions and structure of the settlement.

This study has added new data to the corpus of Iron Age animal burial deposits and helped illuminate the complex character and the versatility of ritual practices in these societies. More studies on such deposits are still needed, however, towards a better understanding of spatial and chronological changes, and ultimately of the culture of the people occupying this territory.

Acknowledgements

I am grateful to my colleagues Dr Alberto Lorrio (Universidad de Alicante INAPH, HAR2010-20479/HAR2013-41447-P, MINECO), D. Josep Maria Segura (Museu Arqueològic Municipal Camil Visedo Moltó, Alcoi), Dr Ignacio Grau (Universidad de Alicante INAPH), and Dr Jaime Vives-Ferrándiz (Museo de Prehistoria de Valencia, HAR2014-54414-P, MINECO) for providing me with the bone assemblages and for their useful comments.

Thanks to Alexandra Livarda, Richard Madgwick and Santiago Riera for organising the session at the 20th EAA annual meeting where we discussed this subject.

References

Albizuri, S. (2011) Depósitos de ovicaprinos en ámbito doméstico ibérico. El ejemplo del poblado de Can'Oliver (Cerdanyola del Vallès, Barcelona). *Archaeofauna* 20, 85–101.

Albizuri, S. and Nadal, J. (1999) Aprovechamiento y Producción animal en época Ibérica. Consideraciones generales económicas. *Limes* 6/7, 41–52.

Alfayé Villa, S. (2007) Rituales relacionados con las murallas en el mundo Celtíbero. *Palaeohispania* 7, 9–41.

Allegro, N., Cosetino, V., Leggio, L., Masala, S. and Svabera, S. (2008) Lo scario del Themophorion di Gortina. In C. A. Di Stefano (ed.) *Demetra: la Divinità, i Santuari, il Culto, la Legenda. Atti del Congresso Internazionale* (Enna 2004), 107–21. Pisa, Biblioteca di 'Sicilia Antiqva' 2.

Andrews, P. and Evans, E. M. N. (1983) Small mammals bone accumulations produced by mammalian carnivores. *Paleobiology* 9, 289–307.

Andrews, P. and Fernández-Jalvo, Y. (1997) Surface modifications of the Sima de los Huesos fossil humans. *Journal of Human Evolution* 33, 191–217.

Aranegui Gascó, C., Bonet Rosado, H., Martí, M. A., Mata Parreño, C., Pérez Ballester, J. (1997) *La Cerámica con Decoración Figurada y Vegetal del Tossal de Sant Miquel (Llíria, València): una Nueva Propuesta Metodológica.* Iconografía itálica. Propuestas de interpretación y lectura, Serie Varia 3. Madrid, Actas del Coloquio Internacional Iconografía Ibérica.

Bartosiewicz, L. (1997/1998) A Systemic Review of Astragalus Finds from Archaeological Sites. *Anthaeus* 24, 37–44.

Belarte, M. C. and Sanmartí, J. (1997) Espais de culte i pràctiques rituals a la Catalunya protohistòrica. *Quaderns de Prehistòria i Arqueologia de Castelló* 18, 7–32.

Belarte, M. C. and Valenzuela, S. (2013) Zooarchaeological evidence for domestic rituals in Iron Age communities of northeastern Iberia (Present day Catalonia) (Sixth– Second century BC). *Oxford Journal of Archaeology* 32(2), 163–86.

Belmonte, C., Albizuri, S., Nadal, J. and Garcés, I. (2013) Équids i gossos en l'economia i en els rituals. Resultats de l'estudi dels materials dipositats en el sitjar iberoromà del Serrat dels Espinyers (Isona, Pallars Jussà). *Revista d'Arqueologia de Ponent* 23, 201–22.

Binford, L. R. (1981) *Bones: Ancient Men and Modern Myths.* Chicago IL, University of Chicago Press.

Boessneck, J. A. (1969) Osteological differences between sheep (*Ovis aries* L.) and goat (*Capra hircus* L.). In D. R. Brothwell and E. S. Higgs (eds.) *Science in Archaeology: A Comprehensive Survey of Progress and Research*, 331–58. London, Thames and Hudson.

Bonet, H. and Vives-Ferrándiz, J. (2011) *La Bastida de les Alcusses. 1928–2010.* València, Diputación Provincial de València.

Bookides, N. and Stroud, R. S. (1997) *The Sanctuary of Demeter and Kore. Topography and Architecture. Corinth XVIII.3.* Princeton NJ, American School of Classical Studies.

Buchsenschutz, O. and Ralston, I. (2007) Dépôts et fortifications à l'âge du Fer. In P. Barral, A. Daubigney, C. Dunnig, G. Kaenel et M. J. Roulière-Lambert (eds.) *L'Âge du Fer dans l'Arc Jurassien et ses Marges. Dépôts, Lieux Sacrés et Territorialité à l'Âge du Fer*, 757–76. Besançon, Actes du XXIXe colloque international de l'AFEAF (Bienne, 5–8 mai 2005) 2.

Burkert, W. (2013) *Homo Necans. Interpretaciones de Ritos Sacrificiales y Mitos de la Antigua Grecia*, 2nd edn. Barcelona, Acantilado.

Bradley, R. (2003) A life less ordinary: the ritualization of the domestic sphere in later prehistoric Europe. *Cambridge Archaeological Journal* 13(1), 5–23.

Bradley, R. (2005) *Ritual and Domestic Life in Prehistoric Europe.* London–New York, Routledge.

Brück, J. (1999) Ritual and rationality: some problems of Interpretation in European archaeology. *European Journal of Archaeology* 2(3), 313–44.

Casellas, S. (1995) Dipòsits faunístics no susbsistencials a la Catalunya prehistòrica. *Cota Zero. Revista d'Arqueologia i Ciència* 11, 89–93.

Celestino, S. and Cabrera, A. (2008) El banquete privado y el banquete comunal en el santuario de Cancho Roano. *Cuadernos de Prehistoria y Arqueología* 18, 189–215.

Coldstream, J. N. (1977) *Geometric Greece.* London–New York, Routledge.

Collins, B. J. (2006) Pigs at the Gate: Hittite Sacrifice in its eastern Mediterranean context. *Journal of Ancient Near Eastern Religions* 6(1), 155–88.

Columela (nd) *De los Trabajos del Campo.* Transl. A. Holgado [1988]. Madrid, Siglo XXI de España.

Desbiez, A. L. J. and Keuroghlian, A. (2011) Ageing feral pigs (*Sus scrofa*) through tooth eruption and wear. *Suiform Soundings* 9(1), 48–55.

De Grossi Mazzorin, J. and Minniti, C. (2013) Ancient use of the knuckle-bone for rituals and gaming piece. *Anthropozoologica* 48(2), 371–80.

Dietler, M. (1996) Feast and commensal politics in the political economy: food, power and status in prehistoric Europe. In P. Wiessener and W. Schiefenhövel (eds.) *Food and the Status Quest*, 87–125. Oxford, Berghahn.

Domínguez-Rodrigo, M. and Piqueras, A. (2003) The use of tooth pits to identify carnivore taxa in tooth-marked archaeofaunas and their relevance to reconstruct hominid carcass processing behaviors. *Journal of Archaeological Science* 30, 1385–91.

Di Stefano, C. A. (ed.) (2008) *Demetra. La Divinità, i Santuari, il Culto, la Leggenda. Atti del I Congresso Internazionale (Enna 2004).* Pisa, Biblioteca di 'Sicilia Antiqva' 2.

Dumézil, G. (1977) *Les Dieux Souverains des Indo-Européens.* Paris, Éditions Gallimard.

Durand, J. L. (1987) Sacrifice et découpe en Grèce ancienne. In J.-D. Vigne (ed.) *La Découpe et le Partage du Corps a Travers le Temps et l'Espace. Anthropozoologica Premier Numéro Special 1*, 59–68. Paris: Éditions du Centre National de la Recherche Scientifique.

Fernández-Jalvo, Y. and Andrews, P. (2011) When humans chew bones. *Journal of Human Evolution* 60, 117–23.

Gierow, P. G. (1966) *The Iron Age Culture of Latium. I. Classification and Analysis.* Acta Instituti Romani Regni Sueciae, Series 4/24, 1. Lund, Gleerups.

Gilmour, G. H. (1997) The Nature and Function of Astragalus Bones from Archaeological Contexts in the Levant and

Eastern Mediterranean. *Oxford Journal of Archaeology* 16(2), 167–75.

González Prats, A. (2011) *La Fonteta. Excavaciones de 1996–2002 en la Colonia Fenicia de la Actual Desembocadura del Río Segura (Guardamar del Segura, Alicante)* 1, 674. Alicante, Área de Prehistoria, Universidad de Alicante.

Grant, A. (1982) The use of tooth wear as a guide to the age of domestic ungulates. In B. Wilson, C. Grigson and S. Payne (eds.) *Ageing and Sexing Animal Bones from Archaeological Sites*, BAR British Series 109, 91–108. Oxford, Archaeopress.

Grant, A. (1984) Animal husbandry. In B. Cunliffe (ed.) *Danebury: An Iron Age Hillfort in Hampshire. Volume 2. The Excavations 1969–1978: The Finds*, 496–548, Council for British Archaeology Research Report 52. London, Council for British Archaeology.

Grant, A. (1989) Animals and ritual in early Britain: the visible and the invisible. In P. Meniel (ed.) *Animal et Pratiques Religieuses: Les Manifestations Materielles. Anthropozoologica* Special number 3, 79–86. Paris, Laboratoire d'Anatomie Comparée.

Graells, R. (2008) Mistophoroi ilergetes en el siglo iv aC: el ejemplo de las tumbas de caballo de la necrópolis de la Pedrera (Vallfogona de Balaguer-Térmens, Catalunya, España). *Jahrbuch des Römisch-Germanischen Zentralmuseums* 55, 81–158.

Grau Mira, I. (2002) *La organización del territorio en el área central de la Contestania Ibérica*. Alicante, Museo Arqueológico de Alcoi.

Grau Mira, I. and Segura Martí, J. M. (2013) *El oppidum ibérico de El Puig d'Alcoi. Asentamiento y Paisaje en las Montañas de la Contestania*. Alcoi, Ajuntament d'Alcoi.

Grau Mira, I., Amorós, I., De Miguel, P., Iborra, M. P. and Segura Martí, J. M. (2015) Fundar la casa: prácticas rituales y espacio doméstico en el oppidum ibérico de el Puig D'alcoi (Alacant). *Archivo Español de Arqueología* 88, 67–84. Madrid, Instituto de Historia Centro Superior Investigación Científica.

Gusi, F. and Muriel, S. (2008) Panorama actual de la investigación de las inhumaciones infantilesen la Protohistoria del Sudoeste Mediterráneo europeo. In F. Gusi, S. Muriel and C. Olaria (eds.) *Nasciturus, Infans, Puerulus, Vobis Mater Terra. Sèrie de Prehistòria i Arqueologia, Servei d'investigacions arqueològiques i prehistòriques*, 257–329. Castelló, Diputació Provincial de Castelló.

Halstead, P., Collins, P. and Isaakidou, V. (2002) Sorting the sheep from the goats: morphological distinctions between the mandibles and mandibular teeth of adult *Ovis* and *Capra*. *Journal of Archaeological Science* 29, 545–53.

Hamilakis, Y and Konsolaki, E. (2004) Pigs for the Gods: Burnt Animal Sacrifices as Embodied Rituals at a Mycenaean Sanctuary. *Oxford Journal of Archaeology* 23(2), 135–51.

Hesse, B. (1995) Husbandry, Dietary Taboos and the Bones of the Ancient Near East: Zooarchaeology in the Post-Processual World. In D. B. Small (ed.) *Methods in the Mediterranean*, 197–232. Leiden, E. J. Brill.

Helmer, D. (2000) Discrimination des genres Ovis et Capra à l'aide des prémolaires inférieures 3 et 4 et interprétation des âges d'abattage; l'exemple de Dikili Tash. *Anthropozoologica* 31, 29–38.

Hill, J. D. (1995) *Ritual and Rubbish in the Iron Age of Wessex. A Study on the Formation of a Specific Archaeological Record*, BAR British Series 242. Oxford, Archaeopress.

Iborra, M. P. (2004) *La Ganadería y la Caza del Bronce Final al Ibérico Final en el Territorio Valenciano. Servicio de Investigación Prehistórica. Serie Trabajos Varios del SIP 103*. València, Diputación Provincial de València.

Iborra Eres, M. P. (2012) El material faunístico recuperado en el depósito votivo de Libisosa. In H. Uzor (ed). *Prácticas rituales, iconografía vascular y cultura material en Libisosa (Lezuza, Albacete). Nuevas aportaciones al ibérico final del Sudeste*, 199–213. Alicante, Universidad de Alicante.

Iborra, M. P. and Pérez Jordà, G. (2013) Three systems of agrarian exploitation in the Valencian region of Spain (400–300 BC). In M. Groot, D. Lentjes and J. Zeiler (eds.) *Barely Surviving or More than Enough? The Environmental Archaeology of Subsistence, Specialisation and Surplus Food Production*, 131–50. Leiden, Sidestone Press.

Landt, M. J. (2007) Tooth marks and human consumption ethnoarchaeological mastication research among foragers of the Central African Republic. *Journal of Archaeological Science* 34, 1629–40.

Levine, M. (1982) The use of crown height measurements and eruption-wear sequences to age horse teeth. In B. Wilson, C. Grigson and S. Payne (eds.) *Ageing and Sexing Animal Bones from Archaeological Sites*, BAR British Series 109, 223–50. Oxford, Archaeopress.

López Bertran, M. and Vives-Ferrándiz, J. (2009) Menjar, beure, cremar. Els rituals com a pràctiques relacionals durant l'Edat del Ferro. *Citerior* 5, 163–88.

Lorrio, A., de Miguel, M. P., Moneo, T. and Sánchez de Prado, M. D. (2010) Enterramientos infantiles en el oppidum de El Molón (Camporrobles, Valencia). *Cuadernos de Arqueología de la Universidad de Navarra* 18(2), 201–62.

Lorrio, A., Iborra, M. P., Sánchez de Prado, M. D. (2014) Depósitos rituales de fauna en el *oppidum* prerromano de El Molón (Camporrobles, Valencia). *Archivo de Prehistoria Levantina* 30, 213–38.

Lyman, R. L. (1994) *Vertebrate Taphonomy*. Cambridge, Cambridge University Press.

Machause, S and Sanchis, A. (2015) La ofrenda animal como práctica ritual en Época ibérica: La Cueva del sapo (Chiva, Valencia). In *Presses petites i grups humans en el passt. II Jornades d'arqueologia*, 261–86. València, Museu de Prehistoria de València, Diputació Provincial de València.

Martínez, G. (2009) Human Chewing Bone Surface Modification and Processing of Small and Medium Prey Amongst the Nukak (Foragers of the Colombian Amazon). *Journal of Taphonomy* 7, 1–20.

Merrifield, R. (1987) *The Archaeology of Ritual and Magic*. London, Batsford.

Miró, C. and Molist, N. (1990) Elements de ritual domèstic al poblat ibèric de La Penya del Moro (Barcelona). *Zephyrus* 43, 311–9.

Monraval, M. and López, M. (1984) Restos de un silicernio en la necrópolis ibérica de El Molar. *Saguntum* 18, 145–62.

Morales, A. and Liseau, C. (1994) Primer estudio sobre la fauna del yacimiento de Capote (Higuera La Real, Badajoz). In L. Berrocal-Rangel (ed.) *El Altar Prerromano de Capote. Ensayo Etno-Arqueológico de un Ritual Céltico en el Suroeste Peninsular*, 289–97. Madrid, Universidad Autónoma de Madrid.

Morris, J. (2011) *Investigating Animal Burials: Ritual, Mundane and Beyond*, BAR British Series 535. Oxford, Archaeopress.

Nieto, A. (2013) Porcs, cavalls, ovelles i infants. Noves aportacions a les pràctiques rituals de la fortalesa dels Vilars (Arbeca, les Garrigues). *Revista d'Arqueologia de Ponent* 23, 127–62.

Olcina Doménech, M., Guilabert Mas, A. and Tendero, E. (2010) Lectura púnica del Tossal de Manises. *Mainake* 32(1), 229–49.

Payne, S. (1973) Kill-off patterns in sheep and goats: the mandibles from Asvan Kale. *Anatolian Studies* 23, 281–303.

Payne, S. and Munson, P. J. (1985) Ruby and how many squirrels? The destruction of bones by dogs. In N. G. R. Fieller, D. D. Gilbertson and N. G. A. Ralph (eds.) *Palaeobiological Investigations: Research Design Methods and Data Analysis*, BAR International Series 266, 31–9. Oxford, Archaeopress.

Pérez Jordà, G., Iborra, M. P., Grau, E., Bonet, H. and Mata, C. (2000) La explotación agraria del territorio en época ibérica: Los casos de Edeta y Kelin. In R. Buxó and E. Pons (eds.) *Els Productes Alimentaris d'Origen Vegetal a l'Edat del Ferro de l'Europa Occidental: de la Producció al Consum. Sèrie Monogràfica* 18, 151–67. Girona, Museu d'Arqueologia de Catalunya-Girona.

Pérez Jordà, G. und Peña-Chocarro, L. (2013) Agricultural production between the 6th and the 3rd millennium cal BC in the central part of the Valencia region (Spain). In M. Groot, D. Lentjes and J. Zeiler (eds.) *Barely Surviving or More than Enough? The Environmental Archaeology of Subsistence, Specialisation and Surplus Food Production*, 81–99. Leiden, Sidestone Press.

Pickering, T. R., Domínguez-Rodrigo, M., Heaton, J. L., Yravedra, J., Barba, R., Bunn, H. T., Musiba, C., Baquedano, E., Diez- Martín, F., Mabulla, A. and Brain, C. K. (2013) Taphonomy of ungulate ribs and the consumption of meat and bone by 1.2-million-year-old hominins at Olduvai Gorge, Tanzania. *Journal of Archaeological Science* 40, 1295–309.

Prummel, W. (1987) Atlas for the identification of foetal skeletal elements of cattle, horse, sheep and pig, Part 2. *Archaeozoologia* 1(2), 11–41.

Reese, S. (1985) Shells, Ostrich Eggshells and Other Exotic Faunal Remains from Kition. In V. Karageorghis (ed.) *Excavations at Kition V. The Pre-Phoenician Levels, Part II*, 340–415. Nicosia, Department of Antiquities.

Rouillard, P. (2010) La Fonteta/Rábita (Guardamar del Segura, Alicante): las excavaciones hispano-francesas, 1996–2011. In *Guardamar del Segura, Arqueología y Museo*, Catálogo de la Exposición, 80–9. Alicante, MARQ.

Ruiz, A. (2008) Íberos. In F. Gracia Alonso (ed.) *De Iberia a Hispania. La Protohistoria de la Península* Ibérica, 733–844. Barcelona, Ariel.

Ruiz, A., Molinos, M., Fernándes, R., Pérez M. and Rueda, C. (2015) El santuario de la puerta del Sol. In A. Ruiz and M. Molinos (eds.) *Jaén, Tierra Ibera. 40 Años de Investigación y Transferencia*, 93–106. Jaén, Universidad de Jaén.

Saladie, P. (2009) Experimental Chewing and Gnawing of Humans and Other Primates Compared to Canids, Suids and Felids. PhD thesis, University Rovira i Virgili.

Sanchez Belda, A. and Sanchez, M. C. (1986) *Razas Ovinas Españolas*. Madrid, Ministerio de Agricultura, Pesca y Alimentación.

Sanmartí, J. (2009) From the archaic states to romanization: a historical and evolutionary perspective on the Iberians. *Catalan Historical Review* 2, 9–32.

Sarrión Montañana, I. (2003) El caballo de la Regenta (Burriana). Estado de conservación y restauración. In F. Quesada Sanz and M. Zamora Merchán (eds.) *El Caballo en la Antigua Iberia*, 187–90. Madrid, Bibliotheca Archaeologica Hispana 19, Real Academia de la Historia.

Ullrey, D. E., Sprague, J. I., Becker, D. E. and Miller, E. R. (1965) Growth of the Swine Fetus. *Journal of Animal Science* 24, 711–7.

Valenzuela Lamas, S. (2008) Ofrenes animals al jaciment ibèric d'Alorda Park (Calafell, Baix Penedès). In M. Miñarro and S. Valenzuela (eds.) *Actes del I Congrés de Joves Investigadors en Arqueologia dels Països Catalans: la Protohistòria als Països Catalans* (Vilanova del Camí, 18 i 19 de novembre de 2005). *Arqueomediterrània* 10, 151–8.

Vives-Ferrándiz Sánchez, J. (2013) Del espacio doméstico a la estructura social en un *oppidum* ibérico. Refexiones a partir de la Bastida de les Alcusses. In S. Gutiérrez and I. Grau (eds.) *De la Estructura Doméstica al Espacio Social. Lecturas Arqueológicas del Uso Social del Espacio*, 95–110. Alicante, Universidad de Alicante.

Vives-Ferrándiz, J., Bonet, E., Carrión, Y., Guillem, J., Iborra, M. P., Quesada, F., Ferrer, C., Tortajada, G. (2015) Ofrendas para una entrada; Un depósito ritual en la puerta oeste *de la bastida de les Alcusses (Moixent, Valencia). Trabajos de Prehistoria* 72(2), 282–303.

Von Nicolai, C. (2009) Pour une 'contextualisation' des dépôts du deuxième âge du Fer en Europe tempérée. *Revista d'Arqueologia de Ponent* 19, 75–90.

Chapter 9

Animal Biographies in the Iron Age of Wessex: Winnall Down, UK, Revisited

James Morris

During the 3rd to the 1st century BC on a small settlement consisting of a handful of roundhouses set in an area of open country close to arable land, a six-year-old male horse was killed. Following its death, the horse was partially dismembered; its lower jaw was removed, its tail and parts of its skin were cut from it, as were its hind legs and left forelimb, all below the knee. The now bloodied horse carcass was dragged, pulled and handled until it lay centrally within a pit on the outskirts of the settlement. The horse carcass was placed on its right side, the remains of its head resting on its chest, its one remaining complete leg extended out away from the body. Once in position, soil was thrown into the pit, until the horse carcass was no longer visible. There the horse lay, the remaining soft tissue being slowly consumed by bacteria from the horses' own gut and the soil which now enclosed it, until only the bones survived. What remains of the horse was next seen over two millennia later in 1977, when a team of archaeologists led by Peter Fasham excavated the site of Winnall Down, Winchester, ahead of the construction of the M3 motorway. The pit in which the articulated horse remains were discovered was labelled as pit 10161 and the horse bones, excavated from layer 10164, were duly bagged, cleaned, boxed and sent to the Ancient Monuments Laboratory at the University of Southampton. There they were subsequently examined and reported on by Mark Maltby, before finally being archived with Hampshire Museum service, where they remain to this day.

***

The horse from pit 10161 (Fig. 9.1) is part of a trend of Iron Age animal burials, with complete and partial deposits present on many British and European sites. Animal burials are recovered from archaeological sites dating to all time periods; however, they appear to be particularly ubiquitous in the Iron Age (Morris 2010a). The above-suggested transformations that the horse from pit 10161 underwent are only possible because of the wealth of information available to archaeologists examining such phenomena. However, it also highlights the difficulty in interpreting and understanding actions often seen as 'strange' to our modern-day mindsets. Indeed, the above description represents a parsimonious explanation of the activities leading to the creation of deposits. Other equally valid explanations are possible including exposure before the missing elements are removed and redeposition. The equifinality of animal burials is evident in the multitude of offered interpretations, ranging from natural deaths (Bourdillon 1990), to population control (Maltby 1988), to the result of rituals linked to the agricultural cycle (Cunliffe 1992). Yet previous approaches to animal burials often, intentionally or unintentionally, resulted in a single common explanation for the majority of deposits for a site or period. For example, Buckland-Wright (1987) suggested the sheep burials from Poundbury died from disease. This is possibly related to how zooarchaeologists consider 'normal' assemblages, often looking for long-term trends and patterns rather than considering individual animals and events. A biographical approach has proven useful when examining individual archaeological artefacts and deposits (see below), but is not commonly used, often because of the need for zooarchaeologists to examine multi-period large-scale assemblages. This chapter explores how a biographical approach can be used to inform on multiple animal burials from a site, using the early and middle Iron Age case study of Winnall Down settlement (Fasham 1985). The biographies of the individual animal burials have been constructed using the available published data (Fasham 1985; Maltby 1985; nd).

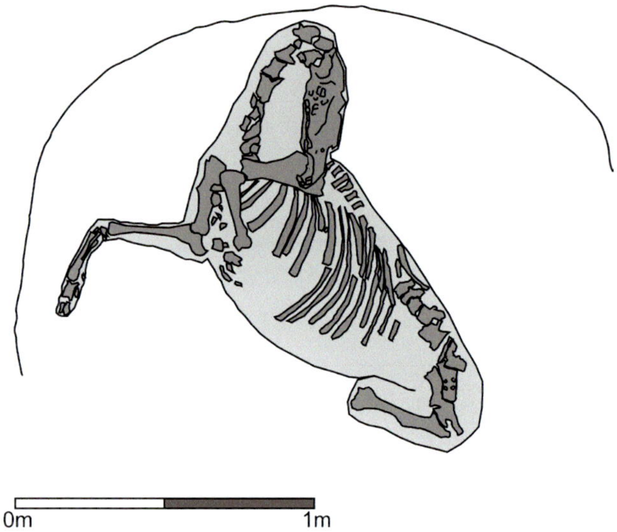

Figure 9.1 Plan of the partial horse burial from pit 10161, Winnall Down. The possible outline of the remaining body has been added in grey. Altered from Hill 1995, fig. 7.1.

To continue the narrative of the horse from pit 10161, we can consider how it has been subsequently interpreted. Maltby (1985) goes into detail, describing the different partial and complete animal skeletons recovered from the Winnall Down excavations, placing the skeletons into three different categories; complete or substantial parts of the carcasses with little evidence of skinning or butchery, bones of neonatal or foetal skeletons, and dumping of parts of the carcass after skinning and dismemberment. However, individual interpretations for remains, such as the horse from 10161, are not discussed. Maltby (1985, 105) points out that the pits with partial and complete skeletons tend to be found on the outskirts of the settlement and it is possible they represent noxious waste. Such interpretations, generally viewing animal skeletons as a product of practical decisions, associated with animal husbandry or settlement organisation were commonly held at the time Winnall Down was published (Morris 2011, 7–9). It should be noted that Maltby would not interpret the animal burials in the same way now; the original faunal report (Maltby nd) was written in 1981 and the publication in 1985 went to press in 1984, before information on the Danebury 'special deposits' (see below) was available (M. Maltby pers. comm.).

In the 1980s, one of the few zooarchaeological researchers suggesting an alternative interpretation was Annie Grant, with her work at Danebury Iron Age hillfort (Grant 1984, 1989). Grant suggested articulated animal remains from Danebury represented 'special animal deposits', proposing that the majority, with some notable exceptions (see below) represented ritual/religious sacrifices. It appears that the zooarchaeological community remained sceptical, a viewpoint best summarised by Wilson (1992). However, ideas that these deposits represent ritual activity were taken forward, with Cunliffe (1992) suggesting an Iron Age 'pit belief system', whereby human and animal remains along with other material were deposited in pits. One of the most influential works on Iron Age pits is J. D. Hill's (1995) consideration of not just the animal remains, but all material deposited within them. Hill (1995, 100) viewed the similarity between the deposition of human and animal remains as key, suggesting that archaeologists who accept the ritual treatment of human remains must extend such interpretations to animal remains, pottery and small finds treated in similar ways. Included in his analysis is the horse skeleton from Winnall Down pit 10161. Hill (1995, 73) suggests that a deliberate choice had been made to exclude large fragments of pottery and bone from the fill of the pit – linked to the presence of the horse skeleton and a hare skeleton in an upper fill.

Hill's work marks a turning point on how complete and partial animal skeletons are interpreted, and by the turn of the millennium most Iron Age deposits were now seen as the result of ritual activity (Morris 2010b; 2011). Such influence also extended beyond the Iron Age, with similar Romano-British (Fulford 2001; Maltby 2012) and Anglo-Saxon (Hamerow 2006; Morris and Jervis 2011) remains also viewed as the results of ritualised activity. Although the subtlety of Hill's arguments is often skimmed over, he does suggest that the deposits are of a ritualised nature. However, he mainly argues that the animal remains have been structured by human action – a human action in which a ritual/functional divide was not present. Rather the ritualised events were part of people's everyday lives, a way of expressing cosmological, social and practical concerns into a specific action.

Despite the considered nature of Hill's arguments there remained a ritual/functional dichotomy in animal burial studies. This is emphasised in the terminology used for such deposits, for example Grant's 'special animal deposits', placing the emphasis on special or 'chosen' for a ritual. Hill suggested using the term Associated Bone Group (ABG), a rather unwieldy term which has also been used in my own previous studies (Morris 2008; 2011), and is now part of Historic England guidance (Baker and Worley 2014). Throughout this chapter the terms ABG and animal burial are used. Regardless of the terminology, however, the important aspect is the separation of a description of a deposit from its interpretation.

Breaking free: animal biographies

The next stage in the biography of the horse from pit 10161 is my own re-examination of ABGs from the Neolithic to late medieval periods (Morris 2008, 2011). In these circumstances, the horse represents one of 746 Iron Age ABGs recorded from 50 archaeological sites from southern England (Dorset, Hampshire and Wiltshire). To break the data down further, 57 horse ABGs were recorded from Middle Iron Age sites, more than cattle (42) and pig (40), but swamped by the number of sheep/goat deposits (150) (for further information see Morris 2011, 41). The horse from pit 10161, therefore, became part of a large dataset concerned with patterns in species composition and deposition over a five-and-a-half-thousand-year time span. The disadvantage of such an approach is that individual deposits become lost in what is effectively a cloud of data. The advantage is that such an approach allows the multitude of deposit types to become apparent, highlighting their inconsistent interpretation at a single site level, let alone a single time period. For example, in Grant's (1984) approach to 'special animal deposits' three types were classified: fully or partially complete animal burials, skulls and horse mandibles, and articulated limbs. The interpretation of the complete and partial animal burials depended upon the age of the animal. When it was possible to calculate a mandible wear stage (MWS, Grant 1982), animals with a MWS of five or less were considered to be natural deaths while those with a MWS of six and over were considered to be 'special animal deposits'. The need to include 'natural deaths' possibly reflected the general undercurrent of economic/environmental determinist explanations zooarchaeologists were using at the time. This also highlights the arbitrary nature behind some interpretations, often driven by a desire to create a limited number of all-encompassing explanations, rather than considering the context and actions behind each deposit.

I argue that the variability in the character of the deposits (see below), combined with the perceived nature of certain time periods – if it is prehistoric and 'weird' it must be ritual – results in *meta-level* descriptions and explanations (Morris 2012). In other words, it results in large-scale categories, such as ritual, being used as description and interpretation, the equivalent of a zooarchaeologist suggesting that mammals were present on the site and given no further information. Despite arguments from authors such as Brück (1999) that we should jettison the term 'ritual', a heavily loaded term to archaeologists, we know that ritual activities do occur in everyday life. These can be for instance secular, religious, class-related, sex-related and personal (Bell 1997; Humphrey and Laidlaw 1994; Kreinath *et al.* 2006). A classic example is the multitude of different feasting events that occur throughout a calendar year in the United Kingdom – birthdays, weddings, funerals, Easter, Christmas, academic conferences and so on. All of these events are ritual activities but as participants we understand their structure, the differences in each, the often-unspoken social rules and the meaning behind them – hence, rather than 'ritual feasts', we refer to them by name. It is useful at this point to define what I mean by ritual as one of the common problems with archaeologists using the term is a lack of clear definition. It has been proposed, for example by Bell (2007), that rituals are exclusively religious, however as suggested above, I see rituals present in both religious and secular activities. Indeed, it has long been argued that to separate the sacred and profane in prehistory is to imply a false Cartesian dualism. The characteristic that most rituals share is that the actions behind them are formulaic (Snoek 2006); there is, in effect, a script. I would also argue another feature of ritual is that there is a specific intended consequence of the formulaic action which most of the participants would understand. Therefore, not all participants in a ritual would know the full script of the event, but they would understand the need for the ritual. The problem we have with the archaeological record is moving beyond the simple meta-level category of 'ritual' or indeed 'functional' towards the specific. For ritual, how do we identify the different scripts or their intended consequences?

I have suggested that a biographical approach towards partial and complete animal burials can help (Morris 2011; 2012). The key consideration in this is human action; in respect of human remains archaeologists have repeatedly emphasised that 'the dead do not bury themselves', and the same is true for animals. When considering animal burials we, therefore, firstly have to consider the human actions that created them, and indeed the wider contexts which created the need for the human actions. This changes the emphasis for these deposits. It is the final act of deposition, the placement within the pit and any associations that archaeologists often try to interpret, at the expense of the above ground transformations animals undergo before deposition.

The most common 'ritual' interpretation of animal burials is that they represent sacrifices (Morris 2011). Yet as archaeologists we are very bad at explaining what we mean by a sacrifice. Ethnographically animal sacrifice appears to be widespread. Animal sacrifices are still an important part of modern day Hinduism (Smith 2015); for example, the festival of Dussehra involves the ritual killing of thousands of animals to the goddess Durga, mother of the universe, in celebration of her victory over evil (DeMello 2012, 317). Animal sacrifice was also an important part of ancient Greek and Roman religious practices (see also Veropoulidou and Nikolaidou this volume). Ogilvie (1986, 41) suggests that traditional Graeco-Roman sacrifices consisted of four phases of activity: preparation, immolation, the slaughter and the final sacred meal, with specific rituals linked to each phase. In principle, all consumed meat in ancient Greece came from sacrificed animals, with the same vocabulary encompassing both sacrifice and butchery (Gilhus 2006). A modern-day analogy would be the use of *dhabihah*, the prescribed method of ritually slaughtering animals in Islamic law (Wheeler 2016). Pointedly these forms of sacrifice would leave no animal burial behind, as the animal is further processed, sold at market and consumed. Although there are numerous archaeological and anthropological definitions of sacrifice (Russell 2012, 89), the killing of the animal is often an important part of a wider ritual. For example, in discussing chicken sacrifice in Africa, Davidson (2015) highlights at least 18 different reasons a person could have for the sacrifice. In many rituals, sacrificed animals are viewed as intermediaries to the gods, able to pass on messages. In the television program 'Around the World in 80 Faiths', broadcast in the United Kingdom on the BBC in 2009, the Anglican vicar Pete Owen-Jones visited the Vodun church of Thron in Cotonou, Benin, Africa (Sheanhan 2009). The service in the church was a mixture of Vodun and Christian practices but primarily involved animal sacrifice to a deity. Domestic animals including cattle, goats, chickens, dogs and cats were all sacrificed and the moment before the animals were killed a message for the deity was whispered into its ear. After the ceremony, the congregation consumes the cow, goat and chicken in a feast, but the dog and cat are discarded in a nearby gutter. The deposition of the animals is not the important part of the ceremony; it is the transition of the animal into a messenger that is of primary importance. Archaeologically the different treatment of the animals post-sacrifice would be identifiable, but the detail of the animals being messengers would certainly be lost. To understand the creation of animal burials/ABGs it is, therefore, important to consider the above ground events behind their creation. It may be the above ground event that was ritualised, not the deposition of the animal's remains.

By considering the biographies of these deposits we can try to ascertain what these above ground events were. Normally, archaeologists look at material culture in what Gell (1998, 11) would describe as a supra-biographical manner, looking beyond the 'life cycle' at longer chronological trends. To explore the life histories of animal burials, however, we need to consider the individual deposits. The majority of archaeological studies employing a biographical approach has been concerned with pottery, metalwork and personal objects (for example Blanco-González 2014 Joy 2009; Kopytoff 1986; Swift 2012). In general, the biographical approach allows artefacts to become 'networks of significance' (Thomas 1996, 159), with artefacts given 'secondary agency'; they do not have the power to initiate happenings, but are objective embodiments of the power society or individuals have given them (Gell 1998, 20–1). The study of the biography of an object, animal or archaeological feature is also the study of moments of transition, be they physical, social or often both. In the aforementioned Vodun ceremony the killing of the dog and cat transforms their spirits into messengers, but their physical bodies then become waste. The process of transformation can also result in the production of new objects with completely different social meanings. For example, the removal of wool from a sheep and its transformation into cloth and clothing, or the working of a cattle horn into a drinking vessel, both examples starting with the transformation of the living animal.

I have argued that zooarchaeology is well placed to adopt a biographical approach to faunal material (Morris 2011; 2012), in part because such a method draws upon an understanding of taphonomic formation processes and transformations. Taphonomic studies have expanded beyond merely considering the biasing effects on an assemblage, and can now be seen as pivotal in understanding faunal assemblages and archaeological formation processes (Madgwick and Broderick 2016; Marín-Arroyo *et al.* 2011). A biographical approach uses taphonomic information to consider the full life history of a deposit. Previous taphonomic work, in particular on the natural disarticulation of animal carcasses (Behrensmeyer and Dechant Boaz 1980; Hill 1979) and butchery (Binford 1978, 386; Lyman 1987; Rixson 1988), can help in considering the above ground actions behind animal burials. The nature of most animal burials recovered from settlement sites suggest they are subjected to human butchery practices rather than natural disarticulation processes. It is, therefore, possible to consider partial animal burials against Rixson's (1988) five stages of butchery; skinning and evisceration, main dismemberment, processing into small portions, utilisation for marrow, and finally bone working. It may also be possible to identify an animal burial that has only undergone the skinning stage but none of the others. Common practices in dismembering animals were also suggested by Binford (1978), drawing on his study of the Nunamiut and other ethnographic works. First, disarticulation of the head from the neck and neck from the rest of the vertebral column, then separation of the front and back legs from the axial skeleton and separation of the lower feet from the legs. Although Binford (1978) was drawing on studies of hunter-gather communities, the butchery tool of choice was the knife. This is also the main tool used throughout most of the later prehistory, with cleavers and saws not in common use until the Roman period in Britain (Maltby 2007). We can, therefore, consider whether an animal burial has undergone some, if not all, of the common butchery processes.

How then does a biographical approach to animal burials work in practice? The account of the horse burial 10161 at Winnall Down, although using deliberately evocative language to paint a picture, is based on a consideration of the deposit's biography (Fig. 9.2). The biography contains a number of assumptions, the first being that the animal was already a part of the community at Winnall Down. Horses, although occasionally eaten, do not appear to be a major contributor to the Iron Age diet (Maltby 1996). It has been suggested that horses in the Iron Age were not bred but rounded up from the wild (Harcourt 1979) or managed feral populations (Hamilton 2000). Alternatively, horses may have only been raised at certain sites and traded, with recent isotopic analysis showing that Iron Age horses from Rooksdown, Hampshire, came to the site from as far away as Wales, Scotland or the continent (Bendrey *et al.* 2009). We, therefore, cannot be certain how the horse reached Winnall Down and indeed it is an assumption that the horse lived around the settlement and was not brought to the site specifically to be killed. In using a biographical approach we are attempting to map out the life history of a specific animal, however, it is never possible to create a complete picture and instead we are examining the animal's life at specific points in its existence. Certain points will be poorly focused, such as its life within the settlement. We can theorise that as a male horse of six years old at time of death it may have been ridden, and may have sired young, but our biography has to rely on wider background knowledge of the time period, rather than specifics. This biography is being constructed from the available records alone and re-examination of the bones might reveal further information. For example, Maltby (nd) does not mention the presence of bit wear on the horses' teeth, which if present would suggest it was ridden. Re-examination would confirm the presence or absence of bit wear, adding to the biography. Additionally, if isotopic analysis were conducted this would help identify whether the horse was raised in the local environment.

When it comes to the animal's death, further assumptions are required – that it was chosen to be killed, rather than died of natural causes, for example – and in making such assumptions we can use the age of the animal. We also do not know how the animal was killed. I have suggested that an artery was cut, so the horse bled to death, a common way of killing animals, and there is no evidence of poleaxing on the skull. We then know for certain that parts of the horse were removed, as they were not present with the rest of the skeleton within the pit. There is also no sign of weathering or gnawing on the remains, which would be possibly present if the missing elements were removed during a period of exposure. At this point, with detailed zooarchaeological and contextual information we can form a more tightly focused picture of the likely activities.

A biographical approach allows us to consider different points in the animal's life, death, and beyond, but with variable focus and chronological resolution. We know little of the horse's life on the settlement but we know some details of its manipulation after death. For other animal burials the situation may well be reversed. By considering the horse's biography we can also group events into moments of physical and social transition and I would suggest the horse went through at least seven major transformations or actions; the first of these is the horse being brought to or raised on the settlement and the last being its placement within the pit (Fig. 9.2). Some may be instantaneous and not involve a physical transformation, such as the choosing of the animal to be killed. At this point, a conscious human choice occurred, possibly changing the way the horse was viewed and certainly its role within the society. Other transformations, such as the removal of the limbs and jaw, may result in both physical and social transformations. What we can be certain of is that such actions were guided by human thought and may represent an event of as much, or perhaps greater, significance than the deposition of the horse's remaining body in the pit. What

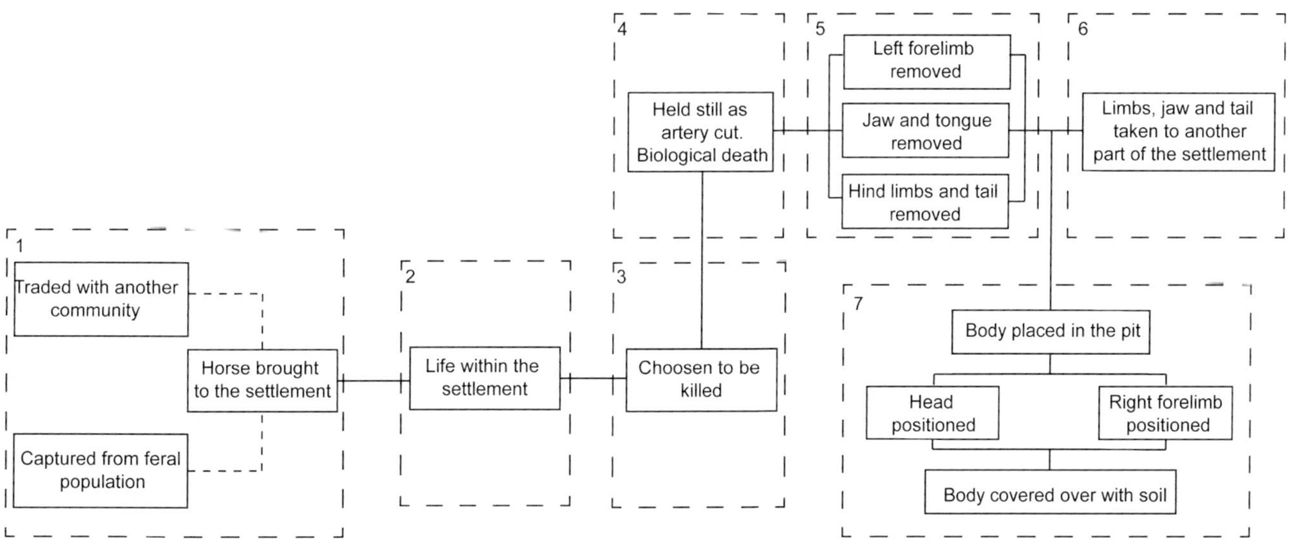

Figure 9.2 A simplified schematic biography of the partial horse burial from pit 10161, Winnall Down. The dashed numbered boxes represent theorized movements of physical and social transition.

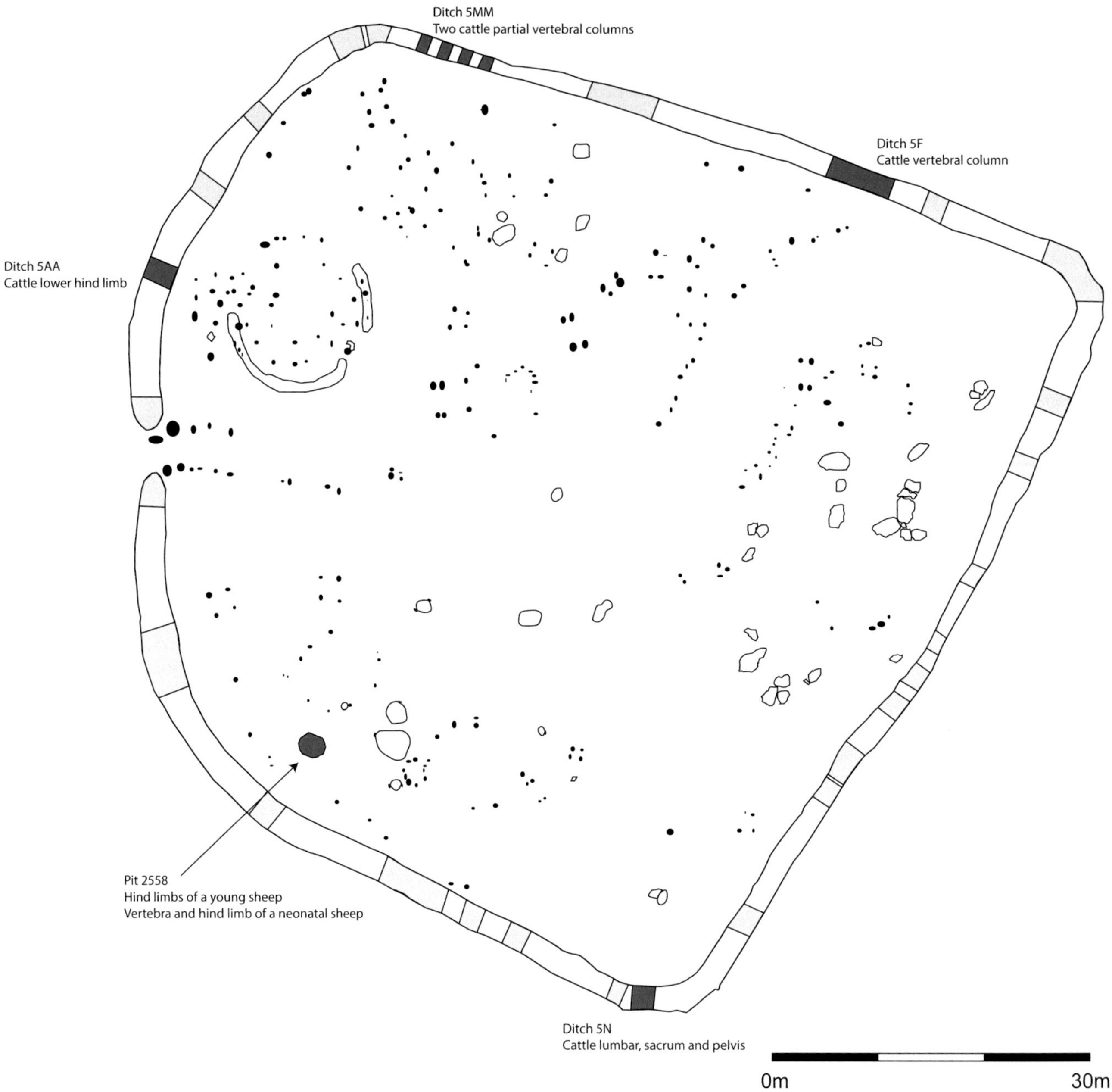

Figure 9.3 Plan of early Iron Age Winnall Down. The darkly shaded ditch sections and pit 2558 contain an ABG. The lightly shaded areas show the location of excavated ditch sections. Altered from Fasham 1985, fig. 9.

is also important to bear in mind is that with the exception of the final transformation, all consisted of 'above ground' pre-depositional events.

The described biography may add further detail to the horse burial, in particular highlighting the above ground events behind the deposits, but does it help us interpret it? The original interpretation is not clear but the horse appears to be seen as an unwanted carcass (Maltby 1985). Hill (1995) suggests the horse is a structured deposit and as such can be viewed as a ritual action, along with the choice to place less pottery in the pit's fill and a hare skeleton in a further fill. The biographical approach highlights that the deposition of the horse was not a stand-alone event but part of a series of actions. It is possible that each one of these actions represented a distinct ritualised activity, and indeed the proposed partial dismemberment of the horse may have had particular significance (see below).

Biographies in action: Winnall Down

So far only one animal burial has been discussed in detail, and a biographical approach would seem to lend itself to detailed narratives of individual deposits. However, the approach can be expanded to consider the animal burials from a site, or indeed region. Ideally for such an approach the biography of each animal burial would be considered, but the key is to look for difference and similarity – not just in species, composition, context and association, but more importantly in the type of transformations the animals had undergone. What this allows us to do is look for patterns in the multiple events leading to the creation of the deposit, rather than just its final resting place. As an example of this approach the ABGs from the early (6th to 3th century BC) and middle (3rd to 1st century BC) Iron Age settlement of Winnall Down (Fasham 1985) will be considered.

Maltby (1985) identified 55 ABG deposits from these phases, seven from the early Iron Age and 48 from the middle Iron Age. The majority of the early Iron Age ABGs were recovered from sections through the settlement's ditch (Fig. 9.3). One of the notable aspects is the concentration of cattle ABGs from the boundary ditch in comparison to the sheep ABGs, which were all recovered from pit 2558. Maltby (1985) suggested this corresponded with the general trend of large mammal remains being more common at the outskirts of the settlement in both the early and middle Iron Age, perhaps to keep the larger quantities of waste produced away from the settlement. The ditch would offer an ideal depository for large amounts of butchery waste generated by cattle, although the taphonomic conditions may have differentially affected sheep remains. A high proportion of loose sheep teeth were present in the ditch as teeth are the elements most likely to survive in poor taphonomic conditions. This may suggest that the lack of sheep bones is due to preservation factors, which have not substantially affected the larger cattle bones. In the ditch however, it is notable that the articulated cattle remains from sections 5AA, 5MM and 5F are all close to possibly contemporary buildings, and if the aim was to keep waste away from the settlement, as suggested by Maltby (1985), then the southern part of the ditch may have been better suited.

Considering the biographies of all the ABGs from the early Iron Age, the consistency of the cattle ABG assemblage is surprising. With the exception of the lower hind limbs from ditch section 5AA, all consist of articulated vertebral columns, interestingly without articulating ribs present. This suggests a large degree of conformity in the above ground actions immediately prior to their creation, all the cattle ABGs being processed, transformed into small groups of bones, probably with a limited amount of soft tissue present, no longer bearing any resemblance to the living animal.

In comparison, the two sheep ABGs from pit 2558 consist of limb bones as well as vertebrae and come from a neonate, and an immature animal with butchery marks present. Although butchery marks are not present on the neonate it is possible modified elements did not survive the burial process. However, the presence of the lumbar vertebrae and pelvis bones, both of which are highly porous, and thus do not normally preserve well, suggests the elements are missing due to human action. The different species, age and location of these remains suggest different actions and possibly different associated meanings behind the creation of these ABGs compared to the cattle. Given that this is just one deposit, it also suggests a much less common action compared to the processes behind the cattle ABGs, although the possible vulnerability of sheep taphonomic destruction could be a factor.

The Middle Iron Age ABG data are much more variable in terms of species, assemblage composition and transformations. Six different species are deposited, with dog ABGs making up 35% (17) of the assemblage, followed by sheep/goat 29% (14) and horse 16% (8). One of the most dramatic changes between the early and middle Iron Age is the drop in the proportion of cattle ABGs, making up only 12% (6) of the assemblage. This does fit a trend in southern England, with cattle making up an even smaller percentage of the Balksbury Camp and Danebury Middle Iron Age assemblages, corresponding with a general increase in sheep/goat ABGs. Where Winnall Down does differ from contemporary sites is the high proportion of dog ABGs, although there is great variety on middle Iron Age sites with dogs making up 2% of the Suddern Farm assemblage compared to 28% from Old Down Farm (Morris 2011, 43–5).

Another difference is the presence of complete, or nearly complete animal burials. Pit 1490 contained a complete female dog, with no evidence of butchery. Another complete female dog was also present in pit 6595, this time also with a complete pig that may have been partially skinned. The pig was a little over two years old, whereas the female dog had lived well into adulthood. The left femur of the dog had a misaligned healed fracture, which would have given it a limp for the remainder of its adult life. The dog is positioned tucked under the overhang of the beehive pit, over a meter deep, its back against the pit wall and legs close to its body. The pig is positioned in the middle of the pit, its legs splayed outwards. It is notable that both these deposits, 1490 and 6595, are in pits very close to contemporary buildings.

Neonatal dog skeletons were present in pits 4006, 6038 and 7257, the latter two pits also being close to contemporary buildings and in the case of 6038, within a roundhouse (Fig. 9.4). These deposits together with the partial horse and complete hare from pit 10161 represent very different above ground events and human choices compared to the remains from the early Iron Age contexts. This is not to say that all dog remains were treated this way, a skull and mandible from an adult dog and articulated thoracic vertebrae were also recovered from pit 7257, hind limbs of dogs were present in pits 7372 and 1055, with 7372 also containing caudal vertebrae (tailbones). It is unknown if the tailbones were still in articulation when

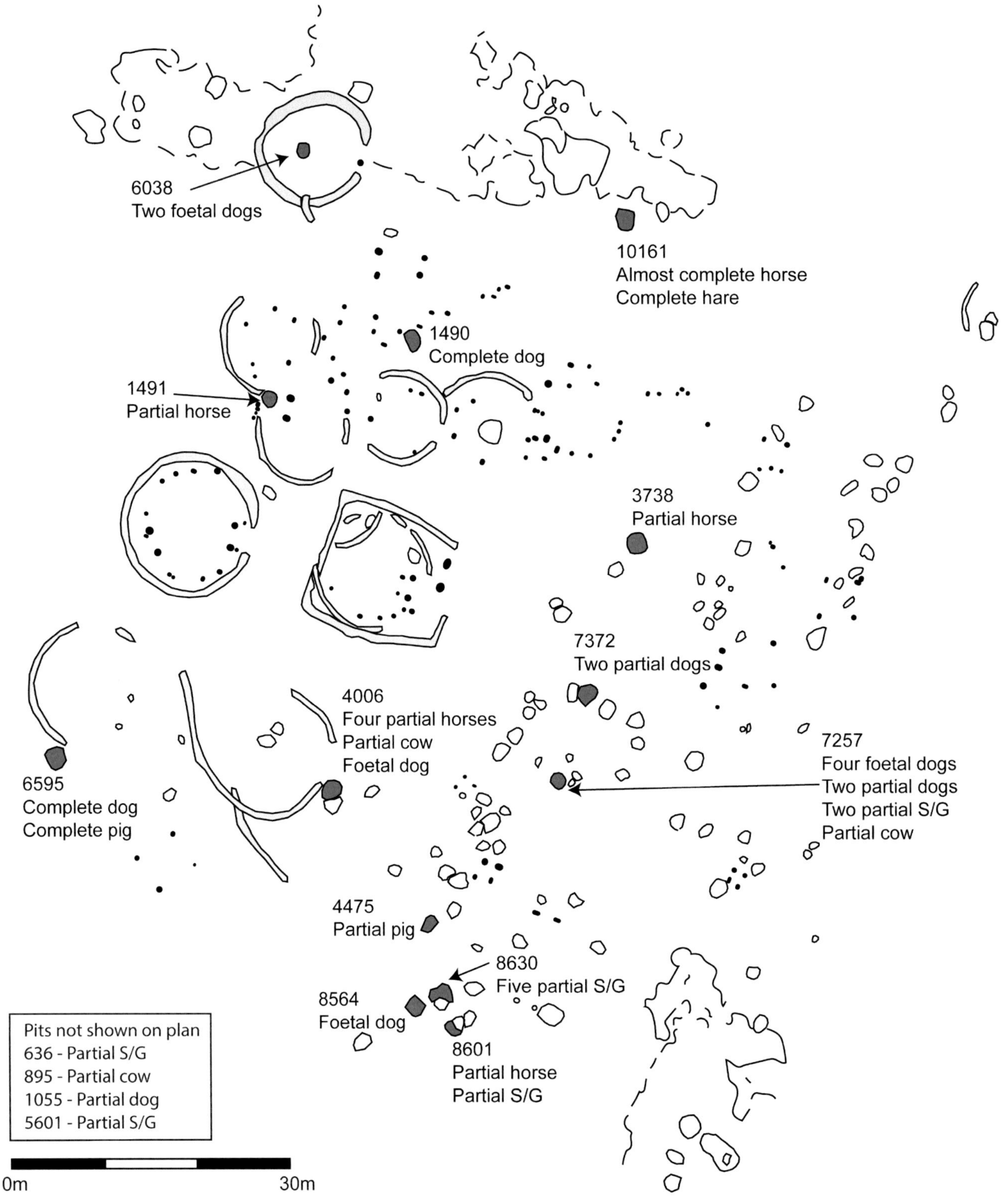

Figure 9.4 Plan of middle Iron Age Winnall Down. The darkly shaded pits contain ABGs. Pits not shown on the plan are due to a lack of information concerning their location. Altered from Fasham 1985, fig. 15.

discovered. However, this shows that dogs were not subject to just one type of transformation resulting in an ABG, but there appears to be at least three different activities. These comprise the burial of complete adult animals, the deposition of neonatal dogs, sometimes in groups, and then the dismemberment of adult dogs and the deposition of small articulated parts. Given that both complete dogs are female it is unfortunate that sexing information is not available for the dismembered adult dogs. Each activity represented very different above ground activities and choices.

In the early Iron Age, all the sheep ABGs were from the same pit and the trend of multiple deposits of the same species does continue in the middle Iron Age. Pit 1941 contains four sheep/goat ABGs, two hind limbs, one forelimb and one vertebral column. Interestingly this was almost mirrored by Pit 8630 that contained two hind limbs, one forelimb, a vertebral column (although made up of just thoracic vertebrae this time) and a skull and mandible. Many of the remains have butchery marks present associated with dismemberment, and for both pits some of the ABGs may be from the same sheep/goat. It is therefore possible that both pits represent very similar transformation events, with the purposeful dismemberment of two or more sheep and the deposition of small partially articulated parts of the animal within a specific pit. However, there is an age difference in the selected sheep; the Early Iron Age pit contained immature and neonate remains, whereas the elements from the Middle Iron Age pit 1941 are from adult animals. There would be practical differences in the deposition of adult compared to neonate remains and if the animals were processed for meat, then fewer people would be fed by the neonate remains. Therefore, this could represent two similar events, but perhaps at different scales.

Other pits contain multiple ABGs from different species. As already mentioned, pit 7257 contains a number of dog remains. Also present are cattle lower hind feet from the same animal, the rear end of a sheep/goat consisting of the hind limbs, pelvis, sacrum and lumbar vertebrae and the left upper forelimb of a sheep/goat, possibly from the same animal. The nine ABGs recovered from this pit can be seen as the result of a number of different human actions: the deposition of dog neonatal remains; the dismemberment of an adult dog and deposition of a recognisable skull and largely defleshed thoracic vertebra; the dismemberment of a cow/bull (the sex is unknown) and deposition of fleshed lower hind feet; the dismemberment of a sheep and deposition of a partially fleshed back end and a partially fleshed upper forelimb.

Pit 4006 also contains a mix of species, but this time mainly large mammals, with the exception of a foetal dog. Four cattle ABGs all consisting of forelimbs were present along with four horse deposits, an upper forelimb, a lower hind limb, a sacrum with lumbar vertebrae and an ABG consisting of both femora, pelvis, sacrum, lumbar and thoracic vertebrae and ribs. Again the composition suggests a number of different above ground events. For the horses, one animal appears to have been highly processed and the forelimb and hind limb deposits may have had little flesh present on them. By contrast, the ABG consisting of the back half of an animal may well have still been recognisable as a horse, with the vertebrae, ribs and upper legs still present. It is notable that of the ABGs consisting of vertebral remains, this and the horse from 10161 are the only ones with the ribs still present, suggesting the animals may have been eviscerated but not fully dismembered. It is difficult to identify the specific ABGs on the plan of the pit, but different groupings of bones can be clearly seen, suggesting that these remains were all deposited at the same time (Fig. 9.5). It is interesting that neonatal human remains are also present within the same fill of the pit. Unfortunately, the detailed stratigraphic information for this pit is not published, so it is only possible to ascertain whether the deposit was quickly infilled or covered gradually by naturally silting. Again a number of different transformations can be noted on the ABGs, and like pit 7257, if all these remains were deposited at the same time they did not only represent different above ground actions but may also suggest they involved a large part of the community to supply the animals and take part in the event.

Conclusion: from human action to human motives

The aim of this chapter was to apply a biographical approach, using published data to the case study site of Winnall Down. Given the constraints of space and data availability, it has not been possible to discuss the biography of each ABG at Winnall Down in detail. It has also not been possible to complete comprehensive linear biographies from birth to reanalysis, in part due to lack of data, but also as Joy (2009) has argued, partly because object biographies consist of connected jumps as objects become active and inactive in clusters of social relationships. Therefore, we are able to focus on certain clusters, points of activity and transitions with variable resolution. This also means that although this approach lends itself to thick narrative descriptions, key biographical information can be acquired without long-winded narratives, as it is the differences in the points of activity and transition that help us consider different human actions.

What a biographical approach brings to the study of ABGs and animals is an emphasis on the different pathways of treatment, and thus, highlights the variation in human actions. This approach also enables us to investigate why the composition of ABGs may change between time periods. The general uniformity of ABGs from the Early Iron Age suggests that the majority and particularly the cattle remains were created by very similar events. What we see in the Middle Iron Age is reduction in uniformity. Remains similar to those from

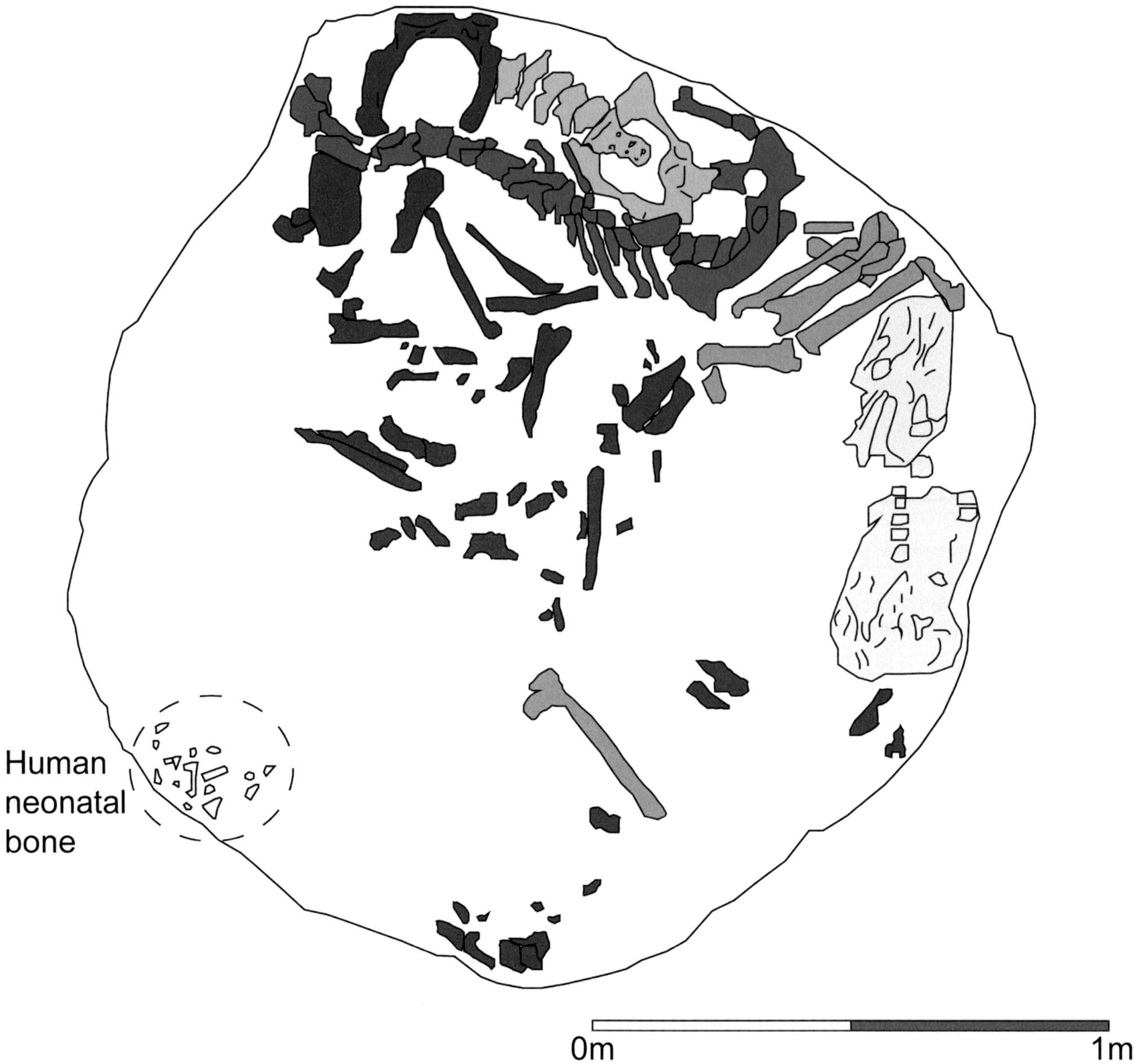

Figure 9.5 Plan of pit 4006. The shading highlights possible ABG groupings, human neonatal bones within the dashed circle. Altered from Hill 1995, fig. 7.1.

the Early Iron Age are still being deposited, although often now as sheep/goat, and with variable biographies, suggested by age differences. The drop in cattle ABGs and the increase in those of sheep/goat mirrors a change in the 'normal' faunal assemblage from the site, although the change for the ABG proportions is dramatic (Fig. 9.6). As discussed above the general increase in the number of sheep/goat ABGs is seen on other sites in southern England, although it is not present on sites in other parts of the country, such as Yorkshire (Morris 2010a). The decrease in cattle ABGs could be linked to a general change in animal husbandry, with more sheep being present on later Iron Age sites. However, the decrease is also part of a pattern where more species are now deposited as ABGs. This suggests that either the animals being used changed but the practices/events behind the creation of the ABGs remained the same, or that there was an expansion in the type of practices/events creating ABGs, or indeed both.

Given the variability in the composition of the ABGs, I would suggest at Winnall Down there was an expansion in the number of pathways to deposition and thus a broadening in the type of above ground events. It is possible that what this represents is an expanding plurality in Iron Age cultural practices. In the Middle Iron Age at Winnall Down at least five different actions appear to exist:

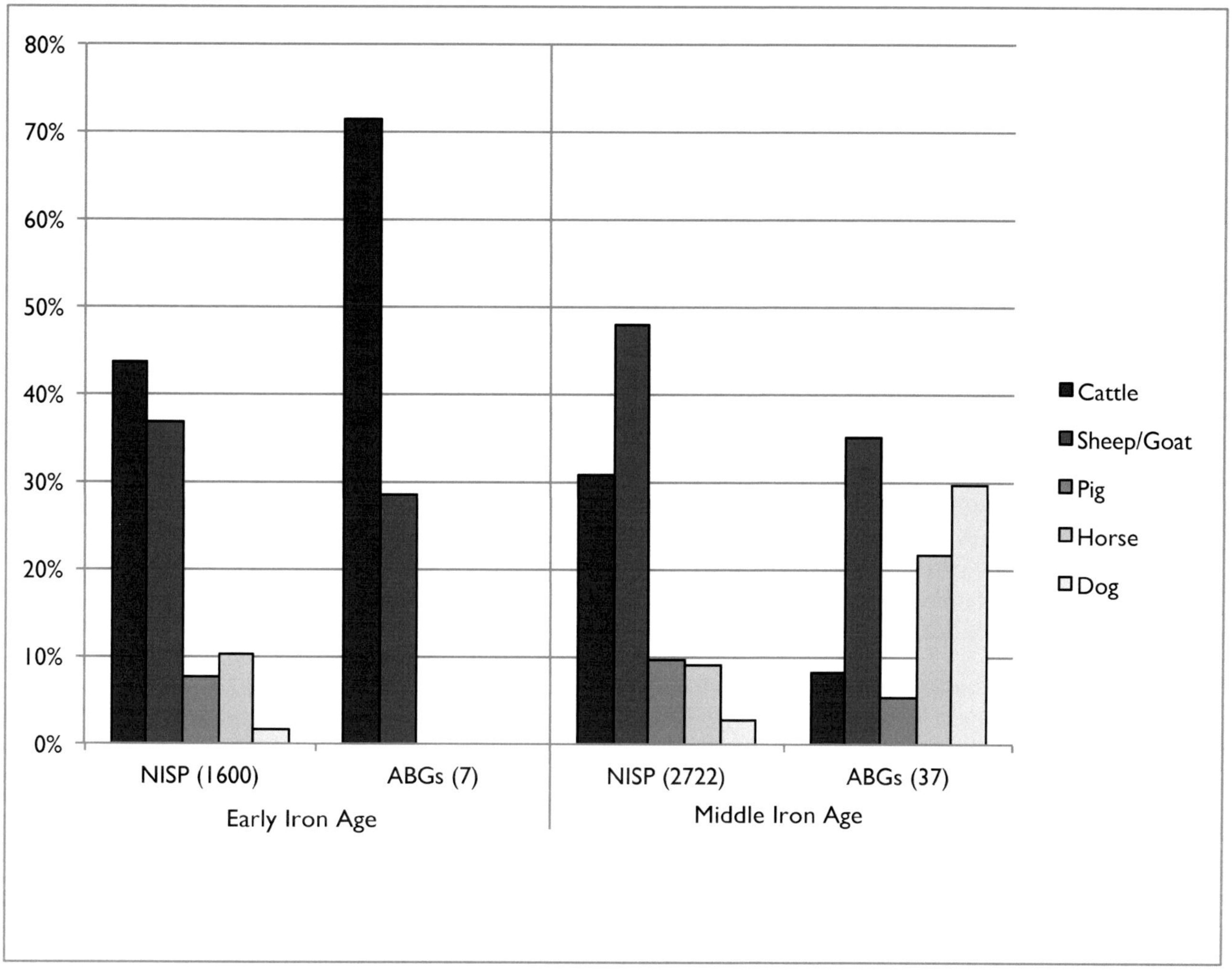

Figure 9.6 Bar chart showing the proportion of cattle, sheep/goat, pig, horse and dog from the disarticulated assemblage NISP (Number of Identified Specimens Present) counts and the ABGs for early and middle Iron Age Winnall Down. Data from Maltby 1985.

- The deposition of complete animals. This appears to be rare and is limited to adult female dogs and a pig deposit. Both deposits are next to houses and no other ABGs are present in the pits.
- The deposition of neonatal dogs, often in groups and with some dismemberment, although the lack of certain elements could be a taphonomic issue. Most of these deposits also contain ABGs from other species that have undergone intensive dismemberment.
- The intensive dismemberment of mainly adult sheep/goat, dog, horse and cattle and their deposition often as a single ABG within a pit.
- The deposition of a number of highly dismembered ABGs together, such as pit 8630 (Fig. 9.4).
- The semi-dismemberment of an animal, for example, horses deposited with a large proportion of the flesh still attached and possibly still in a recognisable form.

In most cases these treatments appear to represent individual events, although occasionally deposits, such as pit 4006 (Fig. 9.4), show a combination. The suggested pathways mainly focus on differences at the end of the animals' lives, but aspects of the whole life cycle are important. For example, in the case of the neonatal dogs, it may be that their lack of an established life history within the community was important, assuming of course they do not represent neonatal natural mortalities. However, one of the biggest issues with a biographical approach is that authors, myself included, are naturally drawn to the data rich clusters of the animals' life histories. For this case study the use of only published records may be part of the problem. A re-examination of the faunal material using the multitude of technical advances since the 1980s may help to add missing parts to these biographies. However, the approach has been successful in moving the focus to the human acts before deposition, although deposition still remains a key part of the biography.

The biographical approach can help us examine human actions, but can it help us elucidate the meaning behind those actions? The first aspect it highlights is the plurality in human actions. The presence of complete animals, purposely placed, close to houses, is intriguing and I have previously suggested pit 6595 (see above) may represent a deliberate, emotive burial with associated offering (Morris 2016). The biography of the dog suggested it had a long life on the settlement, in all likelihood it raised numerous litters and may have received care for its broken leg. It is the only ABG to have pathological evidence and the only ABG to be deposited complete without any alteration (the pig in the same pit has butchery marks on the lower feet).

Its position under the overhang of the beehive pit and over a metre down, suggested human action was involved in its placement. Given the location and space available this would have involved perhaps just two people. I have suggested that such intimate, small-scale contact may be the result of an emotional response to an animal's death (Morris 2016). The dog may not have been a pet in the modern sense, but perhaps was an important part of the community and peoples' lives. It is interesting to note that the dog burial mirrors the treatment of some complete adult human burials at Winnall Down, placed in a crouched position against the edge of a pit. It is also tempting to see horse burial 10161 (Figs. 9.1 and 9.4) in an emotive light, considering its deliberate manipulation and placement. If it does represent a sacrifice, and given the age of the animal this is a possibility, then the removal of certain body parts may have been a key part of a ritualised act. Its deliberate placement with the head clearly visible suggests in this case the act of deposition also had importance. In contrast, the highly dismembered material from pits 1941, 4006 and 8630 (Fig. 9.4) involves a number of different animals and species. Such deposits are not just limited to Winnall Down; pit 197 from Suddern Farm represents a similar deposit. I have suggested that the Suddern Farm pit 197 ABGs represent the 'waste' from a large-scale feasting event (Morris 2011, 174–6). This is due to the level of butchery processing required to create the ABGs, the majority of which consists of vertebral columns with evidence of the meat being stripped from them, perhaps representing uncommon cooking practices. Therefore, they may represent the processing, cooking and consumption of a large amount of meat in a single event, likely to be a feast. In this respect, the deposition of the ABG material may be unimportant and unrelated to the above ground rituals taking place at the feast. It is, therefore, intriguing that human neonatal material is also present in pit 4006. Could this be related to a ritualised feasting event and does its treatment perhaps mirror that of the neonatal dog remains? It is also tempting to see the smaller scale, highly dismembered ABG deposits as the result of smaller scale feasting events requiring just one animal – although Maltby's original argument that they represent 'normal' butchery waste could be just as valid. What is important about this interpretation is that the feasting event may be ritualised, but this does not necessarily mean the deposition of the ABGs was.

A biographical approach does not offer a magic bullet that will explain the human motives behind these deposits. What it does offer is a way to consider the multi-temporal dimensions of animal burials, rather than concentrating on just the final act of deposition. If, as argued, some of these deposits represent an emotive burial, sacrifice or feasting waste at different scales then perhaps it is the animal's deposition, the moment of sacrifice or the human connections and bonds strengthened during the event which were the important aspects behind the ABGs' creation. This case study has used a biographical approach to try and move away from a ritual/functional dichotomy, and although the term ritual has still been used, a ritual involving a horse sacrifice would be very different to the rituals behind a large feast, or the more personal rituals behind a dog burial. However, we must return to the point of equifinality, in that the interpretations I have placed on the deposits are just one possible explanation out of many. The pathways, nevertheless, the biographical approach highlights do help to narrow down our options. It also highlights the futility of suggesting that these deposits can be explained with a single interpretation.

The biographies constructed for the animal burials from Winnall Down were purposely descriptive and separated from the integration of why the human driven transformations were enacted on the animals. However, it is perhaps fitting to return to the biography of the horse from pit 10161 at the end of this chapter, and give just one of its possible full biographies from life to concealment. Some of the following is evidence-based, some conjecture, but what they highlight are the possibilities an animal biographical approach offers for a deposit which in the past was described and interpreted simply as 'waste' and 'ritual'.

***

Sometime during the 3rd to the 1st century BC, a group set out from a small settlement into the surrounding open-country. Riding horses and accompanied by their choice dogs, they headed to an area frequented by a herd of wild horses. Carefully they approached the herd and noticed a young stallion they could separate from the rest of the herd. They knew of the dangers involved, not just to themselves but their animals as well, on the last trip a dog had broken its hind leg. Once captured, the young stallion was taken back to the village, where it was gradually broken and trained to accept a rider. Over the coming years the horse proved useful, carrying members of the community as they visited other settlements and traded items. It was also used on trips back to the place of its birth to collect other young horses for the community. At the age of six, at the height of its powers, something happened in the settlement – a rare event that required an unusual act. The stallion was separated from the other horses it was corralled with and taken to the edge of the settlement. There, a group had gathered, the man who had led the horse now took out a knife that was plunged into the horse's neck, it screamed from the sudden pain, kicking and bucking, before finally collapsing in a pool of its own blood. A group gathered around the horse. The sacrifice had been made but the ritual was not over. Other members of the community gathered around the horse, for it needed to supply important parts for the further rituals to come. Its tail and skin were cut away allowing its hind legs to be removed; its left forelimb was taken off at the elbow, and its jaw and tongue cut away. These parts were carried elsewhere in the settlement, important tokens from the horse's sacrifice. A pit had been prepared and a small group dragged the remaining carcass of the horse to the pit where it was rolled in. People climbed down into the pit and struggled with the weight of the carcass as they positioned it in the center. All who gathered around the edge of the pit could see the horse, now on its right-hand side, the remains of its head resting on its chest, its one remaining complete leg extended out away from the body as though in midstride. The soil and rock from the excavation of the pit was close at hand, this was now thrown into the pit using wicker baskets, until the horse was concealed. But the pit was not completely filled – other deposits would be later placed in it, but that is another story.

Acknowledgements

My whole-hearted thanks to Richard Madgwick, Santi Riera and Alexandra Livarda for inviting me to contribute to this volume, their infinite patience as editors and supporting the rather unusual 'dark and stormy night' beginning and end to this paper. My thanks to the three anonymous reviewers who greatly helped focus and refine the paper. I give thanks to Mark Maltby for his continued support even though I appear to insist on revisiting and reinterpreting many aspects of his earlier work. My thanks to Justine Biddle for her constant support, help and supreme editing skills. All errors of course remain my own.

References

Baker, P. and Worley, F. (eds.) (2014) *Animals Bones and Archaeology. Guidlines for Best Practice*. London, Historic England

Behrensmeyer, A. K. and Dechant Boaz, D. E. (1980) The recent bones of Amboseli National Park Kenya, in relation to East African paleoecology. In A. K. Behrensmeyer and A. Hill (eds.) *Fossils in the Making*, 72–92. Chicago IL, University of Chicago Press.

Bell, C. (1997) *Ritual Perspectives and Dimensions*. Oxford, Oxford University Press.

Bell, C. (2007) Response: Defining the need for a definition. In E Kyriakidis (ed.). *The Archaeology of Ritual*, 277–88. Los Angeles CA, University of California Cotsen Advanced Seminars 3

Bendrey, R., Hayes, T. E. and Palmer, M. R. (2009) Patterns of Iron Age horse supply: An analysis of strontium isotope rations in teeth. *Archaeometry* 51(1), 140–50.

Binford, L. R. (1978) *Nunamiut Ethnoarchaeology*. New York, Academic Press.

Blanco-González, A. (2014) Tracking the social lives of things: biographical insights into Bronze Age pottery in Spain. *Antiquity* 88, 441–55.

Bourdillon, J. (1990) *The animal bones from La Sagesse (The Presbytery) 1988, Romsey, Hampshire*. Portsmouth: Ancient Monuments Laboratory Report 106/90. English Heritage.

Brück, J. (1999) Ritual and rationality: some problems of interpretation in European archaeology. *European Journal of Archaeology* 2(3), 313–44.

Buckland-Wright, J. C. (1987) The animal bones. In C. S. Green (ed.) *Excavations at Poundbury, Dorchester, Dorset 1966–1982. Vol 1: The Settlements*, 4: A1–D14, Monograph Series No 7, 129–32. Dorchester, Dorset Natural History and Archaeological Society.

Cunliffe, B. (1992) Pits, preconceptions and propitiation in the British Iron Age. *Oxford Journal of Archaeology* 11, 69–84.

Davidson, J. M. (2015) 'A Cluster of Sacred Symbols': Interpreting an Act of Animal Sacrifce at Kingsley Plantation, Fort George Island, Florida (1814–39). *International Journal of Historical Archaeology* 19, 76–121.

DeMello, M. (2012) *Animals and Society. An Introduction to Human–Animal Studies*. New York, Columbia Univeristy Press.

Fasham, P. (ed.) (1985) *The Prehistoric Settlement at Winnall Down, Winchester*. Winchester, Hampshire Field Club Monograph 2.

Fulford, M. (2001) Links with the past: pervasive 'ritual' behaviour in Roman Britain. *Britannia* 32, 119–218.

Gell, A. (1998) *Art and Agency. An Anthropological Theory*. Oxford, Clarendon Press.

Gilhus, I. S. (2006) *Animals, Gods and Humans. Changing Attitudes to Animals in Greek, Roman and Early Christian Ideas*. Oxford, Routledge.

Grant, A. (1982) The use of tooth wear as a guide to the age of domestic ungulates. In B. Wilson, C. Grigson and S. Payne (eds.) *Ageing and Sexing Animal Bones from Archaeological Sites*, 91–108. BAR British Series 109. Oxford, Archaeopress.

Grant, A. (1984) Animal husbandry. In B. Cunliffe (ed.) *Danebury: an Iron Age Hillfort in Hampshire. Volume 2. The Excavations 1969–1978: the Finds*, 496–548. CBA Research Report. London, Council for British Archaeology.

Grant, A. (1989) Animals and ritual in Early Britain: the visible and the invisible. *Anthropozoologica* 3, 79–86.

Hamerow, H. (2006) 'Special deposits' in Anglo-Saxon settlements. *Medieval Archaeology* 50, 1–30.

Hamilton, J. (2000) The animal bone. In B. Cunliffe and C. Poole (eds.) *The Danebury Environs Programme. The Prehistory of a Wessex landscape. Volume 2, Part 2. Bury Hill, Upper Clatford, Hants, 1990*, Monograph No. 49, 67–73. Oxford, English Heritage and Oxford University Committee for Archaeology.

Harcourt, R. (1979) The animal bones. In G. J. Wainwright (ed.) *Gussage All Saints*, 150–60. London, Department of the Environment Archaeology Reports No.10.

Hill, A. (1979) Disarticulation and scattering of mammal skeletons. *Paleobiology* 5, 261–74.

Hill, J. D. (1995) *Ritual and Rubbish in the Iron Age of Wessex*. BAR British Series 242. Oxford, Archaeopress.

Humphrey, C. and Laidlaw, J. (1994) *The Archetypal Actions of Ritual*. Oxford, Clarendon Press.

Joy, J. (2009) Reinvigorating object biography: reproducing the drama of object lives. *World Archaeology* 41(4), 540–56.

Kreinath, J., Snoek, J. and Stausberg, M. (2006) *Theorizing Rituals: Issues, Topics, Approaches, Concepts*. Leiden, Brill.

Kopytoff, I. 1986. The cultural biography of things: commoditization as process. In A. Appadurai (ed.) *The Social Life of Things: Commodities in Cultural Perspective*, 64–91. Cambridge, Cambridge University Press.

Lyman, R. L. (1987) Archaeofaunas and butchery studies: a taphonomic perspective. In M. B. Schiffer (ed.) *Advances in Archaeological Method and Theory* 10, 249–337. San Diego CA, Academic Press.

Madgwick, R. and Broderick, L. (2016) Taphonomies of trajectory: the pre- and post-depositional movement of bones. *Archaeological and Anthropological Sciences* 8(2), 223–26.

Maltby, M. (nd) The animal bones from Winnall Down (M3 motorway archaeological rescue committee site R17), Hampshire. Portsmouth, Ancient Monuments Laboratory Report 3453.

Maltby, M. (1985) The animal bones. In P. Fasham (ed.) *The Prehistoric Settlement at Winnall Down, Winchester*, 97–125. Winchester, Hampshire Field Club Monograph 2.

Maltby, M. (1988) The animal bones from a Romano-British well from the 1965–6 excavations at Oakridge II, Basingstoke, Hampshire. Ancient Monuments Laboratory Report 64/1988.

Maltby, M. (1996) The exploitation of animals in the Iron Age: the archaeozoological evidence. In T. Champion and J. Collis (eds.) *The Iron Age in Britain and Ireland: Recent trends*, 17–28. Sheffield, J. R. Collis Publications.

Maltby, M. (2007) Chop and change: specialist cattle carcass processing in Roman Britain. In B. Croxford, N. Ray, R. Roth and N. White (eds.) *TRAC 2006: Proceedings of the 16th Annual Theoretical Archaeological Conference, Cambridge 2006*, 59–76. Oxford, Oxbow Books.

Maltby, M. (2012) Sheep foundation burials in Roman Winchester. In A. Pluskowsk (ed.) *The Ritual Killing and Burial of Animals. European Perspectives*, 152–63. Oxford, Oxbow Books.

Marín-Arroyo, A. B., Madgwick, R., Brugal, J.-P. and Moreno-García, M. (2011) New perspectives on Taphonomy. *International journal of Osteoarchaeology* 22(5), 505–8.

Morris, J. (2008) Re-examining Associated Bone Groups from Southern England and Yorkshire, c.4000 BC to AD 1550. PhD thesis, Bournemouth University.

Morris, J. (2010a) Associated bone groups: beyond the Iron Age. In. J. Morris and M. Maltby. (eds.) *Integrating Social and Environmental Archaeologies: Reconsidering Deposition*, 12–23. BAR International Series 2077. Oxford, Archaeopress.

Morris, J. (2010b) The composition and interpretation of associated bone groups from Wessex. In D. Campana, A. Choyke, P. Crabtree, S. Defrance and J. Lev-Tov (eds.) *Anthropological Approaches to Zooarchaeology: Colonialism, Complexity and Animal Transformations*, 261–71. Oxford, Oxbow Books.

Morris, J. (2011) *Investigating Animal Burials: ritual, mundane and beyond*. BAR British Series 535. Oxford, Archaeopress.

Morris, J. (2012) Animal 'ritual' killing; from remains to meanings. In A. Pluskowski (ed.) *Animal Ritual Killing and Burial: European Perspectives*, 8–21. Oxford, Oxbow Books.

Morris, J. (2016) Mourning the sacrifice. Behaviour and meaning behind animal burials. In M. Demello (ed.) *Mourning Animals: Rituals and Practices Surrounding Animal Death*, 11–20. East Lansing MI, Michigan State University Press.

Morris, J. and Jervis, B. (2011) What's so special? A reinterpretaion of Anglo-Saxon 'special deposits'. *Medieval Archaeology* 55, 65–80.

Ogilvie, R. M. (1986) *The Romans and their Gods*. London, Hogarth Press.

Rixson, D. (1988) Butchery evidence on animal bones. *Circaea* 6(1), 49–62.

Russell, N. (2012) *Social Zooarchaeology. Humans and Animals in Prehistory*. Cambridge, Cambridge University Press.

Sheanhan, T. (2009) Episode Three: Africa. In J-C. Bragard (ed.) *Around the World in 80 Faiths*. BBC Televison.

Snoek, J. (2006) Defining 'Rituals'. In J. Kreinath, J. Snoek and M. Stausberg (eds.) *Theorizing Rituals. Issues, Topics, Approaches, Concepts*, 3–14. Leiden, Brill.

Smith, F. M. (2015) Vedic sacrifice in Modern India. In B. A. Hatcher (ed.) *Hinduism in the Modern World*, 212–26. London, Routledge.

Swift, E. (2012) Object Biography, Re-use and Recycling in the Late to Post-Roman Transition Period and Beyond: Rings made from Romano-British Bracelets. *Britannia* 43, 167–215.

Thomas, J. (1996) *Time, Culture, and Identity: An Interpretive Archaeology*. London, Routledge.

Wheeler, B. M. (2016) Animals, in Law. In K. Fleet, G. Krämer, D. Matringe, J. Nawas and E. Rowson (eds.) *Encyclopaedia of Islam, 3rd edition*. Leiden, Brill Online. Available online: http://referenceworks.brillonline.com/entries/encyclopaedia-of-islam-3/animals-in-law-COM_0169 (accessed 15 March 2016).

Wilson, B. (1992) Considerations for the identification of ritual deposits of animal bones in Iron Age pits. *International Journal of Osteoarchaeology* 2, 341–9.

Chapter 10

Faunal Remains and Ritualisation: Case Studies from Bronze Age Caves in Central Italy

Letizia Silvestri, Mario F. Rolfo, Micaela Angle, Robin Skeates and Leonardo Salari

Introduction

This chapter concerns Middle Bronze Age (MBA) archaeology in Central Italy and focuses on the interpretive potential of faunal remains towards improving our understanding of the human uses of caves in this period. Building on a critical review of all published Bronze Age caves of Central Italy where zooarchaeological research has been carried out, this paper demonstrates how some of the limitations of previous approaches can be overcome by pushing bioarchaeological research towards a more social perspective (Marciniak 2005; Russell 2012). Three new studies carried out on as many cave sites investigated by the authors are presented here. The aims are, firstly to provide an assessment of the relations between bioarchaeology – mostly zooarchaeology – and ritual in Bronze Age Central Italy and, secondly, and more importantly, to propose a more in-depth approach to the study of bioarchaeological remains, which will be beneficial for the wider interpretation of all sites in the area.

The Bronze Age in the Italian peninsula dates to between approximately 2300–200 and 900 BC. Within this timeframe, the MBA has been one of the main subjects of current debates amongst the academic community, for many aspects of this transitional period are still to be clarified. A provisional combination of radiocarbon dates and typologies place the Italian MBA between 1750–1700 and 1350 BC (Cunliffe *et al*. 2009), when the first Mycenaean communities started to approach the Italian coasts (Bietti Sestieri 2010).

The Italian peninsula is crossed by the Apennine Mountains, where limestone complexes are very common. Central Italy, which includes the regions of Tuscany, Marche, Umbria, Lazio and Abruzzo, includes hundreds of caves, rockshelters and crevices, most of which have revealed evidence of prehistoric human use (*e.g.* Barker 1981; Cocchi Genick 2002a; Cremonesi 1968; 1976; Di Fraia and Grifoni Cremonesi 1996; Puglisi 1959; Radi 1981; Radmilli 1974). At least a hundred natural caves in the area are known to have Early-Middle Bronze Age remains dating to the 2nd millennium BC, most interpreted as ritual deposits in sacred spaces (Cocchi Genick 1999; Di Gennaro 1999; Grifoni Cremonesi 1996; 2002; Guidi 1991; Pacciarelli 1997; Skeates 1997; Whitehouse 1992; 2001; 2007), and including human burials. Ritual use of these caves has usually been identified with reference to specific features. First are the location and geo-morphological features of these caves, which are often isolated and difficult to reach, very dark, close to springs, waterfalls, and inner water basins, and characterised by spectacular speleotherms. The second feature consists of the anomalous completeness of artefacts, such as pottery, possibly linked to intentional depositions of offerings or grave goods. This is even more obvious when such artefacts are found in peculiar structures with apparently no practical function (*e.g.* stone circles) or next to features, such as hearths, that do not appear to have served a mundane purpose because of their location (*e.g.* Pastena Cave, see below) and/or contents (*e.g.* Grotta Misa, where accurately separated seed piles of different cereals and legumes were identified lying *in situ* in the centre of a ring-shaped structure, Tongiorgi 1947). Finally, the presence of human bones clearly indicates a mortuary use of the sites or of parts of them.

Caves are the best-known and studied category of archaeological sites of the Central Italian Bronze Age; open-air settlements are less well known. The latter are certainly more difficult to find, as they tend to be located on the shores of lakes and are sometimes submerged, and on defensive hilltops, where medieval villages with a continuity of habitation to present times cover prehistoric features (Bietti Sestieri 2010; di Gennaro 1988). This lack of evidence for open-air settlements has long led to a misconception about settlement patterns in Early and Middle Bronze Age Central Italy: following Puglisi's (1959) extensive study of the Apennine culture, the most credited theory was that communities in this period were formed by pastoralist nomads who lived in caves. This was later disproved by Barker (1981), who inspired by Higgs' school of palaeoeconomy (Higgs 1975), accurately analysed the environment, as well as the faunal and botanical remains of most known prehistoric sites in Central Italy (several of them caves). Among other socio-economic conclusions, his ground-breaking bioarchaeological research demonstrated that agriculture was also practised and sedentism was at least as prevalent as nomadism at various sites.

Since the 1980s the number of known open-air settlements has increased, while most of the features identified in caves were newly interpreted as not indicative of domestic use. The ritualised nature of such contexts became gradually more evident (Cocchi Genick 1999; Skeates 1997; Whitehouse 1992). For example, the Grotte di Belverde (Siena) showed a wide range of different ritual and burial practices, including artefactual depositions, manipulated human skeletons, deposition of swords and building of stone structures; Grotta Morritana (Rome) had a deposit of at least seven bronze axes; Tanaccia di Brisighella (Ravenna) included a large amount of miniature pottery; and Grotta dell'Orso di Sarteano (Siena) is only one example of several caves with pots positioned under water-dripping stalactites (Cocchi Genick 1999 and references therein; Grifoni Cremonesi 1996 and references therein). However, recent soil micromorphology studies (Iaconis and Boschian 2008) indicated that Italian Bronze Age caves with evidence of ritual activities could also have been used for more mundane purposes (*e.g.* Grotta Sant'Angelo, Grotta dei Piccioni di Bolognano). On the basis of soil thin sections, their study suggested that the use of these caves became more intensive and specialised in ovicaprine sheltering throughout the Bronze Age, according to the increased burnt layers of dung found there and the higher amount of coarse ware recorded. Bradley (2005) in fact first claimed that prehistoric societies and communities never distinguished cult from everyday life. The progressive (yet partial) deritualisation (secularisation) of life is only typical of contemporary Western society (*ibid.*). However, it can certainly be that some places were selected as more suitable for more repetitive or intensive activities that can be described as ritual.

Our knowledge of the archaeology of Bronze Age caves is, however, in most cases limited to reports on pottery typology (Cocchi Genick 2002b; Cocchi Genick *et al.* 1995) or to brief reports often published by speleologists and local enthusiasts. Therefore, the majority of these sites remain just dots on a map, sometimes imprecise, with little or no information about their stratigraphy, preservation, features and material classes, other than ceramics and occasional special artefacts (Ricciardi 2015). Zooarchaeological and archaeobotanical aspects have seldom been considered in these contexts and, where they have been included, they have contributed primarily to environmental and subsistence-related interpretations. In fact, given the frequent lack of funding for systematic investigations of open-air sites, assemblages from caves have often been used to reconstruct the palaeoeconomy of broader communities whose representatives frequented those sites (*e.g.* De Grossi Mazzorin 1995). The analysis that follows seeks to provide an alternative approach and interpretation of aspects of life of the communities inhabiting the MBA landscape of Central Italy.

Faunal remains in Central Italy's Early–Middle Bronze Age

Information from only published cave sites that have been both surveyed and excavated at least in a preliminary manner has been collected in order to examine the state of research. The reporting (or absence of reporting) of the following ten variables was noted if a zooarchaeological assemblage was present: preservation of a stratigraphic/chronological distinction of faunal datasets in stratified sites; taxon identification; Number of Identified Specimens (NISP); Minimum Number of Individuals (MNI); age classes/kill-off patterns; body portions and sides; butchery/burning/processing marks; taphonomy/fragmentation; pathologies; and incorporation of zooarchaeological interpretations in a wider, integrated discussion. The results of this review do not only provide useful interpretive data, but also allow for methodological reflection.

The overall results indicated the frequent absence of most of this information from publications of the analysed sites, which can partly explain the problems of zooarchaeological studies of Italian later prehistory. In particular, the investigation first revealed the level of detail in the adopted methodology for each case study, and on what basis conclusions were drawn. Isolating these factors in the analysis of each publication also highlighted the limited comparability of different reports, which do not follow a uniform protocol. However, general inferences can still be made when certain types of information have been covered by more than one report.

Of more than 100 central Italian MBA caves represented in literature by at least a short archaeological report, only 24 were reported to have faunal remains in their archaeological deposits (Fig. 10.1), whereas in only one case (Grotta del Costone di Battifratta, Rieti *et al.* 1985) the almost complete absence of identifiable faunal remains is explicitly mentioned. However, the limited number of faunal reports is not an accurate representation of the original material in

Figure 10.1 Location of central Italian Middle Bronze Age caves with published zooarchaeological reports. Numbers correspond to Table 10.1.

the caves. Firstly, all the latest discoveries and publications always mention the presence of animal bones. Secondly, archival research in museum stores, direct conversations with local archaeological groups and new surveys carried out by the authors of this paper at old and new sites, confirm the existence of faunal remains in every cave that has been examined more closely. This clearly indicates that, especially for the earlier research, bioarchaeological remains have often been overlooked or excluded in favour of more immediately attractive artefacts. Therefore, a significant loss of data has to be acknowledged, as well as a bias in the available information.

Of the published datasets including the three new case-studies (Caves 1, 2 and 3 in Fig. 10.1 and Table 10.1) discussed below, 15 went beyond the mere citation of the identified species (specifying the NISP and maintaining a stratigraphic division at multi-phase sites) and ten at least partially recorded age classes, thus enabling the construction of mortality curves. MNI was calculated in eight cases, providing a complementary taxonomic quantification data. Even in those cases, the exact method used to calculate MNI was rarely specified, making the various samples difficult to compare. Butchery, cut and fire marks were recorded in only six cases. Other aspects, such as body portions and sides, spatial distribution, taphonomy and fragmentation rates, as well as palaeopathology, were very rarely reported, if at all.

Unsurprisingly, the earliest studies generally provide less information. Papers from the 1950s to the early 1990s did not usually record more than three key features (mainly stratigraphy, taxon and NISP), with the only exceptions being the works on Riparo del Lauro (Bigini 1986) and Grotta Beatrice Cenci (Agostini *et al.* 1991). The instigation of palaeoeconomic perspectives (*e.g.* in Italy by Barker 1981) and of more contextual approaches eventually led to an increase in the attention paid to faunal remains, which resulted in more extensive, yet still partial, studies. Apart from the predominance of ovicaprines and other domesticates at all sites (Fig. 10.2) (*e.g.* De Grossi Mazzorin 1995; Wilkens 1991), the only notable identifiable feature recorded in literature has been that of perinatal animals, found in at least one sixth of the caves. More specifically, Grotta Di Carli (Cerilli 2000) and Grotta Sant'Angelo (Wilkens 1996) had lambs/kids, and Grotta dei Cocci (Salari *et al.* 2014) and Grotta Mora Cavorso (Rolfo *et al.* 2013; Silvestri *et al.* 2016) lambs/kids and piglets. The assemblage from Grotta dello Sventatoio (Angle *et al.* 1991) is reported to include only young individuals, but neither the species nor the exact age classes are mentioned. Grotticella 10 di Sorgenti della Nova (Minniti 2012 and references therein), dated to a slightly later phase (the Late Bronze Age), also needs to be mentioned as it contained numerous perinatal piglets. This peculiar occurrence was first noticed by Wilkens (1995), when she analysed ritual animal depositions in central Italian prehistory. She reported depositions of foetuses and newborns of the main domesticates in caves and other structures dating from the Neolithic to the historical times, interpreting them as seasonal sacrifices related to fertility. However, given the heterogeneity of the information provided in the publications examined (see Table 10.1), it is not possible to further investigate the deposition of animals in the caves. Detailed data about age classes combined with MNI are almost always absent. Even more importantly, mention of body portions and spatial distribution, which could aid the contextual interpretation of the datasets, is extremely rare. The problem is accentuated by the fact that integrated discussions, involving the context, the artefacts and the bioarchaeological remains, are completely missing from all publications. Except for the cases of sub-juvenile remains, fauna from caves identified as including ritual features have only been used for wider economic inferences, concluding that the communities who frequented the caves focused on pastoral subsistence strategies. Such economic inferences might, however, be biased by the cultural choices made by the caves' prehistoric occupants.

In the review of the zooarchaeological studies, some of the data provided in the examined texts were not quantified, reducing their potential comparability. For the following six caves (see Figure 10.1 for location/number), the most basic aspect, that of NISP, was lacking and only the range of species identified was provided with some occasional further information: a) Grotta del Costone di Battifratta (20), scarce wild boar (*Sus scrofa*); b) Grotta di Val de' Varri (21), pigs (*Sus domesticus*) and ovicaprines (*Ovis/Capra*) (in order of quantity); c) Grotta Polesini (18), sheep (*Ovis aries*), goat (*Capra hircus*), cattle (*Bos taurus*), fox (*Vulpes vulpes*), domestic cat (*Felis catus*), microfauna, shells; d) Grotta Misa (22), ovicaprines, pig, cattle, dog (*Canis familiaris*), red deer (*Cervus elaphus*), wild boar, hare (*Lepus sp.*), bats and amphibians; e) Buca Tana di Maggiano (24), cattle, sheep, goat, wild boar, (probably) domestic pig, dog, red deer, turtle, badger (*Meles meles*), weasel (*Mustela sp.*), rodents, bats, insectivores, birds, molluscs, reptiles. Abundance of cattle, pig and turtle. Large mammal bones exhibited cut marks; and f) Grotta Nuova (23), domestic and wild fauna.

Finally, in four of the ten datasets that provided kill-off information, the age classes were not specifically quantified, although the publications do state that an unusual percentage of juvenile or sub-juvenile domestic animals were present. These are a) Grotta di Carli (10), sub-juvenile lambs and kids; b) Grotta Sant'Angelo (11), sub-juvenile lambs and kids; c) Grotta dello Sventatoio (19), sub-juvenile domesticates; and d) Grotta Beatrice Cenci (14), 25% of the ovicaprines are 'infant'.

The archaeology of neglected remains

An opportunity to take a step forward in the direction of a more integrated study of faunal remains from central Italian MBA caves has arisen over the last decade. The University of Rome 'Tor Vergata' (Insegnamento di Paletnologia), in agreement with the Soprintendenza Archeologia del Lazio e dell'Etruria Meridionale, joined by the Archaeology Department of Durham University (UK), have recently started or resumed excavations in several caves of the Lazio region. Here we examine the results from three of these sites, which are the most extensively investigated thus far, as well as being the most interesting in terms of the employment

Table 10.1. Caves with published zooarchaeological reports and the type of information included.

Cave no.	*Cave name*	*Location*	*Latest Publication*	*Stratigraphy*	*Species/ taxon*	*NISP*	*Age Classes*	*MNI*	*Measure-ments*	*Body portions*	*Marks*	*Discussion*
1	Grotta Mora Cavorso	Lazio, Jenne (RM)	Rolfo *et al.* 2013	X	X	X	X	X			X	X
2	Grotta del Pertuso	Lazio, Pastena (FR)	Angle *et al.* 2014	X	X	X	X	X				
3	Grotta Regina Margherita	Lazio, Collepardo (FR)	Angle *et al.* 2010	X	X	X	X			X		
4	Grotta del Fontino	Tuscany, Montepescali (GR)	Corridi 2002	X	X	X	X	X	X		X	
5	Grotta del Beato Benincasa	Tuscany, Monticchiello (SI)	Bigini 1981	X	X	X						
6	Grotta dell'Orso di Sarteano	Tuscany, Sarteano (SI)	Cremonesi 1968	X	X	X						
7	Grotta del Mezzogiorno	Marche, Frasassi (AN)	Tongiorgi 1956	X	X	X						
8	Riparo dell'Ambra	Tuscany, Camaiore (LU)	Bigini 1986	X	X	X						
9	Riparo del Lauro	Tuscany, Camaiore (LU)	Bigini 1987	X	X	X	X	X	X		X	
10	Grotta di Carli	Lazio, Ischia di Castro (VT)	Cerilli 2000		X		X	X				
11	Grotta S.Angelo sulla Montagna dei Fiori	Abruzzi, Civitella del Tronto (TE)	Wilkens 1996	X	X	X	X	X	X		X	
12	Grotta a Male	Abruzzi, Assergi (AQ)	Pannuti and Peroni 1969	X	X	X						
13	Grotta dei Piccioni di Bolognano	Abruzzi, Bolognano (PE)	Cremonesi 1976		X	X						
14	Grotta Beatrice Cenci	Abruzzi, Cappadocia (AQ)	Agostini *et al.* 1991	X	X	X	X	X	X	X	X	
15	Grotta La Punta	Abruzzi, Ortucchio (AQ)	Cremonesi 1968		X	X						
16	Grotta Bella	Umbria, Avigliano (TR)	Curci et al. 2014	X	X							

(Continued)

Table 10.1. Caves with published zooarchaeological reports and the type of information included. (Continued)

Cave no.	Cave name	Location	Latest Publication	Stratigraphy	Species/ taxon	NISP	Age Classes	MNI	Measure-ments	Body portions	Marks	Discussion
17	Grotta dei Cocci	Umbria, Narni (TR)	Salari et al. 2014	X	X	X	X	X	X	X	X	
18	Grotta Polesini	Lazio, Tivoli (RM)	Radmilli 1974	X	X							
19	Grotta dello Sventatoio	Lazio, Castel Sant'Angelo Romano (RM)	Angle *et al.* 1991				X					
20	Grotta del Costone di Battifratta	Lazio, Battifratta (RI)	Segre Naldini and Biddittu 1985		X							
21	Grotta di Val de' Varri	Lazio, Pescorocchiano (RI)	Guller and Segre 1949		X							
22	Grotta Misa	Lazio, Ischia di Castro (VT)	Rittatore Vonwiller 1951		X							
23	Grotta Nuova	Lazio, Ischia di Castro (VT)	Rittatore Vonwiller 1951		X							
24	Buca Tana di Maggiano	Tuscany, Maggiano (LU)	Puccioni 1914		X							

of a contextualised social zooarchaeology methodology. Moreover, these three sites, which are located in Southern Lazio, are culturally similar, providing the opportunity to compare and contrast the evidence from each cave interpreted as a ritual context, in regards to faunal remains. The three caves are Grotta Mora Cavorso (Jenne, Rome Province: Rolfo *et al.* 2013), Grotta del Pertuso (Pastena, Frosinone Province: Angle *et al.* 2014) and Grotta Regina Margherita (Collepardo, Frosinone Province: Angle *et al.* 2010) (Fig. 10.1).

Methods of zooarchaeological and archaeobotanical analysis

Faunal assemblages from these three caves have been all subject to a common protocol of investigation, extending from retrieval to post-excavation analysis. The animal bones were numbered and mapped if found *in situ*, or put in separate finds bags identified by square number and context if found during the wet-sieving operations. All excavated soil from all three caves was wet-sieved through 2mm mesh. In addition, a sample of one sixth of soil per context was brought to the laboratory and analysed with a 0.5mm sieve for detailed bioarchaeological analysis. Once in the laboratory, animal bones were examined to determine their preservation status, body part, taxon or size, skeletal element, side, bone fusion or tooth wear, age class (Foetuses and Newborn: 0–3 months; Very young: 3–6 months; Young: 6 months–2 years; Young Adult: 2–3 years; Adult: over 3 years), presence of anthropogenic marks (disarticulation, butchery, cut, exposure to fire) and taphonomic agents (gnawing, roots, patinas, erosion), and evidence of palaeopathology. MNI was also calculated, based on the greatest number of same-sided bones, combined with observations of the dimensions, sex and age of the other skeletal elements, in order to maximise the accuracy of the MNI estimation of the assemblages. A certain degree of older and recent contamination has to be assumed for all caves, as the stratigraphy of these sites is often disturbed by several taphonomic factors including animal burrowing and dens. In addition, animal bones of the Holocene are mostly similar in regards to species and size, making it more difficult to distinguish intrusions compared to artefacts, which can be attributed to specific typologies and periods. Archaeobotanical remains were identified with the aid of a modern reference collection and atlases (*e.g.* Jacomet 2006, Neef *et al.* 2012) and were quantified on the basis of the MNI.

Mora Cavorso

This cave (Fig. 10.3) is located at 715m.a.s.l. in the karst complex of the Simbruini Mountains, in the Upper Aniene River Valley. It is a complex cave with a wide entrance and a sequence of three tunnels followed by inner chambers, extending over a length of at least 60m. It was

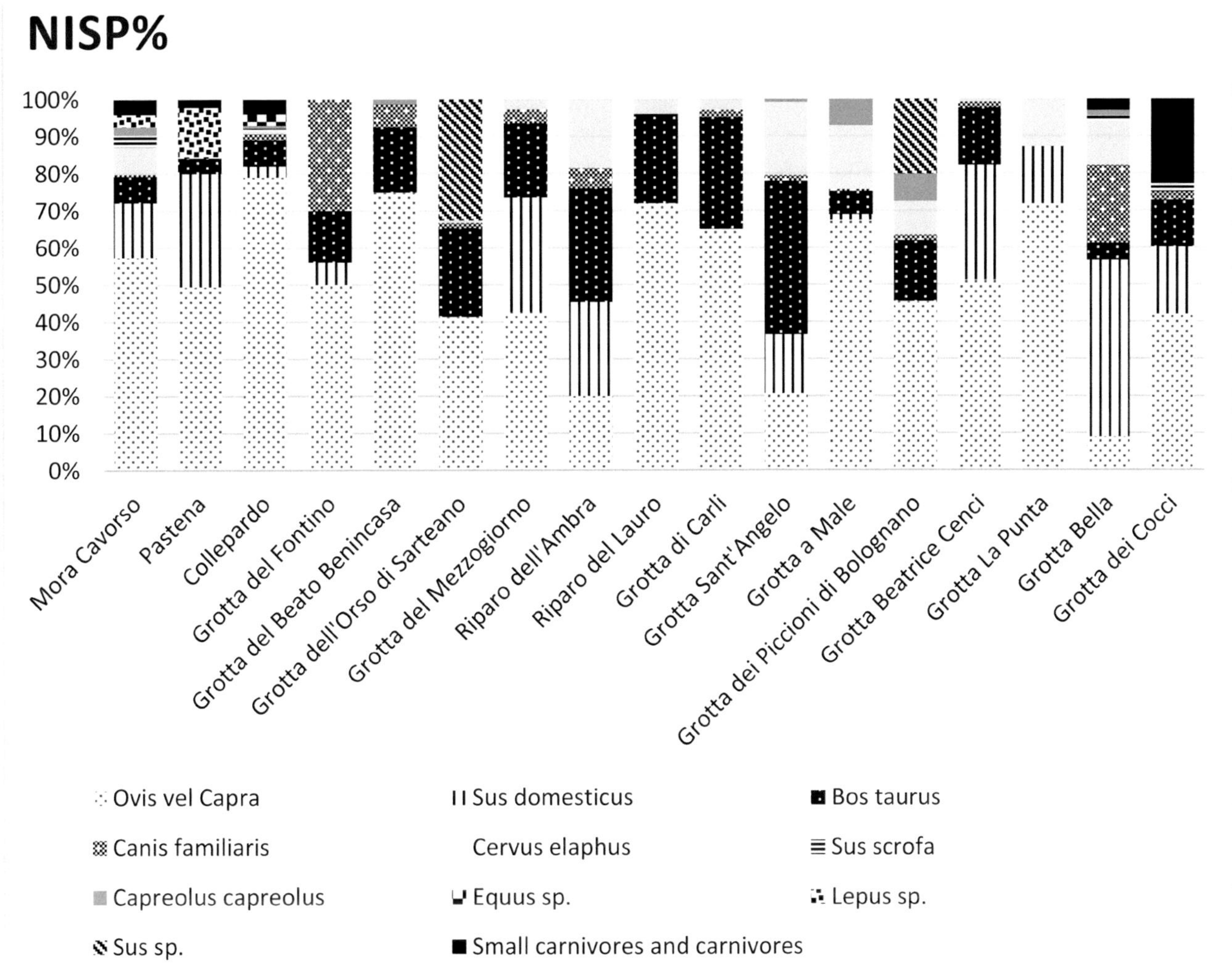

Figure 10.2 Distribution of animal taxa from Early–Middle Bronze Age caves by Number of Identified Specimens (NISP) percentage.

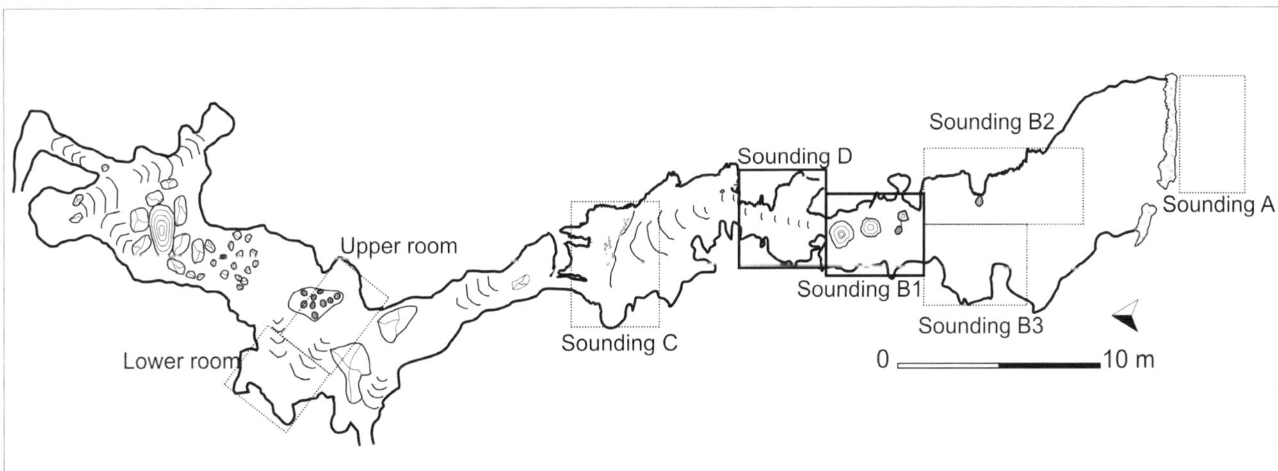

Figure 10.3 Plan of Grotta di Mora Cavorso with location of the archaeological soundings and the Middle Bronze Age deposits.

discovered in the early 2000s by a local speleological group and it has been investigated systematically since 2006 (excavations are still in progress) by the University of Rome 'Tor Vergata'. Although its most significant phase of occupation was the Early Neolithic, when the inner chambers were used for human burial, the cave

Table 10.2 Age classes of the main domestic taxa (MNI) from Early-Middle Bronze Age caves sites and settlements. F/N: Foetus/Newborn; VY: Very young (between 3 and 6 months); Y: Young (between 6 months and 2 years); Y-A: Young adult (between 2 and 3 years); A: Adult (older than 3 years).

MNI	*Mora Cavorso*	*Pastena*	*Collepardo*
Ovis/Capra (Sheep/goat)	37	6	11
Sus domesticus (Pig)	19	4	3
Bos taurus (Cattle)	7	2	3
Canis familiaris (Dog)	2		1
Cervus elaphus (Red deer)	10		1
Sus scrofa (Wild boar)	6		1
Capreolus capreolus (Roe deer)	4		1
Equus caballus (Horse)			1
Lepus sp. (Hare)	5	1	
Small carnivores & *Canis lupus* (Wolf)	13	2	1
Total	103	15	23

was occupied for a long time span, ranging (with some hiatuses) from the Upper Palaeolithic to Late Antiquity, the 17th–18th centuries and the Second World War. A very important period of use was the Early–Middle Bronze Age. Excavations of Soundings D and B2 (Fig. 10.3) revealed the presence of several indications of cult and burial practices, including the presence of a disarticulated but almost complete skeleton of an adult woman, two rather special pits and hundreds of faunal remains (Rolfo *et al.* 2013). The pits' location in the innermost (and darkest) part of the entrance chamber, the presence of an upside-down bowl and a spindle-whorl in one of them, the retrieval of the only two flint arrowheads next to it, their proximity to the woman's commingled human remains, and the presence of many sub-juvenile animals indicated that this context was likely a funerary one. Similar pit features are recurring in several cave sites dated to the Neolithic and the Early–Middle Bronze Age in Central Italy, often associated with burials or other special activities, including water collection cults and depositions of special objects: key such sites are, for example, Grotta Sant'Angelo and Grotta dei Piccioni in Abruzzi (Cremonesi 1976; Di Fraia and Grifoni Cremonesi 1996).

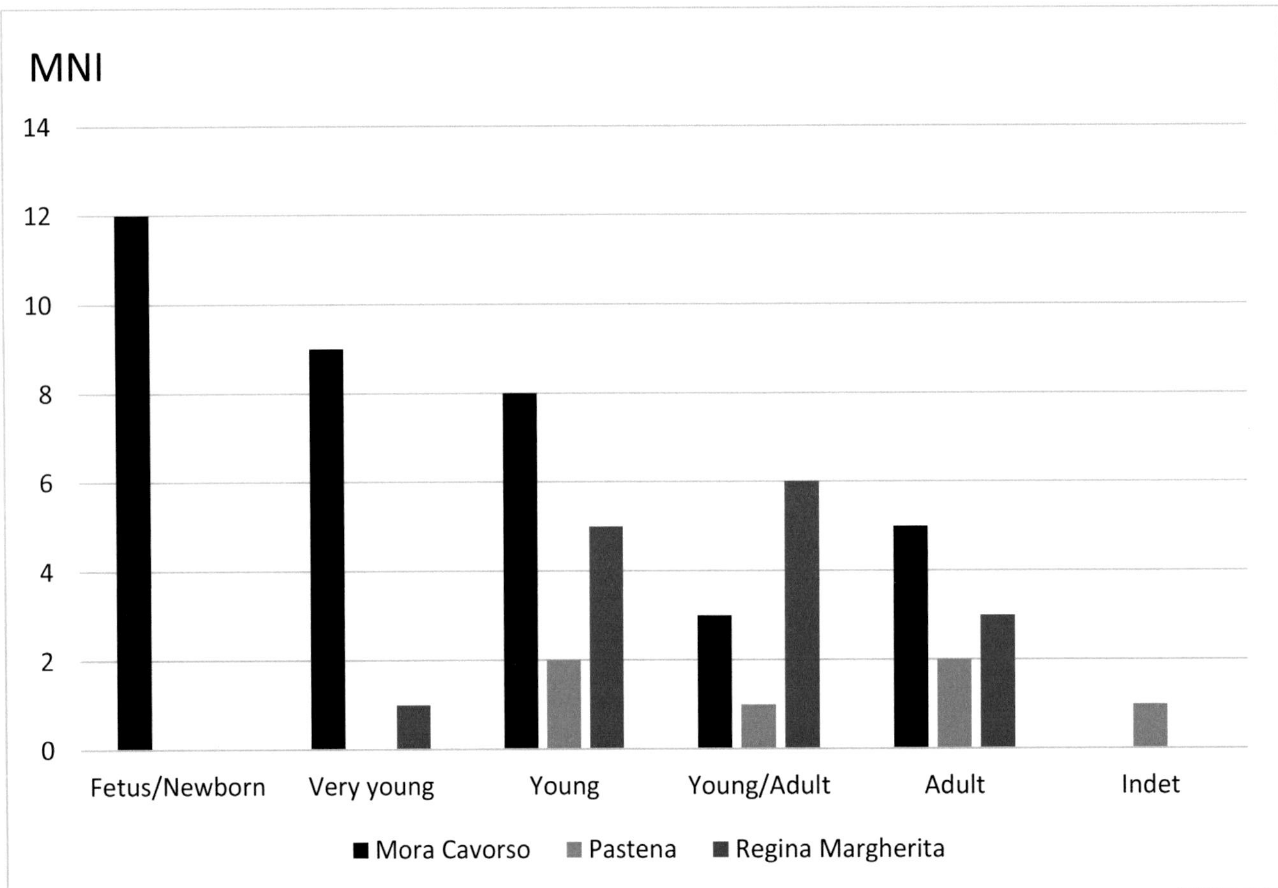

Figure 10.4 Age classes of sheep/goats by Minimum Number of Individuals (MNI) in the three caves under study.

Sub-juvenile lambs/kids and piglets, including some foetuses, dominated the assemblage (60% and 75% respectively of the species, 65% of the whole assemblage) (Figs. 10.4 and 10.5; Table 10.2). The highest concentration of perinatal bones was recorded completely mixed with the human bones of the adult woman, although the distribution of the former occupied a slightly wider area. All body parts were represented and no butchery marks were identified, meaning that the bodies were probably deposited whole, despite the lack of skeletal connections observed at the time of the excavation. The possibility of peri/post-mortem processing is highly unlikely considering that traces of fire were also absent. On the other hand, cut marks were found on wild game (wild boar and red deer) and adult domesticates (cattle and sheep/goat). In addition, remains of these animals were more concentrated on the horizontal floor of the wider and better illuminated part of the entrance chamber. This suggests that adult animals and hunted fauna were treated differently, and deposited in a separate location to the foetuses and newborns, which were probably slaughtered and deposited to become part of the human burial context.

Pertuso (Pastena)

This cave (Fig. 10.6), located at 310m.a.s.l. at the edge of the Pastena Plain, is a very large limestone structure crossed by a seasonal stream, the Rio Mastro. The cave has been known to scholars since the 19th century and became a tourist site in the 20th century. The first archaeological discoveries occurred in the 1980s, when a bronze dagger and axe were found in the inner lake of the cave, modern access to which was only possible via a newly constructed artificial passage realised by drilling a limestone wall. This discovery stimulated research in several small tunnels located at the sides of the main stream bed (Angle *et al.* 2010b; Biddittu *et al.* 2006). Given the frequent floods of the Rio Mastro, which washed out everything situated below a certain level, only elevated tunnels were found to hold archaeological remains. These consisted of Neolithic to MBA artefacts, carbonised plant remains and human and animal bones, all chaotically mixed and not yet radiocarbon dated. Starting from 2012, systematic excavations of the most elevated and widest inner chamber of the cave (Grotticella W2) took place, directed by the Soprintendenza Archeologia del Lazio e dell'Etruria Meridionale and the University of Rome 'Tor Vergata', revealing a relatively well-preserved archaeological deposit. This small and dark space is only reachable by climbing 15 metres of steep rocks. Its stratigraphy was still partly intact and revealed human use typologically dated to the MBA. A sequence of cobble pavings and layers of carbonised legumes and cereals, repeated at least three times, was observed during the stratigraphic excavation of the chamber, leading to the hypothesis of cyclical use of this space. Some *in situ* overturned bowls, in two cases buried in pits surrounded by stones, were also found, along with hearths and ash areas. Faience beads and buttons, a miniature stone axe, two stone pendants, two bronze rings and fragments of a bronze pin, as well as several intact and fragmented spindle-whorls, two small flint arrowheads

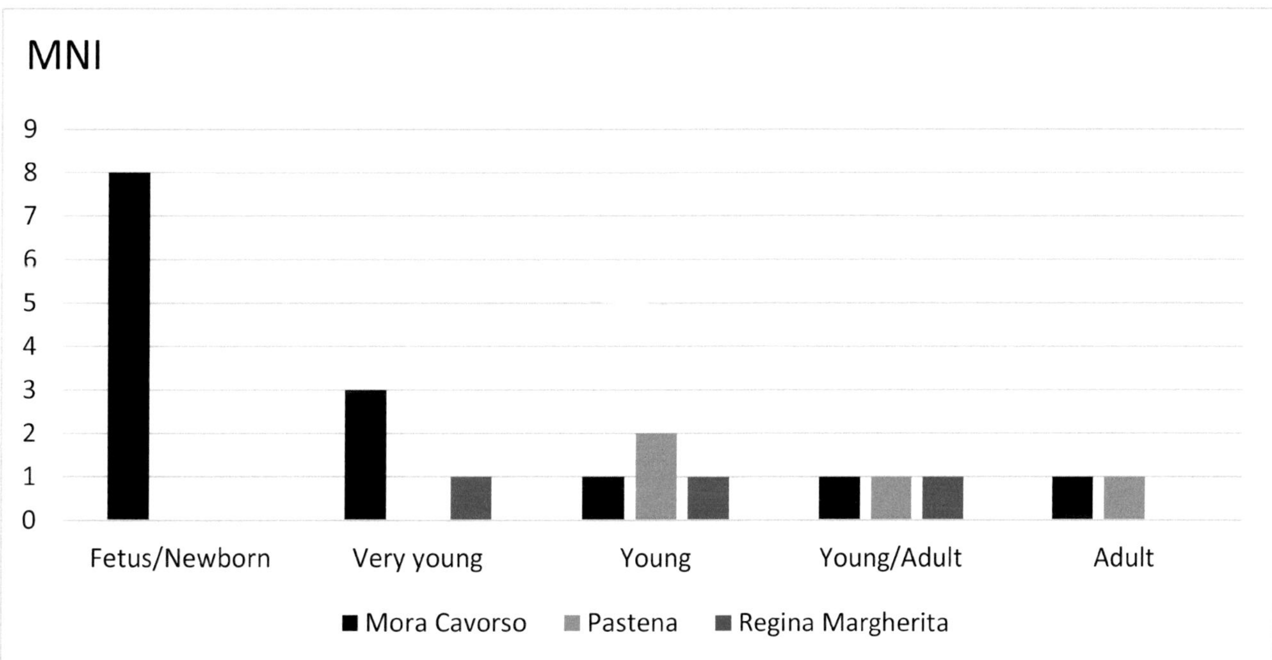

Figure 10.5 Age classes of pigs by Minimum Number of Individuals (MNI) in the three caves under study.

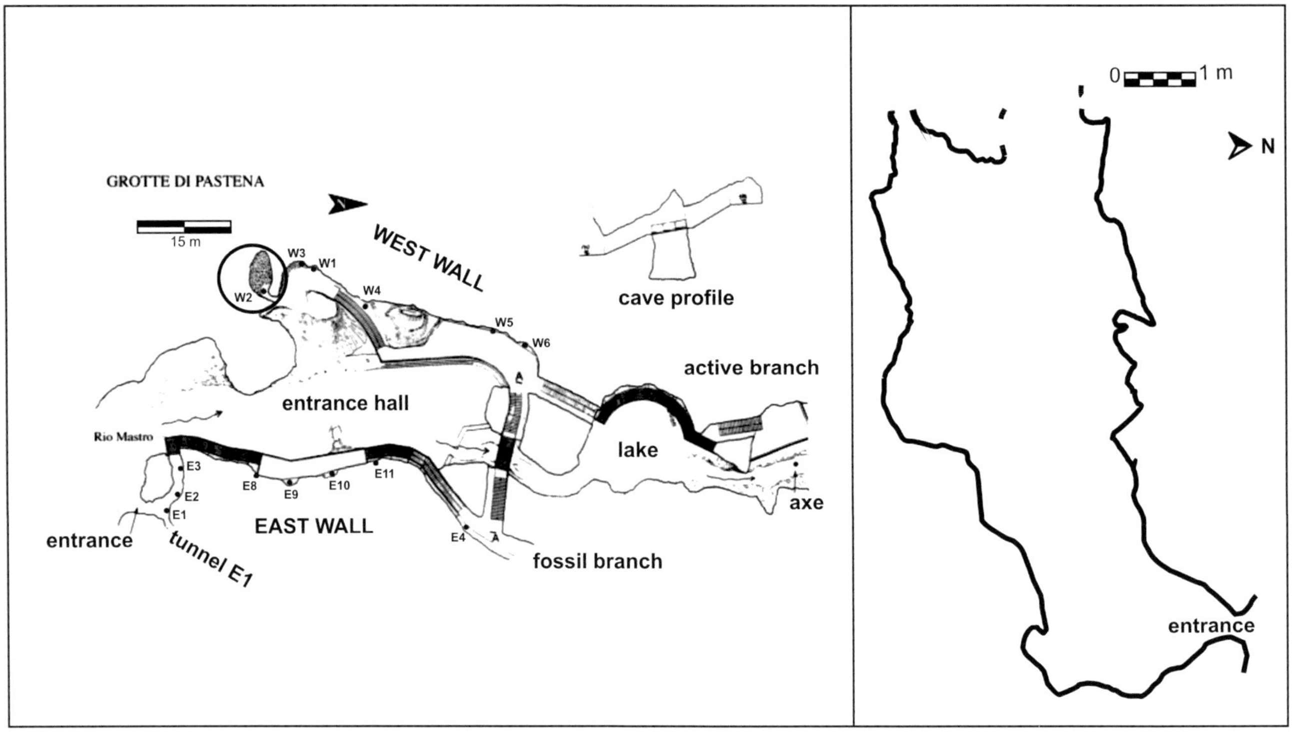

Figure 10.6 Plan of Grotta del Pertuso (left) and Plan of Grotticella W2 (right). Based on Biddittu et al. *(2006) and Angle* et al. *(2014).*

and a bone awl were found lying over the whole excavated area, apparently in a random distribution similar to that of pottery and human and animal bones. The chamber, with its inhospitable geomorphological features, still held the only example of extensive paving structures known in similar contexts in the area, alternated with the above-mentioned layers of carbonised seeds. Plant remains were collected *in situ* and the sampled deposits had 4179 seeds in total and included broad beans (*Vicia faba*), free-threshing wheat *(Triticum aestivum/durum*), glume wheat (*Triticum monococcum/dicoccum/spelta*) and barley (*Hordeum vulgare*), as well as two grape seeds (*Vitis vinifera*) and an olive (*Olea europaea*) or Cornelian cherry (*Cornus mas*) stone (Fig. 10.7).

The discovery of seeds, estimated to number several hundreds of thousands across this space, is exceptional, although the presence of legumes, cereals and fruits has been identified in various caves (Miari 1995), including the nearby Grotta Vittorio Vecchi (Costantini and Costantini Biasini 2007). The last element to add to this picture is the existence of a small natural terrace in the upper part of the chamber, which might have been accessible via a ladder or possibly a collapsed rock path. On this terrace more hearths, seeds (same species and proportions as in the main chamber) and peculiar archaeological finds were found, such as a stone circle covering an u-turned bowl, some human bones, a bronze artefact and faunal remains. All these features appeared intentionally deposited in a special manner, especially when the context of the barely accessible and dark Grotticella W2 is considered.

Faunal remains in this area were scattered amongst hundreds of fragments of coarse pottery and a few human remains. The *c.* 30 human bones belong to at least five individuals from very young to adult age, of both sexes. The faunal remains belong almost exclusively to domesticates. The number of identified specimens is barely more than 100, half of which belong to sheep/goats (Tables 10.3 and 10.4). Of the six sheep/goat present, two were adult, one a young-adult and two around six months old (Figs. 10.4 and 10.5). One third of the remains belong to pigs, represented by four individuals, of which one is adult, two young and one very young. Anatomical elements belonging to all body parts were present. Ten per cent of the bones (especially the radii, humeri, scapulae and tibiae), including ribs of medium-sized mammals, had cut marks. Vertebrae also had butchery marks and 20% of the bones had been exposed to fire (burnt or covered in charcoal). Only four cattle remains, three of which were teeth, were found in the deposit. They belonged to an adult and a young individual. Among the wild fauna, a single bone of wildcat (*Felis silvestris*), one of marten (*Martes martes*) and one of a large bird (with a deep cut mark on it) were recovered. In addition, 13 lower limb bones of an adult hare were found. Most of these hare bones, which did not present skeletal articulation nor anthropogenic marks, were found in one of the most inaccessible areas, the above-mentioned terrace.

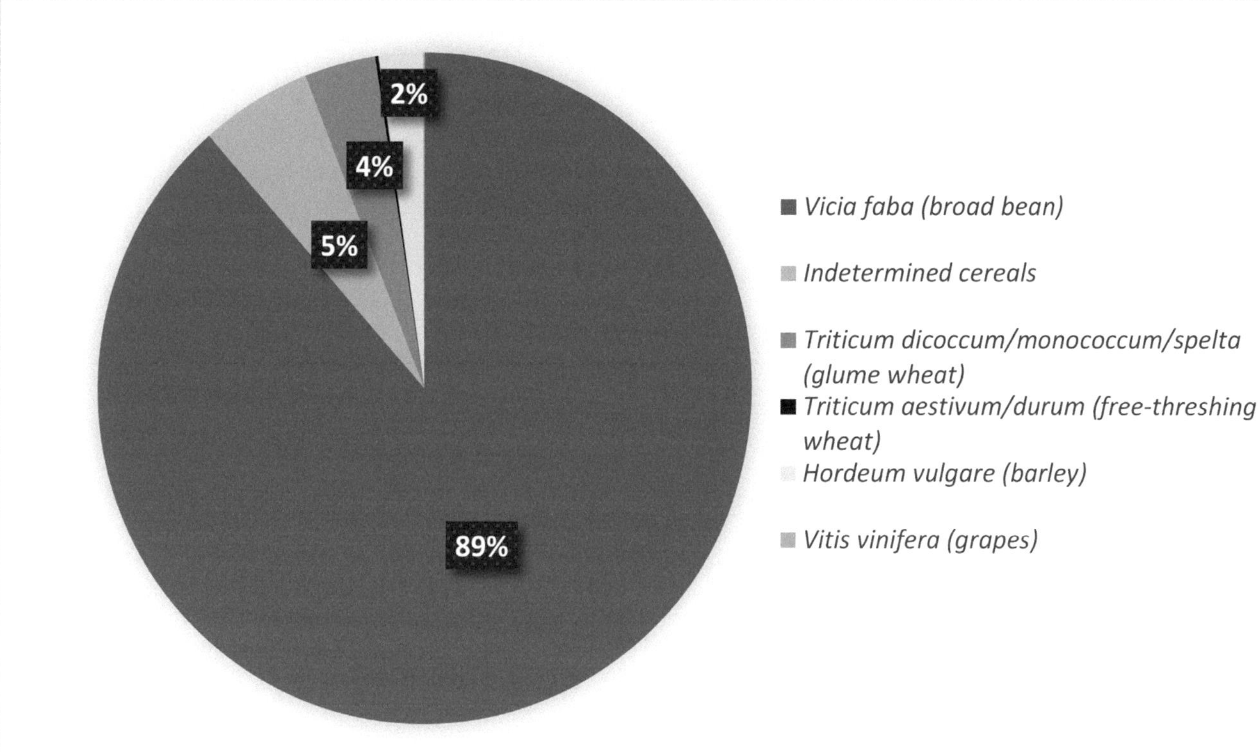

Figure 10.7 Pie chart of botanical taxa found at Mora Cavorso (sample of 4179, c. 5% of estimated total).

The kill-off patterns (following Payne 1973) and the cut and fire marks on the domesticate bones indicate that meat consumption was the main reason for their presence at the site. The animal bones were found lying on the various stone paving structures and next to the hearths or combustion areas. Given the depositional context, it is thus likely that meals were performed as part of some sort of repeated practices, as suggested by the stratified pavings and seeds layers, which involved meat consumption. By contrast, the hare bones seem to have been deposited but not consumed, perhaps as an offering, possibly related to the individuals buried in that space of the cave.

Regina Margherita

Grotta di Regina Margherita (Fig. 10.8), also known as Grotta dei Bambocci and Grotta di Collepardo, is located in a gorge on the edge of the Ernici Mountains at about 400m.a.s.l. It has been known to scholars since the 19th century and soon became a show-cave because of its magnificent karst formations in the huge main hall, thought to resemble human and imaginary figures. This explains the original name of 'Grotta dei Bambocci' (Cave of the Puppets). Systematic excavations at the cave began in summer 2014, under the direction of Durham University and the Soprintendenza per i Beni Archeologici del Lazio, following a rescue campaign carried out in 2008 and sporadic surveys in the 1980s (Angle *et al.* 2010a). This cave is certainly one of the richest cemeteries of the central Italian MBA, including more than 40 individuals of all ages and sexes, deposited in groups in the different sectors of the site, possibly representing family groups. The latter hypothesis was put forward by Cavazzuti who conducted the osteological study and identified similar pathologies or morphological features on skeletal remains of individuals of different ages and sexes that were located in the same area; for instance, some of them had an identical dental malformation, indicative of genetic relations (Cavazzuti pers. comm.).

The faunal sample found in this cave consisted of slightly over a hundred identified specimens and a few hundred indeterminate fragments, whereas no visible burnt layer was present and only four carbonised broad beans were retrieved from two areas (D and G) during wet-sieving. Overall, the faunal and plant assemblages are much smaller than the ceramic and human bone collections, both represented by thousands of finds. Also other types of artefacts, such as bronze spirals, a mother-of-pearl button, faience beads, occasional obsidian and flint flakes, amber bead fragments and a decorated animal bone were found, distributed alongside the human and faunal remains and with no apparent pattern. It has been possible, however, to identify a quantitative difference of the faunal remains between

Table 10.3 Animal taxa from the three new Early-Middle Bronze Age caves by Minimum Number of Individuals (MNI).

NISP	*1*	*2*	*3*	*4*	*5*	*6*	*7*	*8*	*9*	*10*	*11*	*12*	*13*	*14*	*15*	*16*	*17*	*S1*	*S2*	*S3*	*S4*	*S5*	*S6*
Ovis aries vel Capra Hircus (Sheep/goat)	338	47	101	25	60	179	79	15	18	67	27	141	124	66	28	6	37	54	520	103	106	64	144
Sus domesticus (Domestic pig)	87	29	4	3			57	19			21	3		40	6	32	16	23	482	39	65	28	
Bos taurus (Cattle)	42	4	9	7	14	104	37	23	6	31	54	13	45	20		3	11	11	945	123	96	63	35
Canis familiaris (Dog)	3		2	15	5	6	7	4		2	2	1	4	2			2	14	34	7	6	7	10
Cervus elaphus (Red deer)	44		1			3	5	14	1	3	26	36	25	1	5	8		178	4	5	1	6	6
Sus scrofa (Wild boar)	18		1													1	2	10		1	1	1	
Capreolus capreolus (Roe deer)	13		1		1						1	15	20			1		12	3	1			
Equus sp. (Equide)			(.4.)														5		9			18	2
Lepus sp. (Hare)	20	13														1		1	2				1
Sus sp. (Indet. Pigs)						141							55					14					85
Ursus arctos (Cave bear)																1		1					
Small carnivores and *Canis lupus* (Wolf)	24	2	5													13	15	7	11			1	
Turtles																		1	6	62			
Total ID	589	95	128	50	80	433	185	75	25	103	131	209	273	129	39	67	88	326	2016	341	275	188	283

Table 10.4 Animal species identified in the caves and settlements analysed by Number of Identified Specimens (NISP). For site numbers, see Table 10.1.

	F/N	*VY*	*Y*	*Y-A*	*A*	*Total MNI*
Grotta Mora Cavorso	23	14	11	11	10	69
Grotta del Pertuso		1	4	1	4	10
Grotta Regina Margherita		1	9	7	7	24
Grotta del Fontino			1		4	5
Grotta del Mezzogiorno	2	1	5	10	9	27
Grotta dei Cocci	2	2	9	2	10	25
Riparo del Lauro			1	2	6	9
Villaggio delle Macine		4	4	4	11	23
Luni sul Mignone	3		73	92	135	303
La Crocetta			7	1	8	16

the various investigated areas in the cave with the largest number being present at the external areas, close to hearths (Areas A and F) and where the lowest quantity of human bones was recovered, still mingled together. The Bronze Age attribution of the archaeological features identified at the entrance area is hardly debatable, as suggested by the presence of discretely preserved palaeosurfaces, with a relatively undisturbed distribution of artefacts and bioarchaeological remains. The stratigraphy is only few centimetres deep, but it is the only occupation phase at the site. Although in the absence of radiocarbon dating it is currently not possible to make micro-stratigraphic distinctions, the coexistence of human remains and similar artefacts between the entrance area and the inner chambers seems to indicate a contemporaneity between the uses of the different areas of the cave.

Sheep/goat bones made up almost 90% of the assemblage, both in terms of NISP and MNI (Tables 10.3 and 10.4). Most of them were young individuals, killed between six months and one year from birth (Area A) or young-adults (Area F). Detailed analyses of the faunal remains from the other areas did not give any valuable results due to the scarcity and high fragmentation rates of the bones retrieved. The most common anatomical parts were long bones, especially humeri, which showed a recurring fragmentation pattern, having been smashed in the medial portion of the diaphysis, presumably to extract marrow (*e.g.* Outram 2001, 404). Both the kill-off pattern (calculated only for Areas A and F, where the sample was quantitatively more consistent) (Fig. 10.4) and the skeletal element representation (mainly meat-rich long bones) confirm a focus on meat consumption.

Evidence of other domesticates, such as pig and cattle, is much rarer (Fig. 10.2) and includes mostly bones of young individuals. A few metapodials and phalanges of wild boar, red and roe deer (*Capreolus capreolus*), as well as fox might be considered as incidental occurrences brought in by predators such as badgers and martens which still frequent the site, given their extreme rarity and lack of anthropogenic marks. A few adult horse (*Equus caballus*) teeth, metapodials and phalanges found in the deposit were most likely related to a Pleistocene phase, given their advanced state of fossilisation. The existence of Pleistocene fauna at Grotta Regina Margherita was in fact mentioned in an old report by Ponzi (1853). Still, the majority of the faunal remains should be dated to the Bronze Age, which appears to be the only phase of intensive human activity at the site. Areas A and F were certainly used during recent times, as testified by the remains of modern hearths on the surface layers and by recent local sources, such as the cave's tourist guides, who mentioned that the entrance of the cave was used in the 20th century during Christmas for historical re-enactments of Jesus' nativity. However, typologically attributable remains (such as pottery, faience, amber, flint/obsidian and bronze) recovered directly below the superficial contexts of the studied areas dated only to the Bronze Age. The faunal assemblage from Grotta Regina Margherita, then, would seem to be associated with activities linked to funerary practices, although to a certain extent spatially separated from the burials. In fact, hearths and animal bones appear to be more frequent at the entrance of the cave, in a naturally illuminated area, whereas they decrease substantially in the innermost and darkest part of the cave, where human remains and associated artefacts occur almost exclusively.

Discussion: similarities and differences in the ritualisation of caves

The discussion has so far highlighted the variability of practices involving bioarchaeological, and especially faunal remains that can be described as ritual. This potentially also relates to plants, which are a much rarer type of find at these sites, possibly due to unfavourable preservation conditions and the general lack of systematic protocols for their recovery. With regard to the ritual use of plants, variability can be seen at the three caves analysed: at Mora Cavorso, despite the good preservation conditions of the Bronze Age deposit (where several intact foetal bones were retrieved) and of older layers (plant seeds were found in the Neolithic contexts), no plant remains were identified. Methodological biases, in this case, can be excluded since all excavated soil was wet-sieved, including more detailed processing with smaller meshes of a representative selection of samples. On the other hand, the coeval phase of Pastena Cave revealed the presence of entire layers of carbonised seeds (hundreds of thousands) still *in situ*. Finally, Grotta Regina Margherita, where again all the soil was wet-sieved and selected samples were additionally processed with smaller meshes, produced only four carbonised broad

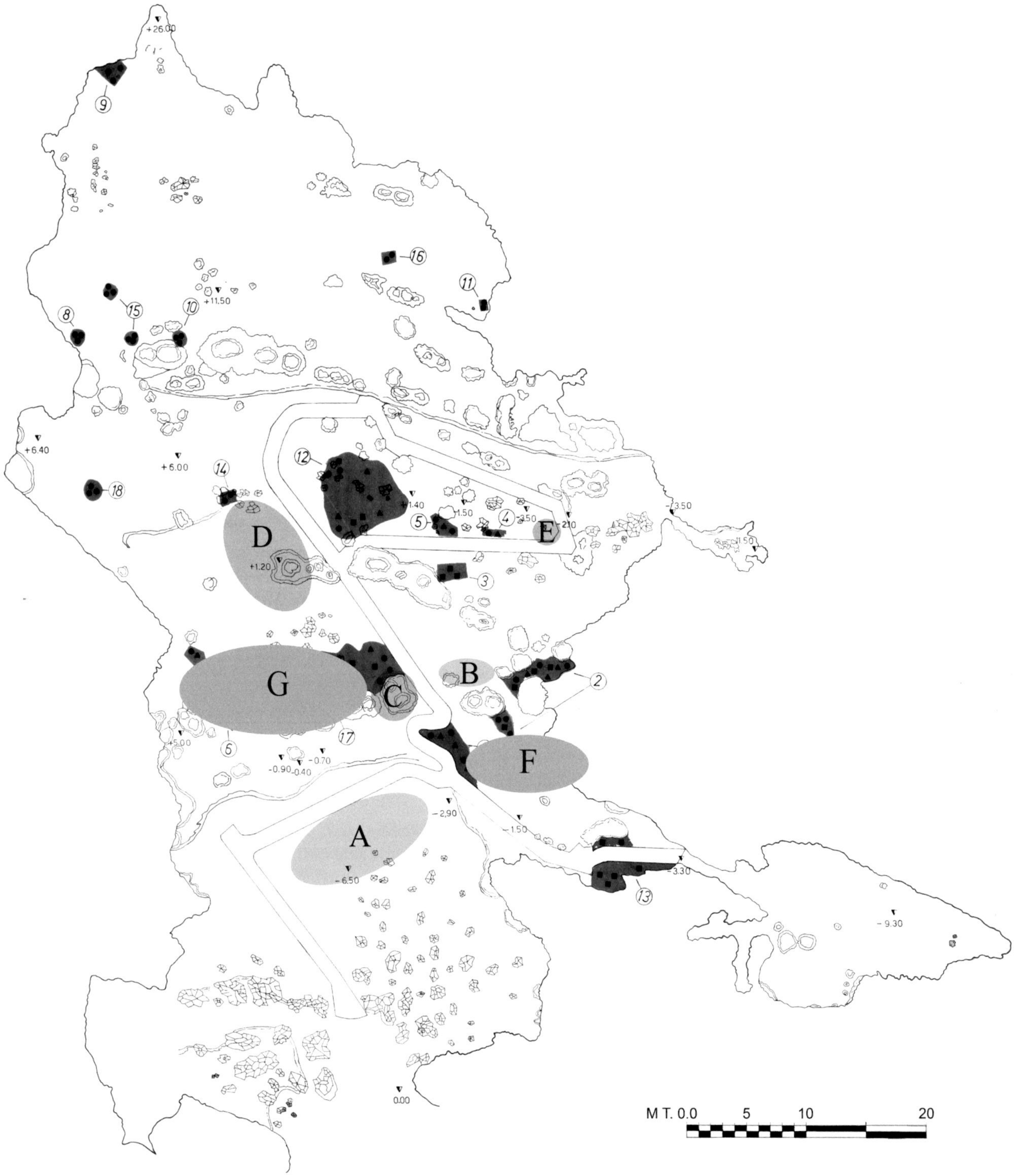

Figure 10.8 Plan of Grotta Regina Margherita with location of the archaeological soundings. Based on Angle et al. *2010.*

beans dispersed in two soundings. Even considering a bias deriving from poor preservation of the vegetal elements or the recovery process during the excavations, it seems evident that the ritual patterns of plant use in the three sites were completely different. The same holds true for the animal assemblages.

Figure 10.2 provides some information on the proportion of animal remains present in various Early to Middle Bronze

Age caves in Central Italy. It would seem that ovicaprines are the most common taxon found, followed by cattle and pig. Hunted and wild animals are less common, but still quite widespread, with a predominance of red deer, wild boar and roe deer. These data do not in themselves have any impact on the current state of knowledge of the Bronze Age in Central Italy. However, there are other features, such as age, body part representation and spatial distribution, that were never previously considered and that can add interesting information, especially when compared to the new case studies presented here and the available data from MBA settlement sites. The only open-air sites for which zooarchaeological analyses are available are Villaggio delle Macine (Rome) (Tagliacozzo *et al.* 2012) (S1; see Table 10.4 and Fig. 10.1 for codes), Coccioli (Teramo) (S2), Cerchio La Ripa (L'Aquila) (S3), Luni sul Mignone (Viterbo) (S4), Castiglione (Rome) (S5) (Minniti 2012) and La Crocetta (Rome) (S6) (authors' ongoing study). However, not all relevant data are available for all of these sites. In terms of taxonomic representation (Fig. 10.2), cattle, which occur in all but three of the caves, are generally more significant in settlements. The occurrence of domestic pigs (or undetermined pigs) is very frequent, in variable amounts, but does not seem to differ between settlements and caves. All but five cave sites show the presence of at least undetermined pigs. Grotta del Fontino has an anomalous 20% of domestic dogs (15 bones), which is otherwise represented by 3–5% of the total specimens at all sites where they are found (all the settlements and 60% of the caves). This might indicate another previously unrecognised ritual dimension, possibly similar to that observed at the Grotte di Belverde, where several dog skulls were reported but for which no quantitative data are available. Another interesting aspect is the greater presence of red deer in caves compared to open sites, with the exception of Villaggio delle Macine. Apart from this case, the percentage of red deer in settlements does not exceed 5%, whereas in seven of the 15 caves that yielded remains of this species, values range between 10 and 20% of the total NISP. This evidence could relate to ritual activities, possibly related to hunting and/or to a connection with the wild, as has been widely suggested in the literature (Baker *et al.* 2015 and references therein). However, it is necessary to stress that these caves were mainly located in highly forested environments, where red deer were more common, and therefore, its presence might also be explained in terms of habitat proximity. Hare is almost absent from most settlements, whereas it is present in small percentages (mostly between 1 and 5%) in at least a third of the caves. Hare appears in one cave (Grotta di Pastena) as a well-represented species (15% of the total animal assemblage). For this cave, its occurrence is unequivocally of ritual nature (as seen, for example, in Glencurran Cave, Dowd 2009), also given its location on the terrace, and body part (lower limbs) selection. Finally, horse is only attested in two settlements (La Crocetta and Cerchio La Ripa), while the horse from Grotta Regina Margherita is most likely an earlier intrusion given its fossilised condition.

In this study age classes were considered only for domesticated animals, as this is the only faunal category whose age trends are usually specified, given their usefulness in reconstructing exploitation strategies. Despite the extremely limited sample, some differences could be recognised between open-air and cave sites (Table 10.2). First, the previously established presence of a relatively constant percentage of adult domesticates (about 50–60% of the total) in the settlements, compares to the more variable and, on average, smaller presence of this age class in the caves. Additionally, there is an almost complete absence of sub-juvenile individuals in the settlements and it can be suggested that the subsistence strategy of settlements was in fact more orientated towards meat and wool exploitation, whereas it might well be the case that milk exploitation was as important an aspect in the utilisation of flocks. In contrast, 55% of the caves include newborns and very young domesticates. Grotta Mora Cavorso and Grotta dei Cocci were already known for this feature along with other caves reported in the literature (*e.g.* Grotta Di Carli and Grotta Sant'Angelo), for which, however, an exact quantification of the remains or of the MNI is not provided. Table 10.2 indicates that at least one more site can be added to this group of caves, namely Grotta del Mezzogiorno. This expands the identification of sub-juvenile animal slaughtering in such contexts also to the Marche region, with all the regions of central Italy now having at least one example of this practice during the MBA. Overall, however, the kill-off trends shown by the caves' datasets are highly variable, especially compared to those of the settlements. Contrary to what could first be inferred by looking at the mortality data, the actual pattern is one of unpredictability of composition. This confirms the limited value of these datasets to palaeoeconomic studies, while testifying to the cultural nature of the human choices that generated the bioarchaeological deposits at these sites.

Body part distribution is the last feature examined here (Fig. 10.9). Overall, the presence of extremities appears more widespread in caves, for ovicaprines and especially for pigs, cattle and dogs. In particular, pig extremities are more frequent in caves where perinatal slaughtering is attested (Mora Cavorso and Grotta dei Cocci), confirming that the carcasses were deposited whole or in large portions. Forelimbs are much more frequent in caves, especially of ovicaprines and cattle, whereas cattle hindlimbs occur much more often in settlements than in caves, with those of pigs being virtually absent from the latter. Red deer does not show particular patterns in body part distribution.

It has to be reiterated that these data are limited to a small group of sites, for which quantitative data (in some cases albeit limited) are available. Interpretations inferred

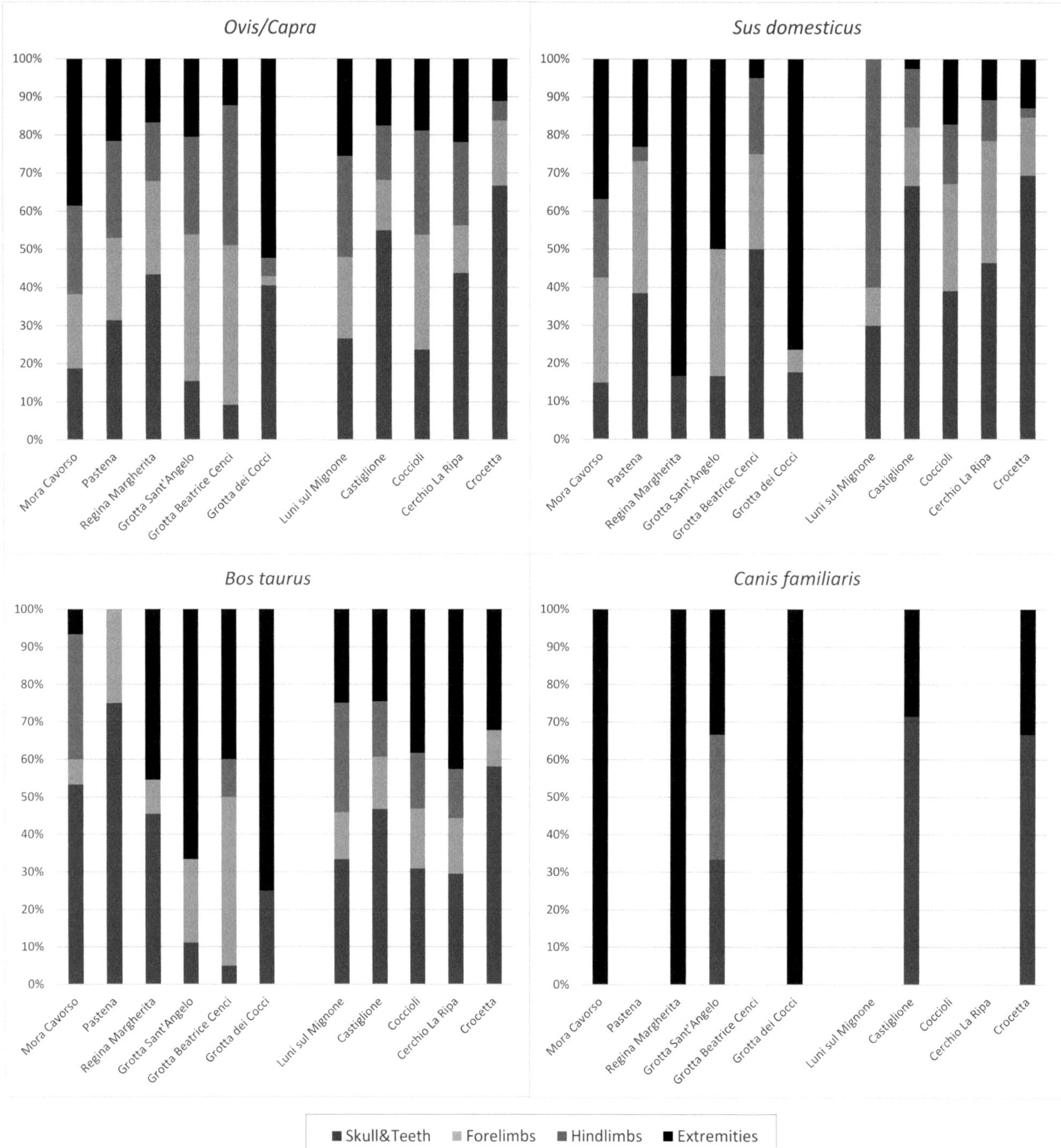

Figure 10.9 Body part distribution (NISP) for the main taxa identified in ritual deposits from cave and settlement sites where information is available and for which intentional selections appear more evident.

from these contexts might well be modified by new data from other sites. However, this study offers a new research perspective and uses partial data in an attempt to move beyond traditional research approaches in the area.

With regard to the new case studies, in the cave Mora Cavorso several perinatal domesticates were deposited close to the body of an adult woman and in association with pits interpreted as ritual in nature. At Pastena Cave meat parts of domestic animals, often bearing traces of butchery and fire, were found in association with alternating layers of paving and carbonised seeds, and the remains of several human individuals of all ages. In the third cave of Collepardo the meat bones of domesticates were generally found concentrated in a specific sector of the site, slightly separated from the main human burial area and close to hearths. Our interpretation of the three patterns varies significantly from case to case. We have a possible example of perinatal slaughtering or sacrifice at the first site, one of a ritual meal in the second, and one of entrance offerings or a meal related to activities taking place further inside the cave in the third. All three show a strong relation with adjacent mortuary contexts, even in the case of spatial distinctions as in Grotta Regina Margherita. Other aspects of the ritual use of the sites analysed, including those used for burial, point to the performance of a more widely-shared set of practices, such as the digging of pits, the *in-situ* up-turned bowls, the presence of recurring artefacts next to human bones (such as personal ornaments or possible grave goods – spindle whorls, arrowheads, beads and bronze jewellery), and the dark and secluded cave locations. However, ritual manifestations related to the bioarchaeological remains show evident variability and, subsequently, a more complex set of practices compared to those previously envisaged of the sub-juvenile deposits and of a few isolated animal burials (Grifoni Cremonesi 1996; Wilkens 1995). In particular, it can be demonstrated how such variability is manifested in contemporary sites which are believed to have been used for similar purposes, all related to the cycle of life and death – both in the human funerary sphere and in the subsistence sphere, linked to the abundance of the harvest and herds (Cocchi Genick 2002a; Grifoni Cremonesi 1996, 2002; Guidi 1991). The diversification of ritual practices involving animals (and plants) in a circumscribed set of sites – in this case the MBA caves of Central Italy – offers a glimpse of the local and conceptual complexity of the symbolic world of these agro-pastoral communities. However, more information needs to be added to this picture through a more focused study of other existing faunal assemblages. This will hopefully be possible in the future with the systematic investigation of new cave sites and with special attention paid to bioarchaeological remains right from the earliest stages of excavation.

Based on the archaeological evidence reported above, tentative parallels with classical literature and ethnographic examples might be drawn, for example between the slaughtering of piglets of Mora Cavorso and at Greek Thesmophoria, where women used to sacrifice these animals and then throw their remains into crevices in honour of Demetra and Kore, archaic deities of the cycle of seasons, fertility and life (Albarella 2014; Versnel 1992). We might recognise the importance of dog as a companion, and imagine that red deer might have had an importance in the affirmation of male identity (Harris 2015) and in rituals of passage of young men (Whitehouse 1992). We might also acknowledge that the hare is a symbol of fertility, death and rebirth in several cultures (Boyle 1973). However, confirmation of a direct link to such later documentation evidence cannot come from the Central Italian Bronze Age, and greater effort therefore needs to be put into interpreting the data that we already have from an archaeological, contextual perspective. Therefore, although interesting, these analogues can serve only as hypotheses to be tested in the future.

A key aim of this paper was to stress that bioarchaeological remains cannot be excluded from the overall interpretation of a site, using them only (if at all) to infer economic strategies. This is especially valid for assemblages that have clearly undergone intentional selections before post-depositional processes altered the contexts. Every site includes a varied use of bioarchaeological remains, which is often helpful in offering a key to understanding the complexity and diversity of the caves themselves. The next challenge for archaeologists working in Central Italy is to establish a methodology through which to identify the presence of animal/plant rituals at all kinds of archaeological sites.

References

Agostini, S., Coubray, S., De Grossi Mazzorin, J., D'Ercole, V. and Remotti, E. (1991) Cappadocia (L'Aquila), località Oveto. Indagini preliminari nella Grotta Beatrice Cenci. *Bollettino di Archeologia* 8, 61–71.

Albarella, U. (2014) *The use of pigs in ritual offerings at Monte Papalucio, Oria (southern Italy*). Paper delivered at the 12th Conference of the International Council of Archaeozoology, San Rafael, Argentina.

Angle, M., Rolfo, M. F., Fusco, I. and Silvestri, L. (2014) New investigations at the Cave of Pastena (Frosinone). Report 2012. In G. Ghini and Z. Mari (eds.) *Lazio e Sabina, Scoperte Scavi e Ricerche, Atti del Decimo Incontro di Studi sul Lazio e la Sabina*, 205–11. Rome, Quasar.

Angle, M., Catracchia, F., Cavazzuti, C., Celletti, P., Malorgio, M. and Mancini, D. (2010a) La Grotta Regina Margherita a Collepardo (Frosinone). In G. Ghini and Z. Mari (eds.) *Lazio e Sabina, Scoperte Scavi e Ricerche, Atti del Sesto Incontro di Studi sul Lazio e la Sabina*, 381–93. Rome, Quasar.

Angle, M., Bruni, N., Cerqua, M., Riva, A., Cavazzuti, C. and Celletti, P. (2010b) *Nuovi scavi nella grotta di Pastena (Frosinone)*. In G. Ghini and Z. Mari (eds.) *Lazio e Sabina, Scoperte Scavi e Ricerche, Atti del Sesto Incontro di Studi sul Lazio e la Sabina*, 369–80. Rome, Quasar.

Angle, M., Gianni, A. and Guidi, A. (1991) La Grotta dello Sventatoio (S. Angelo Romano, Roma). *Rassegna di Archeologia* 10, 720–1.

Baker, K., Carden, R. and Madgwick, R. (eds.) (2015) *Deer and People*. Oxford, Oxbow Books.

Barker, G. (1981) *Landscape and Society: Prehistoric Central Italy*. London, New York, Academic Press.

Biddittu, I., Bruni, N., Cerqua, M. and Riva, A. (2006) La frequentazione delle Grotte di Pastena (FR) dal Neolitico all'Età del Bronzo. In G. L. Carancini (ed.) *Miscellanea Protostorica*, 114–38. Perugia, Ali&no.

Bietti Sestieri, A. M. (2010) *L'Italia nell'età del bronzo e del ferro: dalle palafitte a Romolo (2200–700 a.C.)*. Rome, Carocci.

Bigini, I. (1981) La fauna. In G. Radi (ed.) *La Grotta del Beato Benincasa nel Quadro delle Culture dal Neolitico all'Età del Bronzo in Toscana*, 210. Pisa, Giardini.

Bigini, I. (1986) La fauna. In D. Cocchi Genick, N. Negroni Catacchio, I. Bigini and V. Bertelli (eds.) *Il Riparo dell'Ambra: una Successione Stratigrafica dal Neolitico Tardo al Bronzo Finale*. Viareggio, Assessorato alla Cultura, Museo Preistorico e Archeologico 'Alberto Carlo Blanc'.

Bigini, I. (1987) La fauna. In D. Cocchi Genick (ed.) *Il Riparo del Lauro di Candalla nel Quadro del Bronzo Medio Iniziale dell'Italia Centro-Occidentale*, 189–90. Viareggio, Assessorato alla cultura; Museo preistorico e archeologico 'Alberto Carlo Blanc'.

Boyle, J. A. (1973) The Hare in Myth and Reality: A Review Article. *Folklore* 84(4), 313–26.

Bradley, R. (2005) *Ritual and Domestic Life in Prehistoric Europe*. London, Routledge.

Cerilli, E. (2000) La Grotta Di Carli (Ischia di Castro-VT), campagna di scavo 1996: i primi dati archeozoologici. In N. Negroni Catacchio (ed.) *L'Etruria tra Italia, Europa e Mondo Mediterraneo: Ricerche e Scavi: Atti del Quarto Incontro di Studi, Manciano, Montalto di Castro, Valentano, 12–14 Settembre 1997*, 363–5. Milan, Centro Studi di Preistoria e Archeologia.

Cocchi Genik, D. (1999) La funzione delle grotte e il significato delle acque nelle manifestazioni di culto di epoca protostorica dell'Italia medio-tirrenica. *OCNUS* 7, 167–77.

Cocchi Genick, D. (2002a) Dati relativi ai culti delle acque in età protostorica nell'Italia centrale tirrenica. In N. Negroni Catacchio *Preistoria e Protostoria in Etruria - Paesaggi D'Acque. Atti del Quinto Incontro di Studi, Vol.II*, 405–16. Milan, Centro Studi di Preistoria e Archeologia.

Cocchi Genick, D. (2002b) *Grotta Nuova: la Prima Unità Culturale Attorno all'Etruria Protostorica*. Viareggio, Baroni.

Cocchi Genick, D., Damiani, I., Macchiarola, I., Poggiani Keller, R. and Peroni, R. (1995) *Aspetti culturali della media età del bronzo nell'Italia centro-meridionale*. Florence, Octavo.

Corridi C. (2002) La fauna. In A. Vigliardi and L. Bachechi (eds.) *La Grotta del Fontino: una Cavità Funeraria Eneolitica del Grossetano*, 255–60. Florence, Museo Fiorentino di Preistoria 'Paolo Graziosi'.

Costantini, L. and Costantini Biasini, L. (2007) Economia agricola del Lazio a sud del Tevere tra Bronzo antico e Bronzo medio. *Strategie di insediamento fra Lazio e Campania in età preistorica e protostorica. Atti della XL riunione scientifica: Roma, Napoli, Pompei*, Florence, Istituto Italiano di Preistoria e Protostoria, 787–801.

Cremonesi, G. (1968) Contributo alla conoscenza della preistoria del Fucino: la Grotta di Ortucchio e la Grotta La Punta. *Rivista di Scienze Preistoriche* 23, 145–204.

Cremonesi, G. (1976) *La Grotta dei Piccioni di Bolognano nel Guadro delle Culture dal Neolitico all'Età del Bronzo in Abruzzo*. Vol. 2. Pisa, Giardini.

Cunliffe, B. W., Gosden, C., and Joyce, R. A. (eds.) (2009) *The Oxford Handbook of Archaeology*. Oxford, Oxford University Press.

Curci A., De Angelis M. C., Moroni A., Padoanello S. and Tagliacozzo A. (2014) Grotta Bella (Umbria). Dati per un'analisi economica e paleo ambientale, *Rivista di Studi Liguri* 77–9(2011/13), 143–50.

De Grossi Mazzorin, J. (1995) Economie di allevamento in Italia centrale dalla media età del Bronzo alla fine dell'età del Ferro. In N. Christie (ed.) *Settlement and Economy in Italy 1500 BC to AD 1500. Papers of the Fifth Conference of Italian Archaeology*, Monograph 41, 167–77. Oxford, Oxbow Books.

Di Fraia, T. and Grifoni Cremonesi, R. (1996) *La Grotta Sant'Angelo Sulla Montagna dei Fiori (Teramo). Le testimonianze dal Neolitico all'Età del Bronzo e il problema delle frequentazioni cultuali in grotta*, Collana di studi paletnologici 6. Pisa–Rome, Istituti Editoriali e Poligrafici Internazionali.

Di Gennaro, F. (1988) Il popolamento dell'Etruria meridionale e le caratteristiche degli insediamenti tra l'età del bronzo e l'età del ferro. In G. Bettini, D. Colonnaand and R. Staccioli (eds.) *Etruria meridionale. Conoscenza, conservazione, fruizione. Atti del convegno, Viterbo* 29(30), 59–82.

Di Gennaro, F. (1999) Ipogei artificiali e grotte naturali nell'età del bronzo mediotirrenica. In Gravina, A. (ed.), *Atti 19° Convegno Nazionale sulla Preistoria-Protostoria-Storia della Daunia, Vol. 2*, 135–53. San Severo, Comune di San Severo.

Dowd, M. (2009) *Middle and Late Bronze Age Ritual Activity at Glencurran Cave, Co. Clare*. In N. Finlay, S. McCartan and C. Wickham Jones (eds.) *Bann Flakes to Bushmills: Papers in Honour of Peter C. Woodman*, 86–96. Oxford, Oxbow Book.

Grifoni Cremonesi, R. (1996) Osservazioni sulle buche e sulla problematica dei culti. In T. Di Fraia and R. Grifoni Cremonesi. (eds.) *La Grotta Sant'Angelo Sulla Montagna dei Fiori (Teramo). Le Testimonianze dal Neolitico all'Età del Bronzo e il Problema delle Frequentazioni Culturali in Grotta, Collana di Studi Paletnologici* 6, 305–37. Pisa–Roma, Istituti Editoriali e Poligrafici Internazionali.

Grifoni Cremonesi, R. (2002) I culti e i rituali funerari. In M. A. Fugazzola Delpino and V. Pessina Tiné (eds.) *Le Ceramiche Impresse nel Neolitico Antico. Italia e Mediterraneo*, 209–19. Rome, Istituto Poligrafico della Zecca dello Stato.

Guidi, A. (1991) Recenti ritrovamenti in grotta nel Lazio: un riesame critico del problema dell'utilizzazione delle cavità naturali. *Rassegna di Archeologia* 10, 427–37.

Güller, A. and Segre, A. G. (1949) La Stazione Enea del Grottone di Val de'Varri nell'Appennino Abruzzese. *Rivista di Antropologia* 36, 269–81.

Harris, K. (2015) Cervid exploitation and symbolic significance in prehistoric and early historic periods. Hunting, performance and incorporation: human-deer encounter in Late Bronze Age Crete. In K. Baker, R. Cardenand and R. Madgwick (eds.) *Deer and People*, 48–58. Oxford, Oxbow Books.

Higgs, E. S. (ed.) (1975) *Palaeoeconomy*. Cambridge, Cambridge University Press.

Iaconis, M. A. and Boschian, G. (2008) Geoarchaeology of the deposits of Grotta dei Piccioni and Grotta Sant'Angelo (Abruzzo, Central Italy). *Atti Della Società Toscana di Scienze Naturali, Memorie Serie A* 112, 181–8.

Jacomet, S. (2006) *Identification of cereal remains from archaeological sites*. Basel, Basel University.

Marciniak, A. (2005) *Placing Animals in the Neolithic: Social Zooarchaeology of Prehistoric Farming Communities*. London–Portland OR, Cavendish.

Miari, M. (1995) Offerte votive legate al mondo vegetale e animale nelle cavità naturali dell'Italia protostorica. *Atlante Tematico di Topografia Antica* 1(1), 11–29.

Minniti, C. (2012) *Ambiente, Sussistenza e Articolazione Sociale nell'Italia Centrale tra Bronzo Medio e Primo Ferro*. Oxford, Archaeopress.

Neef, R., Cappers, R. T., Bekker, R. M. and Boulos, L. (2012) *Digital Atlas of Economic Plants in Archaeology*. Groningen, Barkhuis.

Outram, A. K. (2001) A New Approach to Identifying Bone Marrow and Grease Exploitation: Why the "Indeterminate" Fragments should not be Ignored. *Journal of Archaeological Science* 28, 401–10.

Pacciarelli, M. (1997) *Acque, Grotte e dei: 3000 Anni di Culti Preromani in Romagna, Marche e Abruzzo*. Imola, Musei civici di Imola.

Pannuti, S. and Peroni, R. (1969) Gli scavi di Grotta a Male presso l'Aquila, *Bullettino di Paletnologia Italiana* 78(2), 147–247.

Payne, S. (1973) Kill-off patterns in sheep and goats: the mandibles from Aşvan Kale. *Anatolian Studies* 23, 281–303.

Ponzi, G. (1853) Sopra la grotta di Collepardo. Roma, *Atti dell'Accademia Potnificia dei nuovi Lincei* 6, 473–5.

Puccioni, N. (1914) Gli eneolitici della Buca-Tana di Maggiano (Lucca). *Bullettino di Paletnologia Italiana* 40, 1–29.

Puglisi, S. M. (1959) *La Civiltà Appenninica*. Florence, Sansoni.

Radi, G. (1981) *La Grotta del Beato Benincasa nel Quadro delle Culture dal Neolitico all'Età del Bronzo in Toscana*. Pisa, Giardini.

Radmilli, A. M. (1974) *GliSscavi nella Grotta Polesini a Ponte Lucano di Tivoli e la più Antica Arte nel Lazio*. Florence, Sansoni.

Ricciardi, B. A. (2015) Grotte dell'Italia Centrale. Studio della loro utilizzazione nell'Età del Bronzo. In L. Alessandri (ed.) *Speleologia del Lazio. Attraverso il Vuoto. Proceedings of the 6th Meeting of the Federazione Speleologica del Lazio* 8, 98–111. Genzano, Comune di Genzano.

Rittatore Vonwiller, F. (1951) Scoperte di età eneolitica e del Bronzo nella Maremma Tosco-Laziale. *Rivista di Scienze Preistoriche* 6, 3–33.

Rolfo, M. F., Achino, K. F., Fusco, I., Salari L. and Silvestri, L. (2013) La Grotta Mora Cavorso a Jenne (Roma): i livelli dell'antica-media età del Bronzo. *Rivista di Scienze Preistoriche* 63, 95–123.

Russell, N. (2012) *Social Zooarchaeology: Humans and Animals in Prehistory*. Cambridge, Cambridge University Press.

Salari, L., De Angelis, M. C. and Tagliacozzo, A. (2014) La fauna neolitica della Grotta dei Cocci (Narni, Umbria). *Rivista di Studi Liguri* 77/79(2011/13), 105–110.

Segre Naldini, E. and Biddittu, I. (1985) Ceramica dell'Età del Bronzo dal Costone di Battifratta (Rieti). *Archeologia Laziale* 7(1), 26–33.

Silvestri, L., Rolfo, M. F. and Salari, L. (2016) La fauna della media età del Bronzo di Grotta Mora Cavorso (Jenne, Lazio). *Atti 7° Convegno Nazionale di Archeozoologia, Ferrara-Rovigo 22–24 November 2012*.

Skeates, R. (1997) *The Human Uses of Caves in East–Central Italy during the Mesolithic, Neolithic and Copper Age*. In C. Bonsall and C. Tolan-Smith (eds.) *The Human Use of Caves* 79–86. Oxford BAR International Series 667, 79–86. Oxford, Archaeopress.

Tagliacozzo, A, Pino Uria, B. and Angle, M. (2012) La caccia al cervo in un sito dell'Età del Bronzo: Il Villaggio delle Macine (Lago di Albano, Roma). *Atti del 6° Convegno Nazionale di Archeozoologia (Orecchiella, 2009)*, 145–52. Lecce, Associazione Italiana di Archeozoologia.

Tongiorgi, E. (1947) Grano, miglio e fave in un focolare rituale dell'Età del Bronzo a Grotta Misa. *Nuovo Giornale Botanico Italiano* 54, 804.

Tongiorgi, E. (1956) Osservazioni paleontologiche nella Grotta del Mezzogiorno. *Bullettino di Paletnologia Italiana* N. S. X 65(2), 535–40.

Versnel, H. (1992) The Festival for Bona Dea and the Thesmophoria. *Greece and Rome* 39(1), 31–55.

Whitehouse, R. D. (1992) *Underground Religion: Cult and Culture in Prehistoric Italy, Vol. 1*. London, Accordia Research Centre, University of London.

Whitehouse, R. D. (2001) A tale of two caves: The archaeology of religious experience in Mediterranean Europe. In P. F. Biehl, F. Bertemes and H. Meller (eds.) *The Archaeology of Cult and Religion. Archaeolingua* 13, 161–7.

Whitehouse, R. D. (2007) Underground religion revisited. In D. A. Barrowclough and C. Malone (eds.) *Cult in Context: Reconsidering Ritual in Archaeology*, 97–106. Oxford, Oxbow Books.

Wilkens, B. (1991) Il ruolo della pastorizia nelle economie preistoriche dell'Italia centro-meridionale. *Rivista di Studi Liguri* 57(1990–91), 81–94.

Wilkens, B. (1995) Animali da contesti rituali nella preistoria dell'Italia centro-meridionale. *Padusa Quaderni* 1, 201–7.

Wilkens, B. (1996) Le faune. In T. Di Fraia and R. Grifoni Cremonesi (eds.) *La Grotta Sant'Angelo Sulla Montagna dei Fiori (Teramo): le Testimonianze dal Neolitico all'Età del Bronzo e il Problema delle Frequentazioni Culturali in Grotta*, 277–303. Pisa–Rome, Istituti Editoriali e Poligrafici Internazionali.

Chapter 11

Towards an Archaeology of the Social Meanings of the Environment: Plants and Animals at the Prehistoric Ceremonial and Funerary Staggered Turriform of Son Ferrer (Mallorca, Balearic Islands, Spain)

Llorenç Picornell-Gelabert, Manuel Calvo Trias, Jaume Garcia Rosselló, Gabriel Servera-Vives, Giovanna Bosi, Jordi Nadal Lorenzo, Santiago Riera Mora and Ethel Allué

Introduction

In Mediterranean archaeology, environmental datasets are frequently used as secondary sources of information on the available resources. This approach limits a holistic interpretation of the role of the biotic component of the environment in past societies as they are considered passive and monotonous entities with a given material value foreign to social creativity and cultural change. It is from this approach that the study of bioarchaeological remains in so-called ritual sites becomes difficult, as this 'exploitation of natural resources' is not supposed to be performed out of a comprehensible mechanistic and functional rationality, but of a behavioural context ruled by religious or symbolic schemes considered irrational (for criticisms see *e.g.* Bell 1992; Bradley 2005; Brück 1999; Insoll 2004). More recently, however, it has been recognised that both the biotic and abiotic components of an environment actively participate in the construction of a sense of place through social experience, where plants and animals play a social role (*e.g.* Khon 2013), hence they can be envisaged as perishable but active material culture (*e.g.* Livarda *et al.* forthcoming).

This paper, focusing on the Son Ferrer prehistoric ceremonial and funerary site on the island of Mallorca, aims to investigate the social role of plants and animals in a ritual space linked to mortuary practice. The approach adopted does not distinguish between biofacts and artefacts (Morehart and Morell-Hart 2013), but considers that both categories interact in the definition, performance and experience of the social actions that generated the bioarchaeological dataset. We envisage plants and animals as socialised active entities and approach the site as a scene of social performances coherent with the other social spheres and practices of a given community.

Archaeological setting

The site

Son Ferrer is located at the top of a fossil dune (25m.a.s.l.) in the central area of the Calivà peninsula (UTM X 457018; Y 4371845; Fig. 11.1). Its surroundings have been significantly altered in modern times due to urbanisation, although young pine (*Pinus halepensis*) woods with a diverse shrubby stratum still spread in non-urbanised areas. Open sclerophyll Mediterranean maquis grows in areas with the poorest soils, dominated by wild olive tree (*Olea europaea*) and shrubs, such as mastic tree (*Pistacia lentiscus*), Mediterranean heath (*Erica multiflora*), rosemary (*Rosmarinus officinalis*), *Cistus* sp., (rockrose), *Phillyrea* sp (mock privet), Mediterranean buckthorn (*Rhamnus alaternus*), black hawthorn (*Rhamnus oleoides*) and joint-pine (*Ephedra fragilis*).

The site has a quite complex archaeology (Calvo *et al.* 2014; 2015; Garcia *et al.* 2015). The first archaeological evidence corresponds to the Early Bronze Age (EBA, Naviform culture, *c.* 1800–1500 cal BC), when Son Ferrer was part of the Can Vairet necropolis of artificial funerary caves (hypogeum) (Calvo *et al.* 2006), a common type of collective funerary spaces across the island during this period (Veny 1968). The hypogeum of Son Ferrer (Plate 16) was sculpted out from the rock but was intensively reused

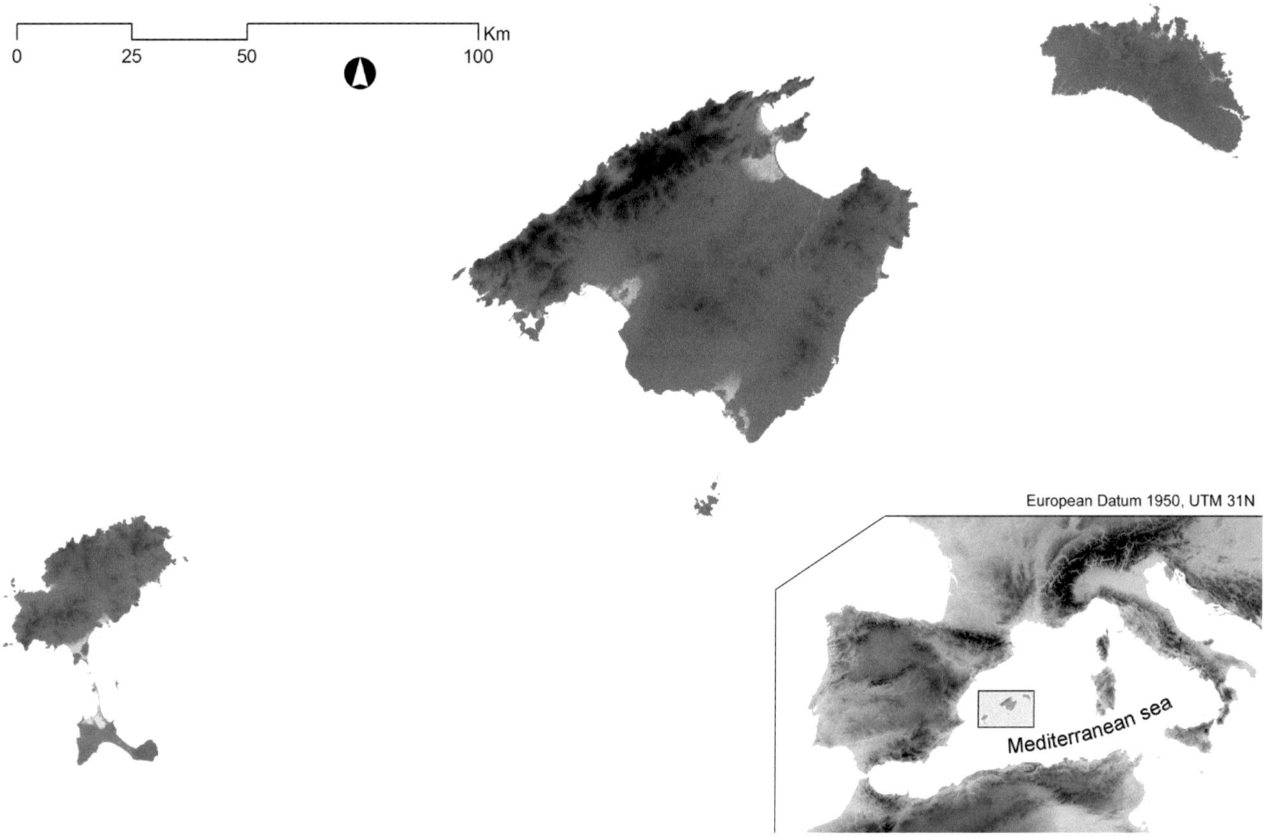

Figure 11.1 Location map of the Son Ferrer site and the Calvià peninsula in the south-west of Mallorca. Drawing by Hector A. Orengo.

in later phases, involving a complete cleaning of the inner space of its cavity.

During the Late Bronze Age (LBA, *c.* 1130–900/850 cal BC) the funerary use of the hypogeum of Son Ferrer was complemented with the construction of an architectural ensemble at the top of the hill where the cave is located, the 'Ritual Context I' (SU17 and 39, Plate 16). This new structure was later altered during the Iron Age, and thus, only part of it is preserved. Therefore, the character of the LBA building cannot be clearly defined beyond identifying that it was a cyclopean staggered construction adapted to the slope of the fossil dune where the EBA hypogeum was excavated.

At the beginning of the Early Iron Age (EIA, Talayotic period, starting from 900/800 cal BC), an important architectural change was carried out. A cyclopean staggered turriform was erected above the Bronze Age hypogeum (Plates 16 and 17). In the Iron Age, the Talayotic culture built a diversity of such cyclopean megalithic buildings, also including the so-called talayots (tower-shaped monuments of circular or square plan), creating a highly visible network through which the social landscape was materially defined and experienced (Calvo 2009; Galmés 2015). At the top of the staggered turriforms there was always a platform that, in some cases, included rooms and/or other structures. All known cases of such monuments are related with ceremonial and ritual activities (Aramburu 1998; Calvo *et al.* 2005). This Talayotic landscape arrangement was always defined around a central settlement, which was interconnected with other monuments placed in important and visible places (Calvo 2009; Calvo *et al.* 2005; Galmés 2015). In the case of Son Ferrer this settlement was Puig de sa Morisca and included several other secondary nodes.

During the Late Iron Age (LIA) (Postalayotic or Balearic culture, 550 cal BC–123 cal BC), Son Ferrer was again re-conceptualised to host a collective necropolis. Different parts of the building were adapted to this new use, among them the EBA hypogeum, which was cleared out, and from this moment onwards and for around two centuries it was used as a collective burial space (Funerary Context I SU 9/101, Plate 16). Both adult and infant corpses were buried inside the cave, with the latter being more abundant. A minimum number of 60 individuals (but possibly up to 101) were documented (Alesan 2007; Alesan and Malgosa 2005). Infantile corpses received a particular funerary treatment, being buried in containers of sandstone or pottery, a common practice in LIA burials of Mallorca. No other material differentiation was identified among the buried individuals.

After the sealing of the cave (354–245 cal BC), the funerary activity focused on the corridor area (Plate 16), where only perinatal inhumations were documented in the space nearer to the cave (331–204 cal BC). Parallel to this funerary activity, in the area closer to the exterior (SU62), a large amount of highly fragmented pottery was found and at the access corridor, a combustion structure was identified (SU64, Plate 16), as well as a rich assemblage of faunal remains.

Finally, later on (*c*. 1 cal BC–1 cal AD) the structural fill of the turriform was excavated to generate spaces to deposit diverse materials and was then sealed (SU44 and 56, Plate 16). The activities at the turriform lasted at least until the end of the 2nd to the beginning of the 3rd century cal AD according to radiocarbon dating and amphorae data (Calvo *et al.* 2014; Quintana 2015).

The (non-perishable) material culture

Due to the intensive reuse of the site, no archaeological remains corresponding to the EBA phase were recorded apart from some potsherds of common types appearing in Bronze Age hypogeum necropolises (Veny 1968). Similarly, no clear pottery assemblage from the LBA phase was found.

Material culture dated to the EIA was limited and was found in the structural fill of the construction between the different rows of the concentric cyclopean walls (SU27, 41, 65, 77, 78, 82) although its origin remains uncertain. It could have originated from previous activities at the site or from the time of construction of the building.

The material culture of the funerary necropolis of the LIA is rather different when compared to other well-known LIA cemeteries of the Balearic Islands. The absence of grave goods and metallic objects is remarkable (especially considering the high number of buried individuals), which are otherwise common in all contemporary necropolises known in the archipelago (Coll 1991; Enseñat 1981; Nadal 2005). In this sense, local hand-made pottery represents the most common non-perishable material culture documented in this context and, as will be argued below, it does not seem to represent grave goods.

LIA pottery distribution patterns have been examined through the combination of photogrammetry, GIS and pottery refitting analysis. The results (Calvo *et al.* 2015) indicated that all sherds from the interior of the cave (SU9/101) in the LIA Funerary Context I had high mobility and the pots identified were found in secondary position. The distribution of pottery fragments is consistent with that of the human remains, which were also found in secondary position and randomly distributed all around the inner space of Funerary Context I (SU9/1901, Alesan 2007). Both pots and human remains were only found in primary position at the entrance of the cave. There, parts of three human bodies with anatomical connections and *in situ* broken pots were found, corresponding to the last three inhumations deposited in the cavity before its final sealing (Calvo *et al.* 2014). The dynamics of space management did not involve specific treatment of the human remains or the pots, as processes of selection, extraction and intentional repositioning were not identified (Alesan 2007; Calvo *et al.* 2015). Thus, this movement of archaeological material has been interpreted as the result of the in-and-out movements during the successive inhumations and related ceremonies in the LIA (Calvo *et al.* 2015).

At the same time as infantile inhumations in urns were taking place at the entrance of the cavity in Funerary Context I, a new area of activity, including the construction of a stone pavement, was established at the corridor, giving access to the cave (SU62, Plate 16). A level of broken ceramic pots was found in this area. The breaking index of this assemblage was particularly remarkable, but the refitting of sherds indicated that almost all pots identified were complete. Moreover, the GIS analysis of pottery fragment dispersion indicated that their movement after breakage was insignificant. This suggests that the pots were used and fragmented *in situ*, and consequently were broken intentionally by the people participating in the ceremonies, possibly once their function was accomplished (Calvo *et al.* 2015).

Parallel to the use of Funerary Context I, different materials were deposited at the exterior, in the northwestern slope of the turriform dated to the LIA and Roman periods. These were located in small cavities created by emptying parts of the structural filling between two different lines of wall belonging to the turriform (Plate 16). They were not directly associated to specific burials, but they were part of the ceremonies that took place there. They included refined dinner service ceramics, such as Roman fine wears, black varnish and small local hand-made cups. Typologically, all of them correspond to fine crockery for food and drink consumption.

Bioarchaeological material and methods

Samples for the recovery of charcoal fragments, seeds/fruits, pollen and faunal remains were selected only from stratigraphically reliable contexts. In total, 13 such units yielded at least one class of these materials (Table 11.1).

Archaeobotanical macroremains

Plant macroremains were systematically recovered during the excavation with a dual strategy. On the one hand, all the dispersed charcoal fragments in the sediment of each stratigraphic unit were systematically handpicked. Additionally, charcoal was collected from hearth SU 64 (Plate 16). In this case, all the sediment was recovered and dry-sieved with a 2mm sieve. On the other hand, during the excavation of the interior of the cave (SU 9/101 of the LIA

Table 11.1 Bioarchaeological records from Son Ferrer considered in this paper.

Phase	*Context & chronology*		*Stratigraphic units (SU)*	*Charcoal*	*Seeds fruits*	*Pollen*	*Animal bones*
Late Bronze Age	Ritual Context I	1130–900/850 BC	17			✓	
			39				✓
Early Iron Age	Construction of Staggered turriform	900–700 BC	27, 41, 65, 77, 78, 82	✓		✓	✓
Late Iron Age	Funerary Context I. Funerary use, interior of the cave	Start: 522–386 cal. BC End: 354–245 cal. BC	9/101	✓	✓	✓	✓
	Funerary Context I Votive use, corridor	331–204 cal. BC	62, 64 (hearth)	✓			✓
	Votive offerings exterior of the turriform	c. 100–1BC c. 1–100 AD	44, 56	✓		✓	✓

Chronology and stratigraphy established after Calvo *et al.* 2014.

Funerary Context I, Plate 16), the totality of the sediment was sampled (9.90 cubic metres) and machine floated using 0.2 to 2mm meshes. From the handpicked samples all available charcoal fragments were analysed. In the case of the floated material from SU 9/101 sub-sampling was necessary due to the enormous amount of >2mm charcoal fragments recovered. All flotation materials were kept for seed/fruit analysis.

The charcoal fragments were manually fractured to provide a clean surface in each section. These were examined with a reflected light microscope with dark/bright field with up to 500x magnifications. The taxonomic identification was based on the observation of the wood anatomical features, the use of identification keys (*e.g.* Schweingruber 1990) and a modern reference collection. Quantification was based on fragment count per taxon.

The seed/fruit remains were examined with a stereomicroscope with up to 80x magnification and identified with the help of a reference collection, atlases and identification keys (*e.g.* Anderberg 1994; Berggren 1981; Cappers *et al.* 2012). Quantification was based on the minimum number of individuals (MNI).

Archaeobotanical microremains

Pollen was recovered in four samples: from the LBA Ritual Context I (SU17), the EIA construction layers of the turriform (SU65) and the LIA Funerary Context I (SU9) and the material at the exterior of the building (SU56) (Table 11.1, Plate 16). From each sample 5g of sediment were treated following standard procedures (Faegri and Iversen, 1989). Samples were dispersed using HCl 35%, followed by digestions with KOH 10%, HF 48% and HCL 35%. A minimum number of 300 pollen grains were counted per sample and the pollen sum includes all identified pollen grains and fern spores, excluding Cichorioideae because of its presumed over-representation. Pollen identification was carried out using published illustrations and morphological keys (*e.g.* Reille 1992–98). Pollen clumps were counted separately as indicators of the local presence of plants (Bui-Thi-Mai and Girard 2002; Clarke 1999). The sum of apophyte taxa includes typical pollen types indicative of human impact (Behre 1981).

Zooarchaeological remains

Faunal remains were the most ubiquitous class of material at Son Ferrer and were retrieved from all contexts (Table 11.1, Plate 16). The bones were macroscopically analysed to identify their taxon, anatomical element, potential fusion in long bones and wear stage in teeth in order to establish the age at death, the side in case of duplicated bones of the skeleton and the level of breakage. Bone measurements were taken following von den Driesch (1976).

Quantification was based on the number of identified specimens (NISP) and the MNI, following Grayson (1984). Tortoise, rabbit and bird remains were considered intrusive specimens due to their taphonomic traces. Rabbit and birds, normally passerines, were found almost complete and without any evidence of human processing (cut marks, fresh fractures). This absence of human marks or breakage, in contrast to other vertebrate, points to a non-human origin of these remains. Moreover, rabbit and tortoise were not introduced to the island until at least the Roman period (Álvarez, 2010; Valenzuela, 2015). The association of passerines with rabbit and tortoise remains reinforces the consideration of this group of taxa as intrusive.

Results

Seeds/fruits

The LIA Funerary Context I assemblage (SU9/101) had almost 500 items per cubic metre. All remains were charred and relatively poorly preserved. Eleven taxa were

Table 11.2 Results of seeds and fruits analysis of the Late Iron Age SU 9/101/Funerary Context I.

Taxon	*Common name*	*Plant part*	*No. remains*
Avena sp.	Oat	Caryopsis	4
Hordeum vulgare	Barley	Caryopsis	32
Triticum aestivum/ durum	Free-threshing wheat	Caryopsis	92
Cerealia	Cereals	Caryopsis	281
Hordeum cf. *murinum*	False barley	Caryopsis	1
Hordeum sp. (wild)	Wild barley	Caryopsis	4
Pisum sativum	Pea	Seed	26
Vicia cf. *ervilia*	Vetches	Seed	2
Fabacaea	Pulses	Seed	16
cf.*Olea europaea*	Olive	Endocarp	2
Polygonum sp.	Knotgrass	Achene	1
TOTAL			457

identified and most of them were cereals, followed by pulses and some olive endocarps (Table 11.2). All cereals were in the form of grains, mostly free-threshing wheat and barley, and no chaff was present. Two types of pulses were present, *Pisum* (pea) (Plate 18) and *Vicia* (vetches), of which peas were predominant, although a third of the pulses could not be identified to genus level. Only five taxa could represent wild species, namely oat (*Avena* sp.), false barley (*Hordeum* cf. *murinum*), wild barley (*Hordeum* sp.), knotgrass (*Polygonum* sp.) and olive (*Olea europaea*), all of which occurred in less than 1% of the total assemblage. Both oat and olive could have also been cultivated, but no such distinction was possible on the basis of the remains available (no oat chaff was present and the olive remains were fragmented). The rest of the possible wild species represented a minor and probably accidental incorporation into the cave flora that accompanied the crops in the cultivated fields.

Charcoal

A total of 930 charcoal fragments belonging to 13 taxa were examined (Table 11.3). The EIA charcoal assemblage included 41 fragments dispersed in the structural fillings of the staggered turrifom. Although the assemblage is not quantitatively reliable, the taxonomical diversity is considerable (five taxa).

The most significant charcoal assemblage corresponds to the LIA Funerary Context I. The entire sample was of charcoal fragments dispersed in the sediment; hearths or charcoal concentrations were not identified. The use of fire inside the cave was not oriented towards product transformation in pots (no cooking pots or hearths were found) and/or corpses' cremation (no evidence of corpse cremation detected, Alesan 2007; Alesan and Malgosa 2005). The evidence points to a successive votive use of fire in relation to the consecutive inhumation of corpses. No evidence of succeeding reuse of the same fire structures were found, as there were no hearths, suggesting that fires were not systematically lit at the same place inside the cave. In contrast, small fires, probably without much preparation, seem to have been made each time that the cave was re-opened to introduce a new deceased.

In terms of taxa, pine (*Pinus* sp.) and mastic (*Pistacia lentiscus*) had very similar values, representing together almost 60% of the whole assemblage. The rest of the taxa were present in less than 10% of the assemblage (Table 11.3). Deciduous and evergreen angiosperms and conifers were identified, but their occurrence was significantly different. Although *Pinus* represents nearly 30% of the assemblage, it is the only conifer identified. Among the angiosperms, deciduous trees (*Quercus*, oak, *Acer* sp., maple) and shrubs (*Pistacia terebinthus*, terebinth) had low values, representing less than 1%. Of the trees, *Pinus* is the best represented, followed by *Olea*, which was less than 10% of the assemblage. Both these trees and the various shrubs identified form part of sclerophyll open forests and shrublands.

Almost all charcoal fragments from hearth (SU64) located at the external part of the corridor after the sealing of the cave were of mastic (Table 11.3). This material represented the firewood that was burnt in the last use/uses of the structure and thus, it can be deduced that almost all fuel consisted of this wood.

Among the deposited materials at the exterior of the turriform 42 charcoal fragments were recovered that belonged to only two species, *Olea* and *Pinus* (Table 11.3), but their exact origin is difficult to establish.

Pollen

Pollen analysis was conducted in samples from ritual and funerary contexts (SU17 and 9) and the construction layers of the turriform (SU 65 and 56) (Table 11.1, Plate 16). The results obtained (Fig. 11.2) showed a very open landscape, with significantly low arboreal pollen values, ranging from 15% of the total in SU9 to 6% in sample SU65. *Quercus ilex*-type (evergreen oak type), *Olea* and *Pinus* appear to be the most common trees represented. Moreover, pollen clumps were recorded in all contexts, indicating the use and deposition of flowers (Fig. 11.2).

The sample from the LBA Ritual Context I (SU17) has low values of arboreal pollen in relation to the total (*c*. 13% of the total pollen). The main taxon present was *Pinus*, while *Quercus ilex*-type and *Olea,* which constitute the expected vegetation of this Mediterranean littoral area in this period, were present in low amounts. Values of shrubby taxa were

Table 11.3 Results of the charcoal analysis at Son Ferrer. Percentages are shown only for Funerary Context I SU 9/101. For the rest of the assemblages the results are shown by fragments count per taxon.

Taxa	*Early Iron Age*	*Late Iron Age*			
	Turriform construction	*Funerary context I SU 9/101*	*Funerary context I SU 9/101*	*Funerary Context I Votive offerings corridor SU 64*	*Votive offerings exterior of turriform*
	No. frags	*No. frags*	*%*	*No. frags*	*No. frags*
Acer sp.		1	0.13		
Arbutus unedo		1	0.13		
Cistus sp.		71	9.23		
Hedera sp.		1	0.13		
Lamiaceae		20	2.60		
Olea europaea	12	54	7.02	3	24
Pinus t. *halepensis*	8	228	29.65		16
Pistacia t. *terebinthus*		4	0.52		
Pistacia lentiscus	6	229	29.78	65	
Deciduous *Quercus*		1	0.13		
Rhamnus alaternus/ Phillyrea sp.	7	17	2.21		
Rhamnus t. *lycioides/ oleoides*	2	2	0.26		
Rosmarinus officinalis		18	2.34		
cf. Arbutus unedo		1	0.13		
cf. Cistus sp.		2	0.26		
cf. Ericaceae		1	0.13		
cf. Lamiaceae		4	0.52		
cf. *Olea europaea*				1	
cf. *Pistacia lentiscus*	2	13	1.69	1	
cf. *Rosmarinus officinalis*		1	0.13		
Undetermined angiosperm	4	65	8.45	9	2
Undetermined conifer		35	4.55		
Total fragments	41	769	100	79	42

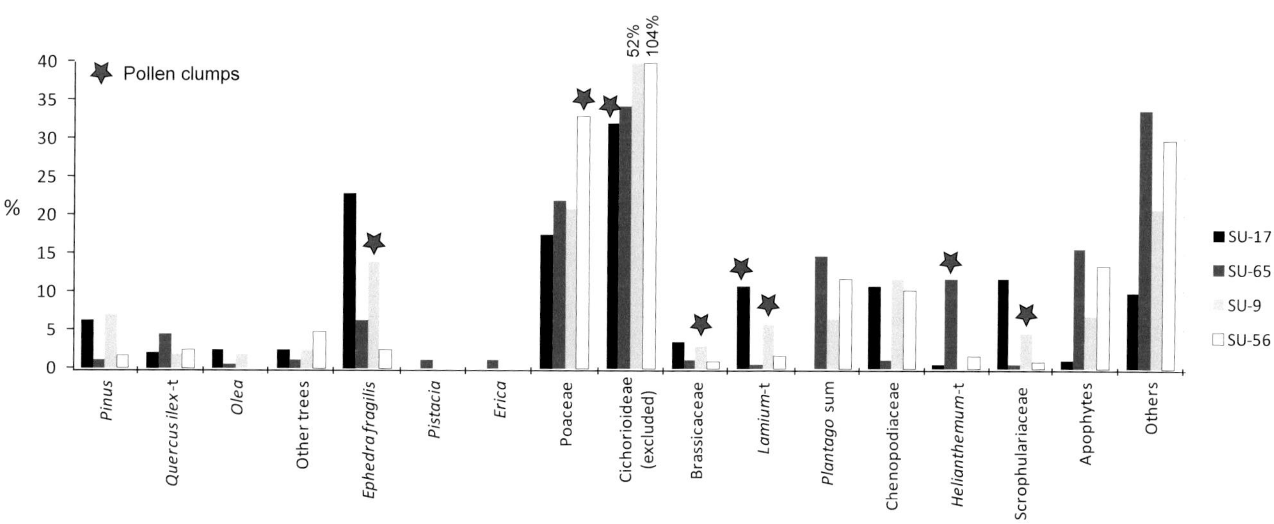

Figure 11.2 Results of pollen analysis. Samples from 4 different contexts: SU 17 = Late Bronze Age ritual context I; SU 65 = Early Iron Age construction of the turriform; SU 9 = Funerary use of the cave during the Late Iron Age, Funerary Context I; and SU 56 = votive offerings at the exterior of the turriform during the Late Iron Age to the 1st century AD.

higher while herb taxa dominated accounting for 62% of total pollen.

The sample of the EIA construction of the staggered turriform (SU65, Table 11.1, Plate 16) had the lowest values of arboreal pollen (6%), suggesting a widely open environment. Within this group, *Quercus ilex*-type was the main tree while *Pinus* and *Olea* were present only as minor components of the assemblage. The total value of shrub taxa reached 10%, including *Ephedra fragilis*-type, *Juniperus* (juniper) and *Pistacia.* The total herb percentage was higher than 80%, including *Helianthemum* (rock rose) pollen clumps.

The sample corresponding to the interior of the funerary cave of the LIA Funerary Context I (SU9, Table 11.1, Plate 16) had arboreal pollen with limited representation of 15%. The main trees were *Pinus*, *Quercus ilex*-type and *Olea.* Pollen clumps of *Ephedra fragilis*-type, Brassicaceae (cabbage/mustard family), *Lamium*-type (dead nettle/henbit type) and Scrophulariaceae (figwort/toadflax family) were recorded, indicating the actual presence of flowers inside the cave during its use as a collective necropolis.

Finally, the sample from the area of votive offerings at the exterior of the turriform (SU-56, Table 11.1, Plate 16) had small quantities of arboreal pollen (10%). The main trees were *Quercus ilex*-t, *Pinus* and *Olea.* The total herb value was the highest reported in the site (86% of the total pollen sum). Also, pollen clumps of grasses (Poaceae) were recorded.

Animal bones

In total, 4125 faunal remains were analysed of which 1423 were identified (34.71%) (Table 11.4). The rest of the assemblage, 2693 fragments (65.29%), was unidentified due to biostratinomic and diagenetic breakage. The mammal remains from the LBA Ritual Context I were few (NISP = 36) but, in contrast, the same taxa were well represented in the EIA and LIA periods. Sheep/goat (*Ovis/Capra*) were by far the most common mammals in all these phases, followed by domestic pig (*Sus domesticus*) and cattle (*Bos taurus*). Cut-marks were very scarce. Only a cut-mark on a caprine radius was identified, but almost all fractures of broken medium/large mammal remains (sheep/goat, cattle and pigs) were fresh, and therefore, they could be related to butchery processes.

The MNI was always smaller for pig and cattle than sheep and goat (Table 11.4). The specimens represented all age stages: infantile, juvenile and adults. A certain predominance of goats over sheep was noticeable. Only one complete individual was found at the exterior of the turriform during the LIA period. This was an adult (fused bones) goat represented by the skull, one complete radium and both complete metacarpals with a very small size (between 50 and 51cm tall according to von den Driesch's greatest length of radius and metacarpal) (Plate 19). Its completeness during deposition is based on its anatomical representation and the models proposed for Western Mediterranean domestic mammals during the 1st millennium BC (Albizuri 2011; Albizuri *et al.* 2015; Miró and Molist 1990).

Dog remains (*Canis familiaris*) were also present in EIA and LIA contexts, as well as Mediterranean monk seal (*Monachus monachus,* a complete post-canine) in an EIA context, which suggests possible hunting activities in relation to marine resources (Table 11.4).

Table 11.4 Results of the zooarchaeological analysis by NISP and MNI.

Taxa	*Late Bronze Age*	*Early Iron Age*	*Late Iron Age*			
	Ritual Context I UE 17/39	*Turriform construction*	*Funerary Context I SU 9/101*	*Votive offerings corridor Funerary Context I SU 62/64*	*Votive offerings exterior turriform SU 56*	*Total Late Iron Age*
	NISP (MNI)	*NISP (MNI)*	*NISP (MNI)*	*NISP (MNI)*	*NISP (MNI)*	*NISP (MNI)*
Ovis aries (sheep)		6	6			6
Capra hircus (goat)	1	33	4		11	15
Ovis/Capra (sheep/goat)	35	593	105	33	97	235
Total sheep and goat	36 (2)	632 (18)	115 (7)	33 (2)	108 (7)	256 (16)
Bos taurus (cattle)		12 (1)	1 (1)	1 (1)	9 (1)	11 (3)
Sus domesticus (pig)		34 (2)	4 (1)	2 (1)	3 (1)	9 (3)
Canis familiaris (dog)		4 (1)	52 (3)	5 (1)		57 (4)
Monachus monachus (Mediterranean monk seal)		1 (1)				
Total NISP	36	683	172	41	120	333

Discussion

Offerings, feasting and foodways at Son Ferrer

Only few remains of the Bronze Age activities carried out at Son Ferrer were preserved due to the intensive reuse and remodelling of the site during later phases. These included a LBA faunal assemblage of sheep/goats (Table 11.4) at Ritual Context I (SU17/39), which was not related to any material culture, and therefore, it is difficult to assess whether these animal bones represent food debris or not. In any case, their context of deposition, a cyclopean staggered ceremonial building related to the funerary use of the hypogeum in the context of the Can Vairet necropolis, was clearly a ceremonial one.

During the EIA, the staggered turriform was built and the site was reconceptualised. In the structural fillings of this cyclopean monument another faunal assemblage was recovered, with the highest taxonomical diversity of all phases. The taxa identified were almost exclusively domestic (with the presence of the one single fragment of monk seal, Table 11.4). The assemblage represents the typical Balearic prehistoric livestock with clear predominance of sheep/goat and minor presence of pig and cattle (Ramis 2006; Valenzuela and Valenzuela 2013). The structural filling of the building, mainly composed of stones and soil, also included bone remains that were present at the surrounding of the site at the moment of construction of the building or before that. It is difficult to disentangle the origin of these faunal remains, as they could have been the residues of different activities that could have taken place at the surroundings of the site before and/or during the construction of the cyclopean monument. In all likelihood, they represent food debris and, one possibility is that this food consumption could have taken place in relation to acts of commensality related to the construction of the turriform, such as feasting related to collective work events. Whatever their original use, what is clear is that, even if these animals were consumed in ritualised eating and drinking ceremonies, they represented the typical meat production documented during the Balearic Bronze and Iron Ages.

The richest bioarchaeological assemblages of Son Ferrer were found in Funerary Context I, where the ceramic pots present were specific products orientated to an aesthetic function linked to food and drink consumption. Two different groups of local handmade pottery items were identified. The first one included medium size ceramic vessels used as funerary containers, most of which included infantile burials (Alesan 2007) and were concentrated in the external part of the Funerary Context I during its later use after the sealing of the cave (Albero 2011; García 2010). The second pottery group was characterised by small size pots. The typology of these small cups was variable, but they all had belt handles. They also exhibited a high degree of variability concerning clay-plastering recipes, with almost no addition of temper to the clay (Albero 2011; García 2010). Therefore, these were technologically vulnerable pots not suitable for direct use with fire. They had polishing of their external surface and no traces of contact with fire. These pots could have been deposited as votive offerings, but considering their fabrics and their functional typology, it seems that they are most likely linked to liquid consumption by the mourners during the funerary activities and left inside the cavity with the corpses of the deceased after the ceremonies (Albero 2007; 2011). In this sense, the pottery assemblage seems to point out to liquid and possibly food consumption during the ceremonies linked to the successive burials and the subsequent deposition of the material culture involved in such funerary ceremonies.

The faunal record of Funerary Context I shows a high taxonomical variability, with the presence of all domestic species. Sheep/goats formed again the most important group and cattle and pigs were found in low numbers (Table 11.4). No significant distinctions among individuals were visible, as sheep/goats, both adult and juvenile, appeared to be the most common remains in all occasions. Although no anatomical connections were identified, the representation of body parts and the number of remains suggested that these animals were not introduced into the cave complete. In most cases, the best-represented animal parts were skull elements (crania and mandibles) and appendicular bones (from fore- and hind limbs). Bones forming the axial skeleton were notably scarce. In this sense, it can be argued that, if they correspond to food offerings, they most likely represent parts of the animal offered by the mourners as food for the deceased. Alternatively, it can be argued that these remains constitute the debris of ceremonial feasts undertaken by the mourners that buried the corpses. Considering that small liquid consumption pots were present, feasts could have occurred as part of the funerary ceremonies. In the same way, it can be also argued that at least part of the food could have been introduced (and thus 'consumed') as offerings to the deceased.

In relation to plant remains, their context of deposition also permits two different interpretations regarding their origin and the causes of their preservation. One hypothesis could be the accidental carbonisation of the plant foods present during the repeated re-openings of the cave. This can be the result of the fact that the mourners were involved in the cooking and eating of already processed food plants inside the cave as part of the funerary ceremony and some grains were accidentally charred during this process. Otherwise, the mourners could have purposely burnt part of the local crop produce as offerings to accompany the dead. So, again, as with the meat products, food consumption could be both related to the ceremonial actions of the mourners during the funerary performances (feasting), and/or to the food offerings for the deceased.

In any case, the characteristics of the plant assemblage, corresponding largely to clean cereals grains and pulses, with a clear emphasis on the former (Table 11.2), fit with the other known archaeobotanical assemblages in the prehistoric Balearic Islands (*e.g.* Aramburu 2010; Arnau *et al.* 2003; Hernández-Gasch and Celma 2011; Lopez Garí *et al.* 2013; Moffet 1992; Pérez 2009a; 2009b; Stika 1999) and Iron Age contexts from the nearby Catalan coast (*e.g.* Alonso *et al.* 2008). These were also mostly composed of cereals, particularly barley and to a lesser extent wheat, some pulses and fruits like olive and grape (*Vitis vinifera*) and only traces of weed taxa. Cultivated pulses at Son Ferrer were found in relatively large amounts in comparison with other prehistoric seed assemblages in the Balearics (Aramburu 2010; Arnau *et al.* 2003; Hernández-Gasch and Celma 2011; Moffett 1992; Pérez 2009a and 2009b; Stika 1999). A predominance of cereals, represented by wheat (mainly free-threshing wheat) and barley, and a low but recurrent presence of pulses appears to be the basis of the prehistoric Balearic agriculture according to the few studies available. Accordingly, similarly to the bone assemblage, there is no evidence of a clear distinction between the consumption of plant products in domestic and funerary contexts. Therefore, the assemblage of Son Ferrer could represent the richness of the agricultural produce and its role in the community's foodways.

Once the Funerary Context I was sealed, infantile burials in urns were taken to the entrance corridor of the hypogeum and at its external part a concentration of purposely broken pottery wares was found *in situ* (SU62, Plate 16). This pottery assemblage is formed by imported crockery (dinner service, plates and vessels/glass) of small size that were produced in different western Mediterranean colonial contexts, including for instance Iberian and Ebusitan commodities. Some bigger plates were also present, all of them handmade local productions with the exception of one Iberian plate. A hearth was also in use in this space during this period (SU64, Plate 16). In this context, thus, the preserved material culture clearly points to a place arranged (stone pavement, hearth) for the performance of acts of commensality, suggested by both the functional typology (serving and consumption vessels for food and drinks) and the imported pottery assemblage. Its direct spatial relation with the last funerary uses of the site is also remarkable, relating such practices of commensality with the funerary ceremonies. In this case, thus, the material culture used in the serving and consumption of food and drinks was differentiated from the local handmade productions dominant in the contemporary domestic contexts. Notwithstanding, among the faunal remains recovered in this feasting context, and thus likely representing food remains, bones of mainly sheep/goat, and also a few cattle and pig, were documented (Table 11.4), suggesting similar modes of meat consumption during these ceremonies to everyday practice.

In the context of the votive deposits at the exterior of the turriform (SU 44 and 56) during its last period of use, domestic animal bones, mainly caprine with some presence of pig and cattle (Table 11.4), together with ceramic pots, were deposited in small cavities created by emptying parts of the structural fillings of the turriform. This material culture included refined dinner service ceramics for food and drink consumption, suggesting that the ceremonies taking place at the exterior of the turriform involved acts of commensality. These feasts would have ended up with the votive deposition in the structure of the monument itself of the perishable and non-perishable material culture involved. This activity does not seem to be directly linked to funerary depositions, but to the significance of the place and the monument itself (Garcia *et al.* 2015). As a place of memory and identity forging, the exterior of the turriform was also the stage of social meetings and celebrations that seem to have included the practice of commensality. In this sense, the faunal remains suggest that after meat was eaten its debris was deposited together with the rest (or a significant part) of the material culture of the ceremonies. Additionally, an adult goat of small size (Plate 19) seems to have been deposited complete at the exterior of the turriform during the LIA period. It is thus possible that goats were votive offerings by themselves, apart from feeding the ceremonial feasts. The meaning of animal depositions in a context like Son Ferrer, as a place where the forefathers of the community had been buried for centuries, may be related to the veneration of the ancestors. Animals could have been offered to the ancestors in a place that materialised the memory and identity of the community.

According to the contextual analysis of the bioarchaeological and pottery data, therefore, it can be suggested that, at least in some cases, both animal and plant remains represent food consumption debris and/or food offerings. If in the LBA and EIA contexts the origin of the animal bones is difficult to assert, in LIA contexts plants and animals seem to have been consumed in refined dinner service ceramics during feastings and/or offered to the deceased in the context of funerary activities and ceremonial rites performed in this significant place for the community. Considering that 'when human beings convert some part of their environment into food, they create a peculiarly powerful semiotic device' (Appadurai 1981, 494) expressed as embodied material culture (Dietler 1996; 2001; Hamilakis 1998; Hamilakis and Sherratt 2012), the consumption of plants and animals at Son Ferrer can be envisaged as a central element in the performance of the ritual and funerary practices. This can be related to both the 'afterlife' of the deceased (food offerings) and the exhibition, negotiation and renegotiation of social identity among the living (feasting) as food constitutes 'an arena for action and social negotiation' (Hastorf and Johannessen 1996, 116).

Feasts are defined as acts of communal consumption of food and drink in a different way from everyday eating and drinking (*e.g.* Dietler 1996, 89; 2001, 66–9; Hamilakis and Sherratt 2012, 194; Pollock 2015). As 'ritualised' consumption practices, feasts constitute especially powerful arenas of 'condensed symbolic representations of social relations' (Dietler 2001, 69). This ritualisation of drinking and eating does not necessarily involve elaborate ceremonies (Dietler 2001, 69) or the consumption of luxury foodstuff. The sensory context of the act of commensality can be also significant. Visual and olfactory stimulation, for example, are important for the creation of the stage of commensality where it acquires its social relevance (Hamilakis 2013) and at Son Ferrer there is evidence that the selection of plants contributed in setting the stage for the ceremonies (see below). The animals and plants involved in the diverse food consumption practices at Son Ferrer were almost exclusively domestic (Tables 11.2 and 11.4). As agricultural and livestock produce is an important part of the social production, their consumption is a key element of social reproduction, so the experience associated to food procurement and production may have played a part in the final act of commensality. The production of domesticated plants and animals involves a set of social actions performed in the environment that plays a role in the creation of a sense of place and time (taskscapes, Ingold 1993), and in the shared experience of dwelling (Ingold 2006). At Son Ferrer, a place with a long-lasting megalithic presence on the landscape where the bodies of the deceased of the community were deposited, feasting could have been an expression of social and public memory.

The fact that the plants and animals offered to the deceased and/or consumed in feasting acts corresponds to the domestic food products suggests that the foodways of the communities that buried and venerated their forefathers at Son Ferrer did not draw up clear boundaries in relation to foodstuff between the domestic and the ritual sphere, between the daily feeding items of the community and those involved in mortuary feasts and/or offerings. There were no luxury foodstuffs identified in contrast with other contemporary necropolises where there was clear individualisation and differentiation in the funerary treatment among the buried. This is the case of the necropolis of Son Real, where a differential consumption of cattle has been identified in relation to domestic zooarchaeological assemblages (Hernández-Gasch and Ramis 2010). Similarly, at the sanctuary of Punta des Patró, very close to Son Real, feasting on bovine meat was identified for the LIA period. Considering that bovines clearly decrease in zooarchaeological assemblages during the Iron Age in Mallorca while there is an increase in the age at which they are slaughtered (possibly indicating use for traction), it has been suggested that these animals became luxury products during this period (Hernández-Gasch and Ramis 2010). Other differences observed at Son Ferrer include the higher presence of infantile burials compared to other contemporary necropolises, and the absence of grave goods and metal objects deposited with the corpses, which is otherwise a common practice during the LIA.

Feasting and commensality at Son Ferrer can be thus envisaged as 'ritualisation' (*sensu* Bradley 2005) of domestic foodways, rather than specific ritual behaviours clearly differentiated from domesticity. Far from materialising luxury and social status, feasting and foodways at Son Ferrer were possibly an arena where social bonds and acts were experienced and performed. Taking into account that feasting is a kind of ritualised social action, in the case of Son Ferrer, this 'complex semiotic relationship to daily consumption patterns' (Dietler 2001, 69) was not evidenced by the vegetal and animal products themselves, but by their context.

Therefore, what is partly observed at Son Ferrer is possibly surplus-based feasting, and in this context, cultivated plants and animals were agents of experiencing and inhabiting the landscape of the community, creating a certain kind of sense of place and identity. This relation with the land and the landscape of the community was also expressed by the place itself. Son Ferrer acted as a persistent place within the landscape for generations and involved the ancestors of the community and their link to the land first with the EBA hypogeum, later with the LBA staggered structure and, finally, from the EIA onwards, with the dynamic and long-lasting use of the megalithic cyclopean staggered turriform.

Plants and ceremonies: the sensory component

In these funerary celebrations, non-food plants played a role in the creation of the visual and olfactory ambience in the different spaces of the turriform. Pollen and charcoal are complementary records that can inform on the choices of plants used. The pollen analysis suggests a largely open landscape in all phases. Pine, oak and olive seem to be the most common trees growing in the surroundings of the site (Fig. 11.2). Of these it is only pine that had significant values in the charcoal assemblage (Table 11.3). In this assemblage, mastic seems to be an important plant, although it is only a minor component of the palynological record. Even though mastic could have played a noticeable role in the Mediterranean maquis of Mallorca (Burjachs *et al.* 1994; forthcoming), the low levels of its pollen at Son Ferrer suggests that this shrub did not occupy the immediate surroundings of the archaeological site. Mastic surface pollen samples in Mediterranean contexts have shown that its pollen reaches about 5% under shrubs and its representation dramatically decrease at 100m from the source area (Erdogan *et al.* 1998; Florenzano *et al.* 2013). Pollen values of mastic at Son Ferrer never reached 1%, suggesting that *Pistacia* scrubland did not spread on the site, even though it is probable that it was a more or less common shrub in the regional vegetation of littoral areas of the island. In Funerary Context I mastic represented almost 30% of the total charcoal assemblage (Table 11.3), which is a significant value considering the high taxonomical variability of the sample (13 taxa). In the hearth related with the feastings at the entrance of the hypogeum (SU 64) this plant represented the almost only fuel used (Table 11.3). These values of mastic differ from its average values in LIA domestic households of the area, where, even being one of the most common taxa, does not normally reach such percentages (Picornell 2012). Therefore, its high values are related with the context of use of its wood as fuel and its cultural value, as the gathering of this wood necessitated reaching beyond the immediate cave environment.

Mastic wood has a high resin content, like pine, which is the other most common taxon in Funerary Context I, also with values close to 30%. The uses of these resinous species, thus, may also be considered in relation to the use of fire inside the hypogeum. In Funerary Context I firewood was burnt in single-use, non-complex hearths that were fired during the successive ceremonies of inhumation. The space where they were lit was a closed one, with no natural light entering and with a large amount of decaying organic matter from the successively deposited corpses. In this context, the olfactory component of these resinous species would have played an important role in the creation of an appropriate ambience for the funerary ceremonies. Other resinous (*e.g. Cistus*) and aromatic plants (rosemary and other Lamiaceae) were also common and together with mastic and pine represented almost 74% of the charcoal assemblage (Table 11.3). Equally, both mastic and pine resinous woods could have provided lighting flames in a dark space like the closed space of the hypogeum. Sensory experiences related to mortuary practices acquire a special significance (Hamilakis 2011, 129–60) and the funerary ceremonies celebrated at Son Ferrer, thus, shared a special sensory component, as experienced by the mourners.

In the case of the LIA hearth, the use of mastic can be also related with its aroma (for other Balearic sites see *e.g.* Javaloyas *et al.* 2009; Riera *et al.* this volume). Additionally, the firewood used here could have been related with the food consumption events and thus its selection could have been also in conjunction with these activities. Overall, the nature of these charcoal assemblages highlights the social significance of mastic among the LIA Balearic society, which was also noted later by the classical geographer Diodorus Siculus (V, 17) when he described how the population of the islands mixed mastic oil and pork lard and smeared the mixture over their bodies (Blanes *et al.* 1990, 32).

The presence of pollen clumps, indicating the deposition of flowers in various contexts is another indication of plants used in relation to sensory aspects. Several clumps were present, including those of deadnettle/henbit type, joint pine type, cabbage/mustard family, figwort/toadflax family, and rock rose (Fig. 11.2). Interestingly, all these taxa have small, colourful and eye-catching flowers of similar shape, which suggests a choice element in the use of flowers in the ceremonies. This use of flowers could have been related with different aspects, such as floral offerings for the deceased or part of the creation of a stage and ambience for the funerary celebrations. Whatever the case, the visual and olfactory characteristics of these wild flowers constituted an important, appealing component of the ritual celebrations, as also noted for other Balearic prehistoric sites (Riera *et al.* this volume; Servera-Vives 2009).

Conclusions

Plants and animals played a major role in the olfactory, taste and visual experiences of the diverse ceremonial celebrations at Son Ferrer. These were interlinked with the creation of social memory and identity during feasting, offerings and funerary practices. The use of fire made out of aromatic woody plants inside the cave of Funerary Context I was an important sensory element, practised and experienced by the mourners that performed the funerary rites, which 'being infused with smoke and smells, and of course with the sensory and embodied presence of others' would also contribute in the production of 'feelings and emotions, time, identities, and personal and collective histories' (Hamilakis 2011, 209). Plants were purposely used to produce light and smells, and provide colourful and eye-catching flowers to the visual component of the funerary celebrations. They were, however, integrated with the other ingredients of the ritualised funerary expressions of group belonging and social identity. Local and imported wears were used during the LIA to consume food in the diverse acts of commensality, in which plants and animals were eaten. Those products were not different from the ones consumed in the households of the community that buried their deceased at Son Ferrer. However, although the foodstuff consumed were the same, the experience of the consumption of these products was completely different, being part of a differentiated sensory environment. This experience of eating the typical everyday food was clearly differentiated (and ritualised) from a sensory point of view. In this process, architecture (monumentality, landscape networks of visibility), a specific sense of place (persistent place, long-term history of ritual and funerary activities and buildings at Son Ferrer from the EBA to LIA) and material culture (non-functional and aesthetic local hand-made pottery, imported wares, absence of metal objects as grave goods) played an important role, together with plants and animals, in shaping the social significance of this site and the activities carried out there.

These ritualised acts and funerary practices in which plant and animals played a role did not involve a distinction and/or specific treatment of individuals among the community buried at the site. Only a differentiation in terms of age was observed with the infantile burials in urns. Accordingly, the social expression of group belonging and communal identity was also experienced by acts of commensality, in which the food consumed was that of the everyday community diet. In this sense, the turriform itself and the long-term history of the place where it stands created a stage where people experienced these sensory, mnemonic and identity expressions during the ceremonies.

Acknowledgements

Llorenç Picornell-Gelabert was awarded a postdoctoral fellowship of the Government of the Balearic Islands – European Social Fund (PD/022/2013) and a Beatriu de Pinós Fellowship (2014 BP_A 00216). Zooarchaeological studies (JNL) received support from the research projects HAR2011-26193 and 2014SGR-; and Gabriel Servera-Vives was supported by a Juan de La Cierva fellowship. The authors are grateful to Dr. Damià Ramis for providing bibliographical references, the two anonymous referees and the editors for helping improve the text. Alejandra Galmés helped with the figures and Susan Gray revised the English, for which we are grateful. This work is part of the knowledge transfer of the R&D projects HAR2012-32620 and HAR2015-67211-P.

References

Albero, D. (2007) Primeras aproximaciones a la tecnología cerámica prehistórica en la península de Calvià (Mallorca). *Arqueología y Territorio* 4, 70–86.

Albero, D. (2011) Caracterización tecnológica, social y adaptación funcional de las cerámicas prehistóricas en el Oeste y Sureste de Mallorca (1700–50 BC). Aproximación sincrónica y diacrónica a partir de estudio arqueométrico de las pastas. PhD thesis, University of Granada.

Albizuri, S. (2011) Depósitos de ovicaprinos en ámbito doméstico ibérico. El ejemplo del poblado de Ca n'Oliver (Cerdanyola del Vallès, Barcelona). *Archaeofauna* 20, 85–101.

Albizuri, S., Subirà, M. E. and Ruiz, J. (2015) Ritual patterns in the deposits of fauna associated with Iberian burials: the necropolis of El Poblado during the fourth to second centuries BC (Archaeological complex of Coimbra del Barranco Ancho, Murcia). *International Journal of Osteoarchaeology* 25, 188–200.

Alesán, A. (2007) Estudi del material antropològic del turriforme de Son Ferrer. Informe inédito. Unpublished report, University of the Balearic Islands.

Alesán, A. and Malgosa, A. (2005) Les inhumacions perinatals del túmul de Son Ferrer (Calvià, Mallorca): estudi antropològic. *Mayurqa* 30, 511–22.

Alonso, N., Junyent, E., Lafuente, A. and Lòpez, J. B. (2008) Plant remains, storage and crop processing inside the Iron Age fort of Els Vilars d'Arbeca (Catalonia, Spain). *Vegetation History and Archaeobotany* 17(1), 149–58.

Álvarez, C. (ed.) (2010) *Seminari sobre Espècies Introduides i Invasores a les Illes Balears*. Sóller, Government of the Balearic Islands.

Anderberg, A. L. (1994) *Atlas of Seeds. Part 4 Resedaceae-Umbelliferae*. Stockholm, Swedish Museum of Natural History.

Appadurai, A. (1981) Gastro-Politics in Hindu South Asia. *American Ethnologist* 8(3), 494–511.

Aramburu, J. (1998) *El patrón de asentamiento de la cultura talaiótica de Mallorca*. Palma, El Tall.

Aramburu, J. (2010) Análisis carpológico. In J. Aramburu (ed.) *Ses Païsses (Artà, Mallorca). Excavaciones en el Recinto* 51, 48–59. Palma, Arqueobalear.

Arnau, P., Gornés, S., Stika, H. P. (2003) Los hipogeos de S'Alblegall (Ferreries) y la agricultura cerealística a mediados del II milenio cal ANE en Menorca. *Trabajos de Prehistoria* 60(2), 117–30.

Bell, C. (1992) *Ritual Theory, Ritual Practice*. New York, Oxford University Press.

Behre, K. E. (1981) The interpretation of anthropogenic indicators in pollen diagrams. *Pollen et spores* 23, 225–45.

Berggren, G. (1981) *Atlas of seeds. Part 3 Salicaceae – Cruciferae*. Stockolm, Swedish Museum of Natural History.

Blanes, C., Bonet, J., Rosselló, A., Font, A. and Rosselló, A. M. (1990) *Les Illes a les Fonts Clàssiques*. Palma, Miquel Font Editor.

Bradley, R. (2005) *Ritual and Domestic Life in Prehistoric Europe*. London, Routledge.

Brück, J. (1999) Ritual and rationality: Some problems of interpretation in European archaeology. *European Journal of Archaeology* 2(3), 313–44.

Burjachs, F., Pérez-Obiol, R., Roure, J. M. and Julià, R. (1994) Dinámica de la vegetación durante el Holocenos en la isla de Mallorca. In M. Durpé, M. Burgaz, I. Mateu and J. Güemes (eds.) *Trabajos de palinología básica y aplicada, X Simposio de Palinologia (APLE)*, 199–210. Valencia, Universitat de València.

Burjachs, F., Pérez-Obiol, R., Picornell-Gelabert, Ll., Revelles, J., Servera-Vives, G., Expósito, I. and Yll, E. I. (2017) Environmental changes and human colonization in the Balearic Islands and their impacts on vegetation: a review. *Journal of Archaeological Science Reports* 12, 845–59.

Bui-Thi-Mai and Girard M. (2002) *L'archeopalynologie, Discipline aux Multiples Facettes. XIe Rencontres Culturelles Interdisciplinaires*. Lévie, Editions Alain Piazzola.

Clavo, M. (2009) Reflexiones en torno a los esquemas de racionalidad espacial reflejados en el paisaje durante la Prehistoria de Mallorca. *Pyrenae* 40(2), 37–78.

Calvo, M., Fornés, J., Garcia, J., Iglesias, M. A. and Juncosa, E. (2005) Condicionantes espaciales en la construcción del turriforme escalonado de Son Ferrer (Calvià, Mallorca). *Mayurqa* 30, 485–508.

Calvo, M., Garcia, J., Iglesias, M. and Juncosa, E. (2006) La necropolis del Bronce Antiguo de Can Vairet/Son Ferrer (Calvià, Mallorca). *Mayurqa* 31, 57–82.

Calvo, M. García, J., Javaloyas, D., Albero, D. and Van Strydonck. M. (2014) Matching data: analyzing the chronological use sequence in the iron Age necropolis of the staggered turriform of Son Ferrer (Balearic Islands, Spain). *Radiocarbon* 56(2), 1–14.

Calvo, M., Garcia, J., Albero, D., and Javaloyas, D. (2015) Análisis de la distribución espacial de la cerámica depositada en el Ámbito Funerario 1 del turriforme escalonado de Son Ferrer (s. V-I a.C.). In A. Matínez and G. Graziani (eds.) *Actes de las VI Jornades d'Arqueologia de les Illes Balears*, 63–72. Formentera, Consell Insular de Formentera.

Cappers R. T. J., Bekker R. M. and Jans J. E. A. (2012) *Digital Seed Atlas of the Netherlands* (2nd edn). Groningen, Barkhuis and Groningen University Library.

Clarke, C. (1999) Palynological Investigations of a Bronze Age Cist Burial from Whitsome, Scottish Borders, Scotland. *Journal of Archaeological Science* 26, 553–60.

Coll, J. (1991) *La evolución del ritual funerario en la cultura talaiótica*. Unpublished PhD thesis, University of the Balearic Islands.

Dietler, M. (1996) Feasts and commensal politics in the political economy. Food, power and status in prehistoric Europe. In P. Wiessner and W. Schiefenhövel (eds.) *Food and the Status Quest: An Interdisciplinary Perspective*, 87–125. Oxford, Bergham Books.

Dietler, M. (2001) Theorizing the feast. Rituals of consumption, commensal politics, and power in African contexts. In M. Dietler and B. Hayden (eds.) *Feasts. Archeological and Ethnographic Perspectives on Food, Politics and Power*, 65–144. Washington DC–London, Smithsonian Institution Press.

Driesch, A. von den (1976) *A Guide to the Measurement of Animal Bones from Archaeological Sites*. Harvard, Peabody Museum Bulletins 1.

Erdogan, V., Kunter, B. and Ayfer, M. (1998) Pollen dispersal in pistachio (Pistacia vera L.). *Acta Horticulrurae* 470, 286–93.

Ensenyat, C. (1981) *Las Cuevas Sepulcrales Mallorquinas de la Edad del Hierro*. Madrid, Excavaciones Arqueológicas en España 111.

Faegri, K. and Iversen, J. (1989) *Textbook of Modern Pollen Analyses*. London, Ejner Munksgaard.

Florenzano, A., Mercuri, A. M. and Carter, J. C. (2013) Economy and environment of the Greek colonial system in southern Italy: pollen and NPPs evidence of grazing from the rural site of Fattoria Fabrizio (6th–4th cent. Bc; Metaponto, Basilicata). *Annali di Botanica* 3, 173–81.

Galmés, A. (2015) Visibilidad y percepción en la construcción de un paisaje prehistórico. El caso de Calvià (Mallorca, Islas Baleares). *Complutum* 26(1), 173–88.

Garcia, J. (2010) Análisis traceológico de la cerámica. Modelado y espacio social durante el Postalayótico (s. V-I AC) en la península de Santa Ponça (Calvià, Mallorca). Unpublished PhD thesis, Universitat de les Illes Balears.

Garcia, J., Calvo, M., Javaloyas, D. and Albero, D. (2015) La secuencia cronológica de uso del turriforme escalonado de Son Ferrer: persistencia de usos e identidad simbòlica. In C. Andreu, C. Ferrando and O. Pons (eds.) *L'Entreteixit del Temps. Miscel•lània d'Estudis en Homanatge a Lluís Plantalamor Massanet*, 188–208. Palma, Government of the Balearic Islands.

Grayson, D. K. (1984) *Quantitative Zooarchaeology: Topics in the Analysis of Archaeological Faunas*. New York, Academic Press.

Hall, V. A. (1988) The role of harvesting techniques in the dispersal of pollen grains of Cerealia. *Pollen Spores* 30, 265–70.

Hamilakis, Y. (1998) Eating and dead: mortuary feasting and the politics of memory in the Aegean Bronze Age societies. In K. Branigan (ed.) *Cemetery and Society in the Aegean Bronze Age Societies*, 115–32. Sheffield, Sheffield Academic Press.

Hamilakis, Y. (2011) Archaeologies of the senses. In T. Insoll (ed.) *The Oxford Handbook of the Archaeology of Ritual and Religion*, 208–25. Oxford, Oxford University Press.

Hamilakis, Y. (2013) *Archaeology and the Senses. Human Experience, Memory, and Effect*. Cambridge, Cambridge University Press.

Hamilakis, Y. and Sherratt, S. (2012) Feasting and the consuming body in Bronze Age Crete and Early Iron Age Cyprus. In G. Cadogan, M. Lacovou, K. Kopaka and J. Whitley (eds.) *Parallel Lives: Ancient Island Societies in Crete and Cyprus*, 187–207. Athens, British School at Athens.

Hastorf, C. A. and Johannessen, S. (1996) Understanding Changing People/Plant Relationship in the Prehispanic Andes. In R. Preucel and I. Hodder (eds.) *Contemporary Archaeology in Theory. A Reader*, 61–78. Oxford, Blackwell.

Hayden, B. (2009) The proof is in the pudding. Feasting and the origins of domestication. *Current Anthropology* 50(5), 597–601.

Hernández-Gasch, J. and Ramis, D. (2010) Economia funerària de la Segona Edat del Ferro de les Illes Balears. L'ús diarcrónic del sacrifici de bòvids en el santuari i les necròpolis de Son Real (Mallorca). *Saguntum PLAV* 42, 71–86.

Hernández-Gasch, J. and Celma, M. (2011) Paleoecologia de l'Edat del Ferro a l'illa de Menorca. Les restes antracològiques i carpològiques procedents del poblat de Biniparratx Petit (Sant Lluís). Unpublished report, Institut d'Estudis Menorquins.

Ingold, T. (1993) The temporality of landscape. *World Archaeology* 25(2), 24–174.

Ingold, T. (2006) *The Perception of the Environment: Essays on Livelihood, Dwelling and Skill*. London, Routledge.

Insoll, T. (2004) *Archaeology, Ritual, Religion*. London, Routledge.

Javaloyas, D., Picornell-Gelabert, Ll. and Servera, G. (2009) PLantas y fenomenología de la muerte durante el Bronce Medio y Final en Menorca. In Organización Jóvenes en Investigación Arqueológica (ed.) *Actas de las I Jornadas de Jóvenes en Investigación Arqueológica: dialogando con la cultura material*, 207–12. Madrid, OrJIA.

Kohn, E. (2013) *How Forests Think*. Los Angeles CA, University of California Press.

Lebreton, V., Messager, E., Marquer, L. and Renault-Miskovsky, J. (2010) A neotaphonomic experiment in pollen oxidation and its implications for archaeopalynology. *Review of Palaeobotany and Palynology* 162, 29–38.

López Garí, J. M., Pérez Jordà, G., Marlasca Martín, R., Farrera Fernandez, V. and Enrich Hoja, J. (2013) La primera agricultura pitiusa y balear: las evidencias de la Cova des Riuets. *Saguntum PLAV* 45, 66–77.

Mercuri, A. M., Florenzano, A., Massamba N'siala, I., Olmi, L., Roubis, D. and Sogliani, F. (2010) Pollen from archaeological layers and cultural landscape reconstruction: case studies from the Bradano Valley (Basilicata, southern Italy). *Plant Biosystems* 144, 888–901.

Miró, C. and Molist, N. (1990) Elements de ritual domestic al poblat ibèric de la Penya del Moro (Barcelona). *Zephyrus* 43, 311–9.

Moffett, L. (1992) Cereals from a Bronze Age Storage Vessel at Torralba D'En Salort, Menorca, Spain. *Vegetation History and Archaeobotany* 1, 87–91.

Morehart, C. and Morell-Hart, S. (2013) Beyond the ecofact: toward a social paleoethnobotany in Mesoamerica. *Journal of Archaeological Method and Theory* 20(3), 483–511.

Nadal, P. (2005) Aproximación cronotipológica a la matreialidad del postalayótico mallorquín: El ajuar funerario no cerámico. Dissertation (MA), Universitat Autònoma de Barcelona.

Pérez, G. (2009a) Estudio carpológico. In J. Aramburu (ed.) *Ses Païsses (Artà, Mallorca). Excavaciones en el Edificio 25 'Guillem Garau'*, 20–35. Palma, Arqueobalear.

Pérez, G. (2009b) Análisis carpológico. In J. Aramburu (ed.) *Excavaciones en el talayot de Son Fred*, 169–79. Palma, Arqueobalear.

Picornell, Ll. (2012) Paisaje vegetal y comunidades prehistóricas y protohistóricas en Mallorca y Menorca: una aproximación desde la antracología. Unpublished PhD thesis, University of Barcelona.

Pollock, S. (ed.) (2015) *Between Feasts and Daily Meals. Towards an Archaeology of Commensal Spaces*. Berlin, Topoi Excellence Cluster.

Ramis, D. (2006) Estudio faunístico de las fases iniciales de la Prehistoria de Mallorca. Unpublished PhD thesis, Universidad Nacional de Educación a Distancia.

Reille, M. (1992–98) *Pollen et Spores d'Europe et d'Afrique du Nord. Atlas Photographique*. Marseille, CNRS.

Servera-Vives, G. (2009) *Usos, simbolisme i significat de les plantes en els rituals funeraris de la Cova des Pas (Ferreries, Menorca), a partir de l'estudi pol•línic i d'altres palinomorfs*. Dissertation (MA), University of the Balearic Islands.

Schweingruber, F. H. (1990) *Anatomy of European Woods. An Atlas for the Identification of European Trees, Shrubs and Dwarf Shrubs*. Stuttgart, Verlag Paul Haupt.

Stika, H. P. (1999) Los macrorestos botánicos de la Cova des Càrritx. In V. Lull, R. Micó, C. Rihuete and R. Risch (eds.) *Ideología y Sociedad en la Prehistoria de Menorca. La Cova des Càrritx y la Cova des Mussol*, 521–31. Barcelona, Consell Insular de Menorca.

Twiss, K. C. (2008) Transformation in an early agricultural society: Feasting I the southern Levantine Pre-Pottery Neolithic. *Journal of Anthropological Archaeology* 27, 418–22.

Valenzuela, L., and Valenzuela, S. (2013) La fauna del Navetiforme I de Closos de Ca'n Gaià (Mallorca). *Arqueología y Territorio* 10, 13–26.

Valenzuela, A. (2015) La gestió dels recursos animals en la integració de les Illes Balears en el món romà. Unpublished Masters dissertation, University of Barcelona.

Van der Veen, M. (2003) When is food a luxury? *World Archaeology* 34(3), 405–27.

Veny, C. (1968) *Las Cuevas Sepulcrales del Bronce Antiguo de Mallorca*. Madrid, Consejo Superior de Investigaciones Científicas.

Vuorela, I. (1973) Relative pollen rain around cultivated fields. *Acta Botanica Fennicia* 102, 1–27.

Chapter 12

Animals and Worldviews: A Diachronic Approach to Tooth and Bone Pendants from the Mesolithic to the Medieval Period in Estonia

Tõnno Jonuks and Eve Rannamäe

Introduction

Animal pendants made of teeth and bones are among the most common ornaments in archaeological contexts worldwide. They are found from the Upper Palaeolithic onwards, from all regions occupied by humans. In spite of this, it is remarkable how rarely these pendants have been subject to wider analyses – both in Estonia and elsewhere – being limited mostly to a brief description and species identification. This can be partly explained by the fact that in many studies, artefacts with very clear anthropogenic origins are the principal focus (*e.g.* traditional archaeological seriation). Therefore, animal pendants together with other natural objects with little anthropogenic modification are often overlooked (see also Gilchrist 2008, 124). Any further discussion is also limited by a common and oversimplified explanation of pendants being merely 'amulets', which, however, occasionally results in overinterpretations of animal cults.

There are several case studies on animal pendants with both religious and social interpretations in the Baltic Sea region (Larsson 2006; O'Shea and Zvelebil 1984; Rainio and Mannermaa 2014; Zagorska and Lõugas 2000), but no large-scale temporal studies have yet been undertaken. This paper aims to provide a diachronic overview of animal tooth and bone pendants in the area that is nowadays Estonia, covering a wide time span from the first human inhabitation in the Mesolithic (*c.* 8900 BC) to the end of the Middle Ages (*c.* AD 1550). As this kind of *longue durée* study is undoubtedly a challenging task, the focus of this paper is not the detailed interpretations but presenting an overview of which kind of tooth and bone pendants have been used and how their distribution has changed through time. By presenting the changing tendencies in species preferences and possible reasons behind them, we aim to examine how animal pendants have not just been decorations and personal amulets, but also socio-cultural markers. Being the most numerous finds related with ideology and worldview, they can be used as sensors for following changes in the symbolic world.

Sources and methods

The main criteria for the inclusion of teeth and bones in this study were either the presence of an anthropogenic modification in the form of an attachment, or contextual information indicating relevant cultural importance. The first criterion was easily applicable, as the attachment of a pendant is usually identifiable as a drilled or carved piercing, or a notch. However, past human-animal relationships could also have been expressed with animal parts lacking any external signs of modification. In these cases, it would be hard to differentiate those from other faunal remains (like food waste), and therefore, evaluation of the archaeological context is needed. This is the case, for example, in the 12th–13th century male burial at Kukruse, northeastern Estonia, where the symbolic meaning of an unmodified eagle claw is indicated only by its location among other gravegoods. For the majority of Estonian pendants, detailed contextual information, including precise dating, is unfortunately limited. This deficiency applies especially to the Stone Age finds that were collected during large excavations in the Soviet period. Furthermore, contrasting scales of the undertaken archaeological fieldwork complicate the comparison between different collections and may bias the interpretations. For instance, a large number of tooth pendants from the extensively studied Stone Age settlement-cemetery of Tamula I and the Late Iron Age hillforts of Lõhavere and Otepää, inevitably dominate in the analysed temporal cohorts.

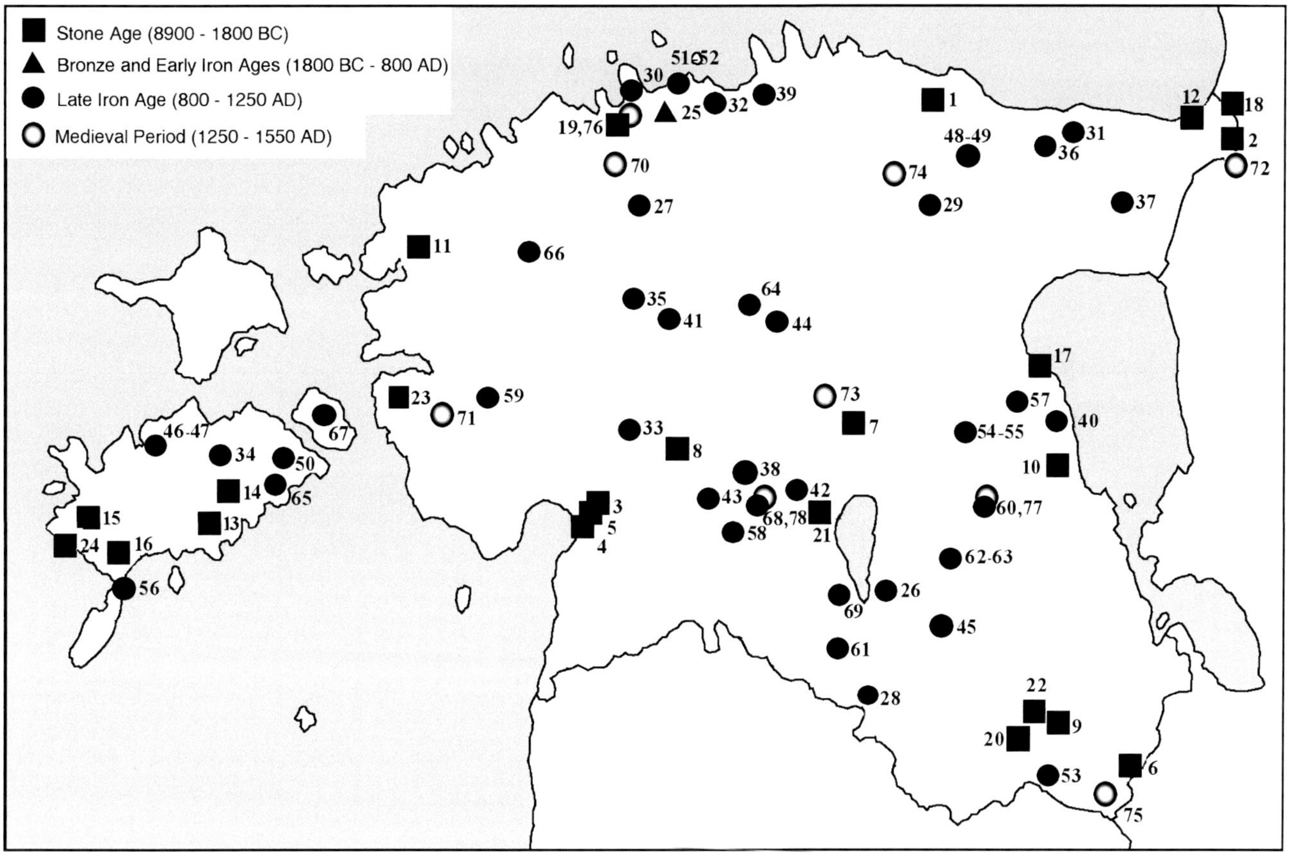

Figure 12.1 Archaeological sites (n = 78) of the analysed tooth and bone pendants (n = 1316) in Estonia. For the names of the sites see Table 12.1. Map by A. Kuperjanov.

The current research is mostly based on actual artefacts in archaeological collections, studied within the project *Materiality of religion: Religion-related artefacts in Estonian archaeological collections* (see Jonuks *et al.* 2013). For the rest of the finds that have gone missing during the decades since their excavation or have unclear documentation – especially those from the Soviet period – available information from previous publications has also been included. Our aim was to record all known pendants, excluding any 'unclear' finds, such as tooth fragments without visible attachment (although documented as artefacts in the collections) or those with an unknown archaeological context. As these 'unclear' finds are not numerous, we consider our dataset of 1316 pendants (1270 teeth and 46 bones – identified by taxon and skeletal element) from 78 archaeological sites all over Estonia comprehensive and sufficient to address the research questions posed above (Fig. 12.1; Table 12.1 for sites mentioned in the text). It must be noted though, that the number of tooth and bone pendants is actually higher, but here we have enlisted only those of a whole unmodified skeletal element (except for the piercing): incisors, canines, molars, astragali, claw bones, bird carpometacarpii and tarsometatarsii, and a sturgeon scute. We have excluded dozens of Early Neolithic beads made of birds' tubular bones (*e.g.*, in the settlement/burial site Tamula I), which have been manufactured by cutting off both ends of a long bone, leaving only a tubular shaft.

A diachronic overview

Stone Age

Stone Age sites are divided between four archaeological periods: Early Mesolithic (8900–5500 BC) – pre-ceramic hunter-gatherer societies; Late Mesolithic (5500–4000 BC) – hunter-gatherer societies with ceramic vessels; Early Neolithic (4000–2500 BC) – hunter-gatherer societies with the first domesticated plants and wider trading networks; Late Neolithic (2500–1800 BC) – societies with small-scale animal husbandry and agriculture used side by side with hunting and gathering.

From the Mesolithic, a large number of pendants are known (n = 119 for the Early and n = 128 for the Late Mesolithic), but the bulk of all pendants in Estonia come from the Early Neolithic (n = 800). The pendant finds from these periods largely show a similar pattern, that of being related to inhumation burials and settlements with mixed inhumation contexts. The most significant assemblage is from the largest Stone Age cemetery in Estonia – Tamula I (*c.* 4600–2600 BC), which alone yields more than a third of the current dataset (n = 541; Fig. 12.2.2, 12.2.3 and Plate 20).

Table 12.1 Tooth and bone pendants (n = 1316) by time period, archaeological site and identified taxon. Archaeological collections are those of the Tallinn University (AI), Estonian History Museum (AM), Estonian National Museum (ERM), Harju County Museum (HMK), Narva Museum (NLM), Pärnu Museum (PäMu), Rakvere Museum (RM), Saaremaa Museum (SM), Tallinn City Museum (TLM), Tartu City Museum (TM), the University of Tartu (TÜ), Valga Museum (VKM), Viljandi Museum (VM), and Võru Museum (VK).

Time period	*No.*	*Site*	*Site type*	*Collection no.*	*elk (Alces alces)*	*wild horse (Equus ferus)*	*dom. horse (Equus caballus)*	*dog (Canis lupus familiaris)*	*wolf (Canis lupus)*	*dog/wolf*	*fox (Vulpes vulpes)*	*canid*	*lynx (Lynx lynx)*	*wild cat (Felis silvestris)*	*carnivore*	*bear (Ursus arctos)*	*wild boar (Sus scrofa)*	*pig (Sus scrofa dom.)*	*wild boar/pig*	*mustelids*[1]	*beaver (Castor fiber)*	*white hare (Lepus timidus)*[2]	*cattle (Bos taurus)*	*aurochs (Bos primigenius)*	*seal (Phocidae)*	*chicken (Gallus gallus dom.)*	*birds of prey*[3]	*sturgeon (Acipenser sturio)*	*unidentified*	*Total pendants*
Early Mesolithic 8900–5500 BC	1	Kunda	settlement	AI 3308, 3359, 3410, 4284	12			1	1						2	2	1												1	20
	2	Narva-Joaoru	settlement/ burial	AI 4264	25			1									11								1				26	64
	3	Pulli	settlement	AI 4441, 4476	18	1			1	1										4	2								1	28
	4	Pärnu	settlement	PäMu 5 A 1596																									1	1
	5	Sindi Lodja	settlement	PäMu 15260 A 2553																									1	1
	6	Tsiistre	settlement	VK 1075	1																									1
	7	Umbusi	settlement	AI 4772											1		3													4
Total Early Mesolithic:																														*119*
Late Mesolithic 5500–4000 BC	8	Kivisaare	settlement/ burial	AI 2421, 4379; PäMu 81 A 2427; TÜ 1113				32	1	5					1		1			3					1					44
	9	Kääpa	settlement	AI 4245; VK A 16; VKM A 18	47	1		3	1							1	17			5					5				4	84
Total Late Mesolithic:																														*128*
Early Neolithic 4000–2500 BC	10	Akali	settlement	AI 4013	1											2	2			1										6
	11	Jalukse	burial	AI 2659				9												3										12
	12	Kudruküla	settlement	NLM 1304	9			1				1				1	2			7	1				2					24
	13	Kõljala	burial	AI-K 35										2						2					1					5
	14	Kõnnu	settlement/ burial	AI 4951	65	1			1	1	2						17							5	22					114
	15	Loona	settlement	AI 4129, 4210	5						1	1			1					6				1	50					65

(Continued)

Table 12.1 Tooth and bone pendants (n = 1316) by time period, archaeological site and identified taxon. Archaeological collections are those of the Tallinn University (AI), Estonian History Museum (AM), Estonian National Museum (ERM), Harju County Museum (HMK), Narva Museum (NLM), Pärnu Museum (PäMu), Rakvere Museum (RM), Saaremaa Museum (SM), Tallinn City Museum (TLM), Tartu City Museum (TM), the University of Tartu (TÜ), Valga Museum (VKM), Viljandi Museum (VM), and Võru Museum (VK). (Continued)

Time period	*No.*	*Site*	*Site type*	*Collection no.*	*elk (Alces alces)*	*wild horse (Equus ferus)*	*dom. horse (Equus caballus)*	*dog (Canis lupus familiaris)*	*wolf (Canis lupus)*	*dog/wolf*	*fox (Vulpes vulpes)*	*canid*	*lynx (Lynx lynx)*	*wild cat (Felis silvestris)*	*carnivore*	*bear (Ursus arctos)*	*wild boar (Sus scrofa)*	*pig (Sus scrofa dom.)*	*wild boar/pig*	*mustelids*[1]	*beaver (Castor fiber)*	*white hare (Lepus timidus)*[2]	*cattle (Bos taurus)*	*aurochs (Bos primigenius)*	*seal (Phocidae)*	*chicken (Gallus gallus dom.)*	*birds of prey*[3]	*sturgeon (Acipenser sturio)*	*unidentified*	*Total pendants*
	16	Naakamäe	settlement/ burial	AI 4211	10			1									3			1					3					18
	17	Omedu	settlement	unknown[4]						1							1													2
	18	Riigiküla III	settlement	AI 4198				1																						1
	19	Tallinn	settlement	AI 6917		1																			1					2
	20	Tamula I	settlement/ burial	AI 3932, 3960, 4118; VK 1088; VK A 3, 7, 8, 14	143			9	3	2	12				5	11	68			220	2	1		3	50		1		11	541
	21	Valma	settlement/ burial	AI 4022	2			2			1						2			2										9
	22	Villa III	settlement	AI 4037	1																									1
Total Early Neolithic																														*800*
Late Neolithic 2500–1800 BC	23	Kaseküla	settlement	AM 807	1															1					1					3
Total Late Neolithic																														*3*
Stone Age (unspecified period)	24	Tagamõisa	burial	AI 1396																					2					2
Total Stone Age (unspecified period):																														*2*
Bronze Age/ Early Iron Age 1800 BC–AD 800	25	Jõelähtme	burial	AI 5306				1																						1
Total Bronze Age/Early Iron Age:																														*1*

(Continued)

Table 12.1 Tooth and bone pendants (n = 1316) by time period, archaeological site and identified taxon. Archaeological collections are those of the Tallinn University (AI), Estonian History Museum (AM), Estonian National Museum (ERM), Harju County Museum (HMK), Narva Museum (NLM), Pärnu Museum (PäMu), Rakvere Museum (RM), Saaremaa Museum (SM), Tallinn City Museum (TLM), Tartu City Museum (TM), the University of Tartu (TÜ), Valga Museum (VKM), Viljandi Museum (VM), and Võru Museum (VK). (Continued)

Time period	*No.*	*Site*	*Site type*	*Collection no.*	*elk (Alces alces)*	*wild horse (Equus ferus)*	*dom. horse (Equus caballus)*	*dog (Canis lupus familiaris)*	*wolf (Canis lupus)*	*dog/wolf*	*fox (Vulpes vulpes)*	*canid*	*lynx (Lynx lynx)*	*wild cat (Felis silvestris)*	*carnivore*	*bear (Ursus arctos)*	*wild boar (Sus scrofa)*	*pig (Sus scrofa dom.)*	*wild boar/pig*	*mustelids*[1]	*beaver (Castor fiber)*	*white hare (Lepus timidus)*[2]	*cattle (Bos taurus)*	*aurochs (Bos primigenius)*	*seal (Phocidae)*	*chicken (Gallus gallus dom.)*	*birds of prey*[3]	*sturgeon (Acipenser sturio)*	*unidentified*	*Total pendants*
Late Iron Age 800–1250 AD	26	Aakre Kivivare	settlement/ hillfort	AI 4726; TÜ 1928; VKM A 29				2													11									13
	27	Angerja	settlement	AI 5368														1												1
	28	Hummuli	burial	AI 1979 V												1														1
	29	Inju	burial	AI 570				1																						1
	30	Iru	hillfort	AI 3428, 4051												2		1												3
	31	Järve	burial	AM 112												1														1
	32	Kaberla	burial	AI 4116				2														2			1				1	6
	33	Kadjaste	burial	AI 2596																									1	1
	34	Karja	burial	AI 4115														1												1
	35	Keava	hillfort	TÜ 1026																		1								1
	36	Kukruse	burial	TÜ 1777																2							1			3
	37	Kuremäe	burial	unknown[5]													2													2
	38	Kuude	burial	unknown[6]																									1	1
	39	Kuusalu	settlement	AI 5043, 5099							1					2				1			1						2	7
	40	Lahepera	burial	AI 4978																									1	1
	41	Linnaaluste III	settlement	TÜ 1115																							1			1
	42	Lõhavere	hillfort	AI 3578, 4133, 4297				5		1						4		11	2	2					1	3			7	36
	43	Mustivere	settlement	AI 3993				2								1	1													4
	44	Nurmsi	burial	AI 2533																									1	1
	45	Otepää	hillfort	AI 4036				2					1		1	8	2	7	1		4					1				27
	46	Paatsa	hillfort	AI 5982																					1					1
	47	Paatsa	settlement	AI 5983														1												1

(Continued)

Table 12.1 Tooth and bone pendants (n = 1316) by time period, archaeological site and identified taxon. Archaeological collections are those of the Tallinn University (AI), Estonian History Museum (AM), Estonian National Museum (ERM), Harju County Museum (HMK), Narva Museum (NLM), Pärnu Museum (PäMu), Rakvere Museum (RM), Saaremaa Museum (SM), Tallinn City Museum (TLM), Tartu City Museum (TM), the University of Tartu (TÜ), Valga Museum (VKM), Viljandi Museum (VM), and Võru Museum (VK). (Continued)

Time period	*No.*	*Site*	*Site type*	*Collection no.*	*elk (Alces alces)*	*wild horse (Equus ferus)*	*dom. horse (Equus caballus)*	*dog (Canis lupus familiaris)*	*wolf (Canis lupus)*	*dog/wolf*	*fox (Vulpes vulpes)*	*canid*	*lynx (Lynx lynx)*	*wild cat (Felis silvestris)*	*carnivore*	*bear (Ursus arctos)*	*wild boar (Sus scrofa)*	*pig (Sus scrofa dom.)*	*wild boar/pig*	*mustelids*[1]	*beaver (Castor fiber)*	*white hare (Lepus timidus)*[2]	*cattle (Bos taurus)*	*aurochs (Bos primigenius)*	*seal (Phocidae)*	*chicken (Gallus gallus dom.)*	*birds of prey*[3]	*sturgeon (Acipenser sturio)*	*unidentified*	*Total pendants*
	48	Pada	settlement	AI 5082				1									1											1		3
	49	Pada	burial	AI 5366; AM A 1036				3	2	1					1	1						2					2		1	13
	50	Pöide	hillfort	SM 1460															6						1					7
	51	Rebala	settlement	AI 5916														1									1			2
	52	Rebala	burial	AI 5490				1																						1
	53	Rõuge	settlement/ hillfort	AI 4040, 4100			1		1							4	1	1	1	1	6								4	20
	54	Saadjärve	settlement	AI 5307																	1									1
	55	Saadjärve	hillfort	AI 5308						1																				1
	56	Salme II	burial	SM 10602												2														2
	57	Savastvere	settlement	AI 1459																1										1
	58	Sinialliku	hillfort	AI 4407									1							1									1	3
	59	Soontagana	hillfort	PäMu 1971, 2434, 2765-2767				2		1	1					1		7	2										2	16
	60	Tartu	hillfort	TM 260																	1									1
	61	Tõrva Tantsumägi	hillfort	VM 9402 A31												1														1
	62	Unipiha	settlement	AI 4474																	1									1
	63	Unipiha	hillfort	AI 4472												1					1									2
	64	Valgma	burial	unknown[7]																									1	1
	65	Valjala	hillfort	AI 4300				1			1							8	1											11
	66	Varbola	hillfort	AI 4084, 4783, 5299; ERM 1941				1										5				1							9	16
	67	Viiraküla	burial	AI 1044					1																					1

(Continued)

Table 12.1 Tooth and bone pendants (n = 1316) by time period, archaeological site and identified taxon. Archaeological collections are those of the Tallinn University (AI), Estonian History Museum (AM), Estonian National Museum (ERM), Harju County Museum (HMK), Narva Museum (NLM), Pärnu Museum (PäMu), Rakvere Museum (RM), Saaremaa Museum (SM), Tallinn City Museum (TLM), Tartu City Museum (TM), the University of Tartu (TÜ), Valga Museum (VKM), Viljandi Museum (VM), and Võru Museum (VK). (Continued)

Time period	*No.*	*Site*	*Site type*	*Collection no.*	*elk (Alces alces)*	*wild horse (Equus ferus)*	*dom. horse (Equus caballus)*	*dog (Canis lupus familiaris)*	*wolf (Canis lupus)*	*dog/wolf*	*fox (Vulpes vulpes)*	*canid*	*lynx (Lynx lynx)*	*wild cat (Felis silvestris)*	*carnivore*	*bear (Ursus arctos)*	*wild boar (Sus scrofa)*	*pig (Sus scrofa dom.)*	*wild boar/pig*	*mustelids*[1]	*beaver (Castor fiber)*	*white hare (Lepus timidus)*[2]	*cattle (Bos taurus)*	*aurochs (Bos primigenius)*	*seal (Phocidae)*	*chicken (Gallus gallus dom.)*	*birds of prey*[3]	*sturgeon (Acipenser sturio)*	*unidentified*	*Total pendants*
	68	Viljandi	hillfort	VM 10741-10742, 10877, 10921, 10952, 11472					1				1			1		2	1		5									11
	69	Vooru	hillfort	VM 331																	1								1	2
Total Late Iron Age:																														*231*
Middle Ages 1250–1550 AD	70	Lehmja	settlement	AI 5310; HMK 6010						1						1		2												4
	71	Lihula	town	AM 583, 808				2											1											3
	72	Narva	town	NLM 2420															1											1
	73	Põltsamaa	castle	TÜ 714															1											1
	74	Rakvere	castle	RM A 4871, 4873												2														2
	75	Siksali	burial	AI 5101				4								3		1		1	1									10
	76	Tallinn	town	AI 6332; TLM 28149															3											3
	77	Tartu	town	TM 2126/ A51; TM A 91, 188, 202							1				1				2		1									5
	78	Viljandi	town	VM 10235, 10258, 10534												1			2											3
Total Middle Ages:																														*32*
				Total tooth and bone pendants:	*340*	*4*	*1*	*90*	*13*	*15*	*20*	*2*	*3*	*2*	*13*	*54*	*135*	*49*	*24*	*264*	*38*	*7*	*1*	*9*	*143*	*4*	*6*	*1*	*78*	***1316***

[1] Including pine marten (*Martes martes*), western polecat (*Mustela putorius*), badger (*Meles meles*) and otter (*Lutra lutra*).

[2] It is most probably the white hare (*Lepus timidus*), because the European hare (*Lepus europaeus*) arrived to Estonian area only as late as in the 16th century (Kirk 2003; Lepiksaar 1986).

[3] Including osprey (*Pandion haliaetus*), white-tailed eagle (*Haliaeetus albicilla*) and golden eagle (*Aquila chrysaetos*).

[4] Data from Roio *et al.* (2016, fig 5).

[5] Data from Ligi (1993, 128).

[6] Data from Luik (2005, tab 1).

[7] Data from Vassar (1943, 81).

During the Stone Age, there were four clearly dominant taxonomic groups of mammals used for pendants – elk, mustelids, seals, and wild boar; followed by canids, bear, aurochs, wild horse, beaver, wild cat, and hare. It seems that pendants were often (or perhaps in most cases) attached to head-gear and clothing, as is indicated by their position in the graves: they have been found frequently in rows or in closed groups on the body, or in connection with limbs and the head (see *e.g.* Larsson 2006; Zagorska and Lõugas 2000). There is also evidence of more complicated configurations of artefacts as is the case in Tamula I, where a bead of a bird's tubular bone was inserted into the hole of an elk tooth (Fig. 12.2.2), hence indicating an ornamentation where teeth alternated with bone beads (see also Brea *et al.* 2010, 137, fig 17). These beads of birds' tubular bones are quite abundant in Tamula I as well as in a few other Estonian Neolithic sites (Naakamäe and Kääpa), but do not seem to appear in any other time period. Besides the beads, only one presumable bird pendant – a claw of an osprey (without a man-made attachment) – is known from the Stone Age, again from Tamula I.

Late Neolithic societies started exploiting a variety of domestic animals and plants, but simultaneously tooth and bone pendants drop almost to nonexistent (n = 3).

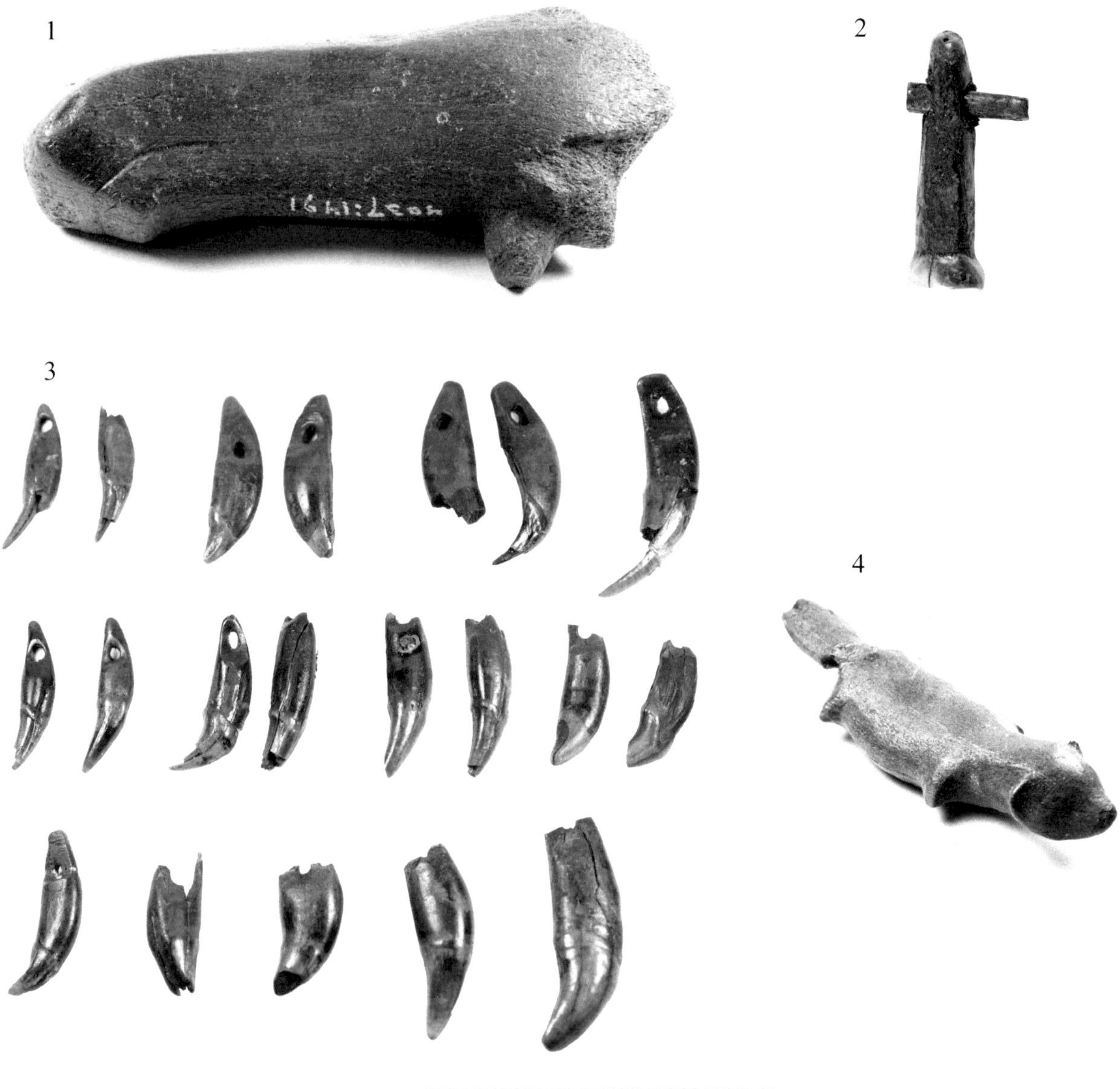

Figure 12.2 Selection of tooth pendants and animal figurines from the Stone Age: 1) antler figurine of an elk head from the settlement of Villa III (AI 4037: 1401), determined by the pintle to have originally been attached to a shaft; 2) elk incisor together with a bead made of a bird's tubular bone from the settlement-cemetery of Tamula I (AI 4118: 1407); 3) set of mustelid canines from the settlement-cemetery of Tamula I (AI 3960: 244); 4) antler figurine of a beaver from the settlement-cemetery of Valma (AI 4022: 5725).

Decrease in the number of pendants, together with changed burial customs and economies seem to have marked wider changes in religion, mentality, and worldview. In the course of such developments animal pendants apparently lost their symbolic meaning and significance (see below; also Jonuks 2009, 155ff). Based on archaeological evidence, this tendency continued into the following period (Bronze Age, 1800–500 BC), from which only a single pendant is known despite the fact that more than a hundred Late Bronze Age (850–500 BC) graves have been investigated. Similarly, there are no pendant finds that can be confidently dated to the following Early Iron Age (500 BC – AD 550) and Pre-Viking Age (AD 550–800) periods either.

Late Iron Age

Usage of animal pendants is introduced again only during the Viking Age (AD 800–1050) and becomes prominent in the Final Iron Age (AD 1050–1250), before Christianisation in the 13th century. Pendants from this period (n = 231) are distributed across the country, but have most commonly been recovered from central hillforts, such as Lõhavere, Otepää, and Rõuge. More than half of the Late Iron Age pendants (n = 139) come from hillforts and from the combined material of hillforts/settlements (n = 33), while settlements alone (n = 22) and burials (n = 37) yield significantly fewer finds.

Compared to the Stone Age, the number of pendants from the Late Iron Age is much smaller, yet the list of species is wider. Previous preferences for elk and mustelids were replaced by domestic pig, bear, beaver, and dog. Other species, like mustelids, wolf and fox, hare, wild boar, lynx, seal, cattle, and horse, were employed more rarely. It is interesting to note that not a single elk-tooth pendant has been found in Estonia after the Stone Age. Among birds, domesticated species (chicken) as well as birds of prey (white-tailed eagle and osprey) seem to have been valued for pendants; and this is also the period when the sturgeon scute originates (Fig. 12.3.7). In contrast to the Stone Age, only some dozens of tooth pendants from the Late Iron Age are known from burials, children's, males' and females', but being more numerous in the latter. There, among other modes, the pendants have been attached to impressive bronze chest decorations (*e.g.* Pada, Kaberla, Kukruse, and Viiraküla; Fig. 12.3.8). The apparent association between female burials and pendants, however, must be interpreted with caution, as most of the Late Iron Age finds come from mixed occupation layers on hillforts and settlements, making it difficult to speculate about their actual usage. Luik (2010a, 50) has stressed the possibility that during the Late Iron Age, wearing pendants could have been more valued during the person's lifetime and not in the afterlife, and that is why pendants are sparse in graves.

Medieval period onwards

Very few animal pendants (n = 32) are known from the medieval period (AD 1250–1550). These occur almost exclusively in towns and castles and only few examples come from rural settlements or burials. One reason for such uneven proportion could be the fact that the majority of studies during the past couple of decades have been carried out in urban contexts. However, those at rural sites have indeed revealed tooth pendants very rarely, and therefore, the overall distribution pattern might be similar to that of the Late Iron Age. With most of the finds coming from central hillforts and settlements (Late Iron Age) or towns and castles (Middle Ages), a connection between animal pendants and power centres may be suggested. Also, the main animals used in the Middle Ages – domestic pig, bear, and dog – remained the same as during the previous period. Regarding burials, the only medieval example comes from Siksali, southeastern Estonia, where ten pendants of pig, bear, dog, beaver, and otter have been found. Still, it must be noted that this site is exceptional in Estonia and represents rather a specific cultural area in southeastern Estonia and northeastern Latvia (Valk and Laul 2014, 180f). This area had more connections with Latvia and Lithuania, where tooth pendants were common in medieval burials until the 17th century (*e.g.* Griciuvienė and Vasiliauskas 2005, 213 and references therein).

From the end of the 16th century onwards, animal pendants are absent in the Estonian archaeological record, and, interestingly, also in ethnographical collections. This does not necessarily exclude the usage of tooth pendants, but considering the large-scale excavations of modern age layers in Estonian towns without any pendant finds, they must have been rare. There are several examples of animal pendants from the recent periods, but these come from outside Estonia. For example, in the early 20th century Finland, folk doctors, foretellers, and witches used pierced canines and bear claws (Stark 2015, 137f; see also Choyke 2010, 202 and references therein).

The animals

Almost every wild species in Estonia has been used for pendants – some taxa are represented more or less continuously throughout the history, while others occur only in certain periods. Domestic animals have been favoured as well, but overall to a lesser extent. This pattern suggests that wild animals could have held more ideological symbolisms derived largely from the wider religious and mythological background, while the ones living close by and controlled by humans had, although not necessarily exclusively, more practical value. To explore this hypothesis, each species will be discussed separately.

Elk: the king of the woods

The chronological distribution of elk-tooth pendants is particularly interesting. They appear first in the Early Mesolithic, become one of the most numerous pendant types during the Late Mesolithic and Early Neolithic, and then disappear completely. It is significant that although

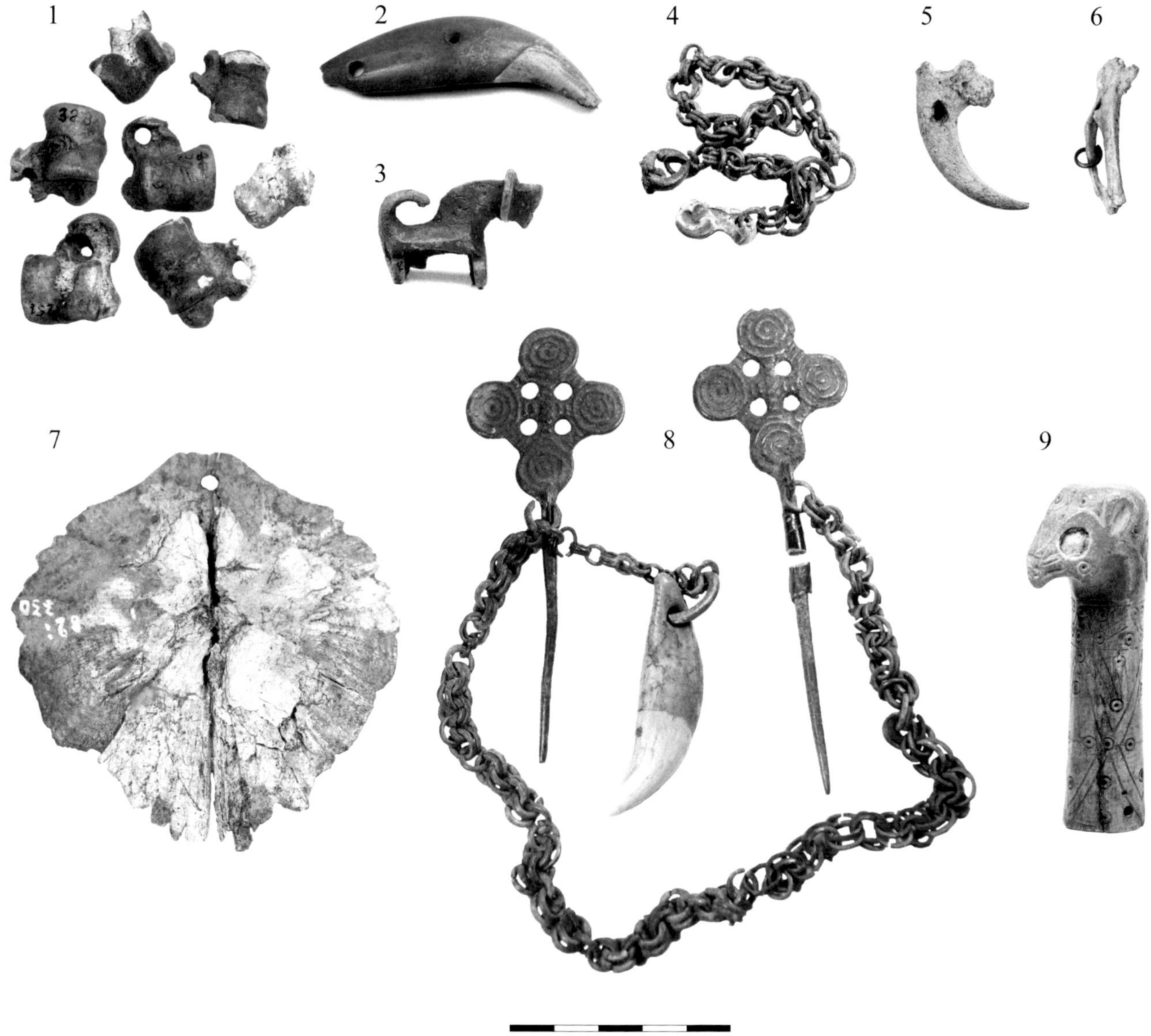

Figure 12.3 Selection of animal pendants from the Late Iron Age: 1) beaver astragali from the settlement of Aakre Kivivare (VKM A 29); 2) bear canine with double piercing from the cemetery of Hummuli (AI 1979 V: 19); 3) dog-shaped figurine from the stone grave of Ehmja (AM A 554: 777); 4) hare astragalus with a bronze chain from the cemetery of Pada (AI 5366 LXXX: 1); 5) eagle claw bone from the settlement of Linnaaluste III (TÜ 1115: 188); 6) chicken wing bone (carpometacarpus) from the hillfort of Lõhavere (AI 4133: 1225); 7) sturgeon scute from the settlement of Pada (AI 5082: 330); 8) female jewellery set together with a wolf canine from the cemetery of Viiraküla (AI 1044: 1); 9) eagle-headed handle from the hillfort of Lohavere (AI 3578. 526).

elk was continuously hunted in the later periods and up to the modern times (Paaver 1965, 257), it was not chosen for pendants anymore.

Large numbers of elk-tooth pendants have given rise to different and complicated interpretations. The popularity of elk has been explained functionally, as the shape of its teeth make it relatively easy to drill or carve a hole in the flat root (Wood 1957, 382). Such utilitarian explanation, however, cannot be universal, as elk pendants are missing in later periods when the animal was still hunted. Pendants could also have been attached to clothing just for decoration with 'no studied care' (Wood 1957, 384). Considering the wider archaeological context though, we still may claim that elk had a very special role in the religion and worldviews of the Stone Age. Elk was one of the three most used creatures (besides human and bear) in Eastern Baltic Stone Age art as is reflected by elk-headed staffs from *c.* 7000–1500 BC (Fig. 12.2.1; Carpelan 1977; Iršenas 2000), and by rock carvings all over northern Europe, where depictions of ritual processions led by a human, holding high the same staff, have been presumed (*e.g.* Helskog 2004, 278). Elk-headed staffs have often been found from settlements but in some occasions also from burials, for example in the Oleni island cemetery at Onega lake, northwest Russia (Gurina 1956, 214). The equivalent to elk in southern areas is the red deer, one of the most common animals in central and southern Europe for tooth pendants (Choyke 2001). It is interesting to note that red deer was present also in Estonia during the climate optimum in the Late Mesolithic, but its teeth were never chosen for pendants. Osteological remains of red deer have been found in several sites together with multiple elk-tooth pendants (*e.g.* Tamula I, Kääpa; Paaver 1965). Both species represent similar phenomena – they are among the largest and most attractive animals in the local ecosystems and at the same time important game animals for meat, fur, bone, and antler. Yet, it seems that in the ideology of the northern forest zone, elk was the one occupying a dominant position.

Primarily, elk pendants could be related to a sort of hunting magic, possibly linked to achieving good hunting success, or they could have been worn as trophies of a successful hunt. However, considering their dominant role and the relative scarcity of other game among pendants, it is likely that elk reflects some particular ideology (for a detailed study about game animals in Estonia see Lõugas and Maldre 2000; see also Paaver 1965). Ethnographic analogies from the early 20th century indicate, for instance, that among the Blackfoot tribes in northern America, elk-tooth pendants were used as means to obtain general health and well-being (Wood 1957, 384). Another potential interpretation could be related to a myth of the master of nature, still narrated in present-day northern Eurasia, whereby a spirit rules the nature and takes the form of a specific animal, such as bear, elk, or raven (Antanaitis 1998; Jonuks 2009; Willerslev 2007; Zvelebil and Jordan 1999). The spirit is responsible for the well-being of nature, but it also guarantees the animal for the hunter as a gift. We do not wish to claim here that this Eurasian myth could originate directly from the Palaeolithic or Mesolithic; we propose that the number of elk pendants, the presence of rock carvings, as well as the probable ritual staff heads would fit rather well into this concept of a nature guarding spirit, represented in the form of an elk.

Bear: depiction of danger and prestige

Bear was as important creature in the Neolithic northern European iconography as elk, and might also be connected to the myth of the master of nature described above. Bear, however, appeared in art considerably later (since *c.* 3000 BC) and Carpelan (1977, 41) has even suggested that the bear 'cult' developed from the elk 'cult'. This idea is in accordance with the appearance of bear-tooth pendants in Stone Age Estonia: they were absent in the Early Mesolithic (even though bears were hunted; Kriiska 2004, 24), then appeared in the Late Mesolithic, and were the most numerous during the Early Neolithic. Distinctly bear-orientated regions are northern Scandinavia, Finland, and Karelia, where most of the bear figurines and bear burials come from (see Fossum 2006; Jennbert 2003; Zachrisson and Iregren 1974). Zvelebil (2003, 12) has suggested a ritual practice of separating bears' crania and jaws from the body, possibly associated with bear funerals. A kind of specific treatment of bears' heads or skulls is very likely at some Estonian sites as well. For example, in the Mesolithic site Kääpa, 60 bear skull fragments have been found, whereas usually in the faunal assemblages there are just a few specimens (Paaver 1965, tab. 21).

During the Late Iron Age, bear was one of the most numerous animals used for pendants (n = 30). While being present in every site type, most of the bear pendants come from hillforts (n = 18) and only few from settlements (n = 3) or burials (n = 5). Some of these pendants carry bronze rings or traces of those, indicating that they were hung on a chain or attached to a decoration. There are two dominant interpretations for the bear pendants. Firstly, as shown in medieval literature and contemporary ethnography, bear has often been associated with aggression and male power (Tolley 2006 and references therein). The second interpretation for bear canines is being a protective magical amulet, according to the presence of such finds in burial contexts, which, notably, are exclusively associated with females and children (Fig. 12.3.2; Asplund 2005; Kivisalo 2008). While these interpretations could be rather gender/ age biased, there is another meaning that should not be ignored – namely, of being a valued trophy. To be able to disentangle the potential meaning of the acquisition of these pendants, their full contextual background, including associations with other finds, needs to be taken into account (see below), but it is likely that the Late Iron Age communities were inspired by a combination of the above-mentioned reasons.

Mustelids and beaver: connecting land and water

Mustelids are the most numerous group of animals used for pendants during the Neolithic. As mustelid bones are rather small and unsuitable for processing, and the meat of carnivores is usually not particularly valued, the main reason for hunting them was probably fur. However, this could not have been the only reason for using their teeth in such high numbers during the Stone Age. Fur and its trade were important also in the Bronze and Early Iron Age, but there are no such pendants known from these periods. Furthermore, compared to elk and bear teeth, those of mustelids are tiny, which renders making the fastening more complicated (Fig. 12.2.3). Hence, there must have been some strong cultural reason for favouring them. One possible cause for ascribing certain symbolic meaning to those animals might have been their specific gait and hiding habits. Mustelids are very agile, and some of them – especially otter, but also polecat – are active both on land and in water. Considering the general importance of rivers, they could have represented a medium between different worlds – a role that has been previously assigned to waterfowl (Zvelebil 2003, 18). Similar ideological significance could also be discussed in relation to badger with its striated face and underground setts. Across Europe, weasels have been related to midwives, female healers and witches, and in some cases even believed to be a reincarnated witch herself (Bettini 2013).

After a long gap since the Neolithic, mustelid canines reappeared as pendants only in the Late Iron Age, but in much lower numbers and with lower species diversity. Most of those are from otter, and were often attached to a bronze chain, obviously being on display. Here the connection with fur trade would be more plausible, as this is also the case with beaver pendants of the same period (see below).

Beaver has been hunted from the Early Mesolithic onwards (Kriiska 2001, fig. 5), but in contrast to mustelids, beaver bones and teeth were rarely used for pendants during the Stone Age (n = 5). It seems that the importance of this animal was demonstrated with figurines instead (Fig. 12.2.4; Ots 2010). Only in the beginning of the Viking Age, beaver gained an important role in terms of adornments. The astragalus pendants from the Late Iron Age settlement sites in Rõuge, Viljandi, and Kivivare (n = 31) are particularly prominent examples of this practice (Fig. 12.3.1). The social importance of beaver during that period has been discussed at length (see Leimus and Kiudsoo 2004; Luik 2010a), including speculations on a beaver cult (Tvauri 2001, 161). The latter is supported by beaver paws made of clay from the Åland Islands and Russia, astragalus pendants made of bronze from Latvia (Luik 2010a, 49 and references therein), and – as for the Stone Age mustelids – a possibility of being a medium between different worlds. Nevertheless, it seems that the Late Iron Age beaver pendants in Estonia could also reflect their utilitarian value, being connected with hunting and trade. The distribution of these pendants – two from settlements, 12 from hillforts and 17 from mixed assemblages of hillforts/settlements – is remarkable, as all examples come from southeastern Estonia and are associated with a large-scale trade of beaver fur and possibly of castoreum (the latter as a mean for increasing virility) during the Viking Age eastern Europe (Leimus and Kiudsoo 2004). The relatively large number of beaver bones among faunal waste from this period further supports the idea of being an economically valued mammal (Paaver 1965, 62; Rannamäe and Lõugas in press). Nonetheless, in spite of the economic importance, such a focused hunt might still have increased the visibility of this animal and created specific connections and beliefs, which were materialised in beaver pendants or replicas of beaver parts (see also Callmer 1994, 31; Luik 2010a, 52). It is interesting to note that unlike otters, beaver pendants do not have any signs of contact with bronze ornaments, which indicates a different manner of wear. Differences in nuances are well illustrated by pendants made of beaver astragali in lower reaches of Daugava River in Latvia. The majority of the pendants there are marked with green bronze oxide and originate mostly from burial contexts (Zarina 2006). Thus, despite using and valuing the same animal, the details of its regional usage are significantly different.

Dogs and wolves: the confrontation of wild and domestic

The interpretations of the use of dog and wolf must be diametrically different: one was a domesticate that probably hunted with humans, and the other was its wild counterpart, potentially representing danger from the woods. However, in many instances the two cannot be confidently identified to species level, thus detrimentally affecting the interpretative potential of such pendants. Dog was the first domesticate in Estonia and it is possible that its symbolic role was primarily connected with the phenomenon of domestication – dog as the closest animal to a dangerous predator in the forest could have been perceived as a sort of medium between human environment and the wilderness (Jonuks 2006, 41). This could be the reason for 60 dog-tooth pendants recvovered from Stone Age sites, whereas wolf is represented with only eight. It must be noted though, that 32 of these dog teeth are from a single site – the settlement and cemetery of Kivisaare.

During the Late Iron Age, dog and wolf pendants (n = 32) form quite a large group, with only pig being more numerous. Most of those which could be identified are, again, dog (n = 23), and only five are wolf. Of the latter, three are from the Pada and Viiraküla cemeteries, and similarly to bear, occur only in female burials (Fig. 12.3.8). Dog, on the other hand, represents a more complicated phenomenon, as this species has a more diverse appearance in the material world. In addition to tooth pendants, there are also dog burials (Mägi *et al.* 1998), in which dogs were executed and buried together with their owner (see the description of a Viking burial by ibn Fadlan, Abu-Chacra 2004). Other dog-symbolism is seen in small bronze figurines suggested to be dogs (Fig. 12.3.3). The figurines with their large ears and back-arching tail (Jonuks 2006) are mostly the same weight, and have thus been interpreted as weight scales (Kiudsoo and Russow 2011). On the other hand, some of them were hung on a bronze chain, which indicates their usage as decorations and not functional objects. In sum, dog definitely had a particular role in the Late Iron Age mentality and value system, including being part of an important elite tradition of hunting and noble lifestyle (Sørensen 2013).

Other game

Wild boar pendants – of canines and lower incisors – are prevalent in the Stone Age (n = 128), whereas very few occur in the Late Iron Age (n = 7). Most of those from the Stone Age are lower incisors and only a dozen are canines. The relative abundance and similar shape of the incisors suggest that they were worn in the same manner as elk incisors and possibly carried a similar meaning.

Seals were mainly important during the Stone Age. As expected, most of the seal-tooth pendants come from the coastal sites in Saaremaa Island – Loona (n = 50) and Kõnnu (n = 22) – but there is also a considerable number (n = 50) from the southeastern inland site Tamula I, more than 150km from the sea. Only four pendants have been found from the Late Iron Age burials and hillforts.

With only single examples are represented lynx (n = 3) from the Late Iron Age hillforts and fox (n = 16 from the Stone Age and n = 4 from the Late Iron Age/medieval sites). Particularly the Late Iron Age examples associate well with the general choices of the period, when predators, like bear, were preferred.

Hare is not an animal to which rich symbolism is usually ascribed. One pendant made of a hare incisor has been found at Tamula I, but the exact context is unknown. Surprisingly, although few in number, hare astragali were used for pendants during the Late Iron Age. It is noteworthy that four of those derive from the breast ornaments of two burials of 30–40-year-old females in Kaberla and Pada cemeteries (Fig. 12.3.4). Another two are from the hillforts of Keava and Varbola, the one from the latter attached to a short bronze chain together with a dog canine (Tamla and Maldre 2001). Clearly these astragali were on display – the pattern seen here is similar to that of the Late Iron Age burials, where otter canines were attached to clothing ornaments.

Other animals like wild horse and especially aurochs were hunted during the Stone Age (Paaver 1965, 180, 309), but for some reason not favoured for pendants. There are only four horse incisiors from the entire Stone Age, and nine incisors of an aurochs from the Early Neolithic. Roe deer is also worth mentioning as it has been hunted in every period (though not in great numbers; Paaver 1965, 218ff), but not a single pendant of this species is known. Both examples illustrate well that the symbolic meanings were associated only with selected species and not with every utilised animal.

There are also unique cases that are difficult to interpret and most probably represent a single event. This could apply to a Viking Age sturgeon scute (Luik and Maldre 2005, 272), as catching such a large fish must have been a great challenge, which was later materialised as a pendant (Fig. 12.3.7). The only two known pendants of a wild cat, found at the Stone Age Kõljala burial site, could also indicate a personal connection with that particular animal.

Familiar domesticates

Domesticated animals have rarely been chosen for pendants: there is only one horse-tooth pendant from the Late Iron Age site of Rõuge and one probable cattle incisor from the Late Iron Age Kuusalu settlement. Besides dog (see above), the most numerous domesticate used for pendants was pig. Although the differentiation between the teeth of wild boar and domestic pig can be complicated, the latter was clearly favoured during the Late Iron Age and in the medieval period (n = 49 for pig; n = 7 for wild boar; n = 24 for wild boar/pig). It is interesting to note that while during the Stone Age incisors were preferred, in the later periods canines were favoured. It is also remarkable that most of the Late Iron Age pig canines come from hillforts and only single samples are known from burials and settlements – a distribution that resembles the one of bear canines. Pig and boar pendants have been rarely discussed in Estonia. Arukask (1999) has briefly considered the connection between modern Estonian folk tradition and the ritual role of wild boar mentioned in the description of *aestii*s in Tacitus's *Germania*, but the symbolic role of domestic pig has largely been ignored. It seems likely that the symbolic meaning of wild boar passed to the domestic pig, as suggested for southern and western regions (Kajkowski 2012 and references therein).

Birds: chicken wings and grand eagles

In addition to the beads of birds' tubular bones mentioned above, only one bird find relevant to this study – an osprey's claw bone – is known from the Stone Age. It is without any man-made attachment, but the protuberance of the claw would allow hanging it without special piercing. In the current state of research, it is difficult to prove whether this claw was a pendant or not, but as it comes from the Tamula settlement/burial site, it might have had at least some symbolic meaning.

Late Iron Age bird pendants can be divided into two groups. The first includes the only known domesticated bird of that period in Estonia – chicken: three carpometacarpal bones (of a wing) come from Lõhavere hillfort and had clearly been on display once, because they were hung on a bronze chain (Fig. 12.3.6); and the fourth find is a tarsometatarsus with a spur (presumably of a male individual) from the Otepää hillfort. The second group is that of eagles, and as demonstrated by several contemporaneous eagle-headed handles (Fig. 12.3.9; Luik 2010b), these birds obviously had some wider symbolic meaning. While three eagle claws have piercings (Fig. 12.3.5), two are without any attachment, but the fact that they are grave goods, supports their interpretation as pendants. In our dataset, all eagle claws come from male burials, which is in contrast to the previously discussed female burials with bear and wolf teeth. This male focused concept concurs with two hawk skeletons found in a mass grave of Scandinavian warriors in Salme, Saaremaa Island, dated to *c.* AD 750 and interpreted as evidence of falconry (Maldre *et al.* in press).

Final discussion and conclusions

Despite the apparent variability in practices and diverse selection of animals, our analysis has shown that certain patterns emerge, highlighting significant differences between time periods. In particular, there are two clearly distinct groups of animal pendants based on chronology and species preferences: the Stone Age with elks, mustelids, seals, and wild boars; and the Late Iron Age and, to a lesser extent, medieval period with pigs, bears, beavers, and dogs. These groupings support the hypothesis that the choice of animals for pendants was ideological and that not all animals were suitable for that purpose. As many species were hunted, but not all of them were used for pendants or they were used only on single occasions, the practice cannot be explained as solely related to general hunting magic. Considering the distribution of pendants in two chronological groups with a clearly different selection of species, the universal interpretation that humans wore pendants to attain animal powers and abilities, can be largely discounted as well. Of course, neither hunting nor the desire for animal attributes should be entirely excluded, especially considering unique and single cases. But, for example, while the mustelids and elk pendants have a dominating role among Stone Age material, it is highly unlikely that a large part of society wanted to gain the properties of such a narrow selection of animals. Instead, more variability in the list of species would be expected.

So, how can these groupings be explained? In northern Eurasian historic hunter-gatherer societies, pendants – both teeth and the ones shaped like humans or animals – have been often associated with a sort of shamanism-like religion; pendants represented the helping spirits and were attached to the clothing of a ritual leader (Devlet 2001, 51; Jonuks 2009; Zvelebil 2003, 11 and references therein). Those spirits followed the leader (shaman) during the spiritual travels and helped him or her with their 'animal' powers (see *e.g.* descriptions of Nganasan shaman journeys in Lintrop 1996). This kind of use of pendants is characteristic of Eurasian shamanism, for which good analogies are available from the 19th–20th-century shamanistic societies (Hultkranz 1978). Considering that one of the purposes of these adornments was to assist the shaman to enter into a trance, often the entire face was covered with different pendants and beads to prevent vision and by that to predispose for hallucinations (Strassburg 2000). This could explain some rich head sets of tooth pendants, for example, from the Mesolithic Zvejnieki cemetery in northern Latvia (see also Larsson 2006; Zagorska and Lõugas 2000). But it does not mean that using tooth pendants was limited to ritual specialists only. In many anthropological studies (*e.g.* Willerslev 2007), the difference between a shaman and a common member of society has shown to be vague, meaning that anyone could have access to natural powers. This kind of orientation towards communication with nature's spirits changed significantly during the Late Neolithic. As indicated by zooarchaeological evidence, the importance of hunting reduced significantly, although did not lose its role entirely (*e.g.* Lang 2007, 110; Paaver 1965). Therefore, mastering hunting and wilderness was no longer as important and ideologies related to domestication started to prevail instead. Together with new ideologies, the role of shamanism as a belief system seems to have disappeared, aiding spirits lost their importance, and tooth pendants fell out of use (see Jonuks 2009 for a detailed study on this subject).

In the Late Iron Age cemeteries, animal pendants have been often found in the burials of women and children, leading to a mere suggestion of them needing more protection by animals than men (Kivisalo 2008 and references therein). However, other pendants and decorations also occur mainly in the burials of women and children (Kurisoo 2012, 228), and thus, the occurrence of animal pendants in these contexts comes as no surprise. This does not necessarily exclude interpretations related to magical protection, but demonstrates that animal pendants should not be considered separately from other finds. Only by considering both the temporal and spatial context, binary dichotomies and gender-/age-biased interpretations can be avoided.

The comparison of the Late Iron Age and the medieval period provides another good example against simplistic interpretations. The number of pendants decreased dramatically during the Middle Ages and they were not used as grave goods (except in Siksali, which, as stated above, represents a culturally distinct area). This observation could be associated with changes in religion and worldviews, as the 13th century marked the transition to Christianity in Estonia, and thus, such 'pagan' pendants no longer fitted into a Christian world. However, this may be erroneous, as most medieval animal pendants come from Christian contexts (towns, castles) and not from rural settlements. The actual reason for the disappearance of animal pendants is more likely to be a significant change in the tradition of decorations, because pendants in general vanished at that time, and those remaining became more stylised and geometric (see Valk 2001, 52).

Religion-based interpretations have traditionally dominated in discussions concerning animal pendants (*e.g.* Asplund 2005). However, most of these studies have focused on single species and have not considered any chronological variability. Even though the Late Iron Age and medieval pendants have been interpreted here in a socio-cultural framework, religious meanings and protective qualities should not be entirely excluded. Although we have argued against the classical 'amulet' interpretation on the basis of a diachronic analysis, on the vernacular stage (see Primiano 1995) any of the pendants could have functioned as an amulet. For instance, beaver astragali from the Late Iron Age southeastern Estonia have been first of all connected with fur (and castoreum?) trade of a profitable animal, indicating a broader, ideological importance of one of the sources for wealth and prosperity. At the same time, they could have signified capable hunters and been used as amulets to secure lucky hunts in the future. However, in archaeology, such vernacular meanings are difficult to identify and can be inferred only for specimens from particularly good contexts. So, first of all, we should interpret animal pendants on the basis of wider social and/or economic patterns and ideological symbolism, and only then add potential personal values and personal symbolism.

The minor role of magic in pendant production, usage, and deposition during the Late Iron Age is also indicated by their archaeological context. As it has been demonstrated, compared to the Stone Age, Late Iron Age pendants occur in burial contexts to a lesser extent, and available examples show that pendants were proudly on display. Such contextual differences lead to an interpretation of pendants being not that closely associated with the individual, but rather representing general cultural or ideological values of that period. The more complex meaning of some species, for example elk, eagle, and dog, is stressed by their representation in art or in special burials. Still, the connection between pendant and iconographic use of an animal is not necessarily direct and comprehensive, as several important animals like bear or wild boar have only been depicted in single cases during the Stone Age.

Our study has shown that the essence of animal pendants is more multifaceted than only being amulets or simply linking humans with the wilderness. Instead, interpretations vary depending on the context, and therefore, the wider socio-cultural, economic, and ideological background should be considered when discussing human-animal relations. Here we have given a basic framework for interpretations of Estonian animal pendants, but more comprehensive interpretations are beyond the remit of this paper. One topic that was excluded from this study was the production of pendants and use-wear analysis, which would be worthy of future research, as different ways of attaching and modifying animal teeth and bones could shed light on the actual ways the pendants were used. Furthermore, there are many enigmatic examples waiting to be deciphered.

Acknowledgements

The authors are grateful to Liina Maldre, Lembi Lõugas, and Teresa Tomek for their assistance with the osteological analysis, Tuuli Kurisoo, Heidi Luik and Silvia-Kristiin Kask for additional information on the finds, and Aivar Kriiska for providing information on Stone Age sites and chronology. We also thank the two anonymous reviewers and editors of this book who significantly helped to improve the original manuscript. This study was supported by institutional research funding IUT20-7 and IUT22-5 of the Estonian Ministry of Education and Research, the European Union through the European Regional Development Fund (Centre of Excellence in Cultural Theory and Centre of Excellence of Estonian Studies), and the European Research Council grant OVinE H2020-MSCA-IF-2016 749226.

References

Abbreviations

AI	Archaeology collections at the Tallinn University
AM	Archaeology collections at the Estonian History Museum
TÜ	Archaeology collections at the Univeristy of Tartu
VKM	Archaeology collections at Valga museum

Abu-Chacra, F. (2004) *Vikings through Arab Eyes. 'Ibn Fadlan' A.D. 922*. Helsinki, Ammatour-Press.

Antanaitis, I. R. (1998) Interpreting the meaning of East Baltic Neolithic symbols. *Cambridge Archaeological Journal* 8(1), 55–68.

Arukask, M. (1999) Oletusi eestlaste uskumustest ja maailmapildist Tacituse teadete taustal. *Keel ja Kirjandus* 10, 665–72.

Asplund, H. (2005) The bear and the female. Bear-tooth pendants in Late Iron Age Finland. In S. Mäntylä (ed.) *Rituals and relations. Studies on society and material culture of the Baltic Finns*. Suomalaisen Tiedeakatemian toimituksia. Humaniora 336, 13–30. Helsinki, Finnish Academy of Science and Letters.

Bettini, M. (2013) *Women & Weasels. Mythologies of birth in Ancient Greece and Rome*. Transl. E. Eisenach. Chicago IL–London, The University of Chicago Press.

Brea, M. B., Mazzieri, P. and Micheli, R. (2010) People, dogs and wild game: evidence of human-animal relations from Middle Neolithic burials and personal ornaments in Northern Italy. *Documenta Praehistorica* 37, 125–45.

Callmer, J. (1994) The clay paw burial rite of the Åland Islands and Central Russia: a symbol in action. *Current Swedish Archaeology* 2, 13–46.

Carpelan, C. (1977) Älg och björnhuvudföremål från Europas nordliga delar. *Finskt Museum 1975*, 5–67.

Choyke, A. (2001) Late Neolithic Red Deer Canine Beads and Their Imitations. In A. Choyke and L. Bartosiewicz (eds.) *Crafting Bone: Skeletal Technologies through Time and Space. Proceedings of the 2nd Meeting of the (ICAZ) Worked Bone Research Group Budapest, 31 August–5 September 1999*. British Archaeological Reports International Series 937, 251–66. Oxford, Archaeopress.

Choyke, A. (2010) The bone is the beast: animal amulets and ornaments in power and magic. In D. Campana, P. Crabtree, S. D. deFrance, J. Lev-Tov and A. Choyke (eds.) *Anthropological Approaches to Zooarchaeology. Complexity, Colonialism, and Animal Transformations*, 197–209. Oxford, Oxbow Books.

Devlet, E. (2001) Rock art and the material culture of Siberian and Central Asian shamanism. In N. S. Price (ed.) *The Archaeology of Shamanism*, 43–55. London–New York, Routledge.

Fossum, B. (2006) *Förfädernas land. En arkeologisk studie av rituella lämningar i Sápmi, 300 f. Kr. – 1600 e.Kr.* Studia Archaeologica Universitatis Umensis 22. Umeå, Univeristy of Umeå.

Gilchrist, R. (2008) Magic for the dead? The archaeology of magic in Later Medieval burials. *Medieval Archaeology* 52, 119–59.

Griciuvienė, E. and Vasiliauskas, E. (2005) *Žiemgaliai: baltų archeologijos paroda; katalogas*. Vilnius, Lietuvos nacionalinis muziejus.

Gurina, N. (1956) = Гурина, Н. Н. (1956) *Оленеостровский могильник*. Москва-Ленинград. Издательство академии наук СССР.

Helskog, K. (2004) Landscapes in rock-art: rock-carving and ritual in the old European North. In C. Chippendale and G. Nash (eds.) *The Figured Landscapes of Rock-Art. Looking at Pictures in Place*, 265–88. Cambridge, Cambridge University Press.

Hultkrantz, Å. (1978). Ecological and phenomenological aspects of shamanism. In V. Diószegi and M. Hoppál (eds.) *Shamanism in Siberia*, 27–58. Budapest, Akadémiai Kiadó.

Iršėnas, M. (2000) Elk figurines in the Stone Age art of the Baltic Area. In A. Butrimas (ed.) *Prehistoric Art in the Baltic Region*, 7–29. Vilnius, Vilnius Academy of Fine Art.

Jennbert, K. (2003) Animal graves. Dog, horse and bear. In K. Cassel and A. Gustafson (eds.) *Current Swedish Archaeology, 11. Water – Resource, Regenerator, Communicator*, 139–52. Stockholm, Swedish Archaeological Society.

Jonuks, T. (2006) Koerad Eesti asukate viikingiaja maailmapildis. *Mäetagused* 31, 29–48.

Jonuks, T. (2009) *Eesti muinasusund*. Dissertationes archaeologiae Universitatis Tartuensis 2. Tartu, University of Tartu Press.

Jonuks, T., Oras, E. and Johanson, K. (2013) Materiality of Religion: Religion-Related Artefacts in Estonian Archaeological Collections. *Folklore. Electronic Journal of Folklore* 55, 151–76.

Kajkowski, K. (2012) The boar in the symbolic and religious system of Baltic Slavs in the Early Middle Ages. *Studia Mythologica Slavica* 15, 201–5.

Kirk, A. (2003) Halljänes on meil uustulnuk. *Eesti Loodus* 2/3, 68–9.

Kiudsoo, M. and Russow, E. (2011) Hoards discovered in 2010. In E. Oras and E. Russow (eds.) *Archaeological Fieldwork in Estonia 2010*, 225–32. Tallinn, Muinsuskaitseamet.

Kivisalo, N. (2008) The Late Iron Age Bear-Tooth Pendants in Finland: symbolic mediators between women, bears and wilderness. *Temenos* 44(2), 263–91.

Kriiska, A. (2001) *Stone Age Settlement and Economic Processes in the Estonian Coastal Area and Islands*. Academic Dissertation. Helsinki, University of Helsinki. Available at: http://ethesis.helsinki.fi/julkaisut/hum/kultt/vk/kriiska/stoneage.html (accessed 10/10/2017).

Kriiska, A. (2004) *Aegade alguses. 15 kirjutist kaugemast minevikust*. Tallinn, Tallinna Raamatutrükikoda.

Kurisoo, T. (2012) Ristripatsid Eesti 12.–13. sajandi laibakalmistutes: kas ehted või usu tunnused? In H. Valk (ed.) *Õpetatud Eesti Seltsi Aastaraamat 2011*, 213–34. Tartu, Õpetatud Eesti Selts.

Lang, V. (2007) *The Bronze and Early Iron Ages in Estonia*. Tartu, University of Tartu Press.

Larsson, L. (2006) Tooth for a Tooth for a Grave: Tooth Ornaments from the Graves at the Cemetery of Zvejnieki. In L. Larsson and I. Zagorska (eds.) *Back to the Origin: New Research in the Mesolithic-Neolithic Zvejnieki Cemetery and Environment, Northern Latvia*, 153–288. Stockholm, Almqvist and Wiksell.

Leimus, I. and Kiudsoo, M. (2004) Koprad ja hõbe. *Tuna* 4, 31–47.

Lepiksaar, J. (1986) The Holocene history of theriofauna in Fennoscandia and Baltic countries. Nordic Late Quaternary Biology and Ecology. *Striae* 24, 51–70.

Ligi, P. (1993) Vadjapärased kalmed Kirde-Eestis (9.–16. sajand). In V. Lang (ed.) *Vadjapärased kalmed Eestis 9.–16. sajandil*. Muinasaja teadus, 2, 7–175. Tallinn, Eesti Teaduste Akadeemia, Ajaloo Instituut.

Lintrop, A. (1996) The Incantations of Tubyaku Kosterkin. *Folklore. An Electronic Journal of Folklore* 2, 9–28.

Luik, H. (2005) *Luu- ja sarvesemed Eesti arheoloogilises leiumaterjalis viikingiajast keskajani*. Dissertationes Archaeologiae Universitatis Tartuensis. Tartu, University of Tartu Press.

Luik, H. (2010a) Beaver in the economy and social communication of the inhabitants of South Estonia in the Viking Age (800–1050 AD). In M. Kucera and G. K. Kunst (eds.) *VIAVIAS Vienna Institue for Archaeological Science*, 46–54. Vienna, Universität Wien.

Luik, H. (2010b) Linnupeakujuline sarvest käepide Lõhavere linnamäelt. In Ü. Tamla (ed.) *Ilusad asjad. Tähelepanuväärseid leide Eesti arheoloogiakogudest*. Muinasaja Teadus 21, 127–38. Tallinn, Ajaloo Instituut.

Luik, H. and Maldre, L. (2005) Bone and antler artefacts from the settlement site and cemetery of Pada in North Estonia. In H. Luik, A. Choyke, C. Batey and L. Lõugas (eds.) *From Hooves to Horns, from Mollusc to Mammoth: Manufacture and Use of Bone Artefacts from Prehistoric Times to the Present*. Muinasaja teadus 15, 263–76. Tallinn, Teaduste Akadeemia Kirjastus.

Lõugas, L. and Maldre, L. (2000) The History of theriofauna in the Eastern Baltic region. *Folia Theriologica Estonica* 5, 86–100.

Maldre, L., Tomek, T. and Peets, J. (in press) Birds of prey from Vendel Age ship burials of Salme (c. 750 AD) and in Estonian archaeological material. In K.-H. Gersmann and O. Grimm (eds.) *Raptor and Human – Bird Symbolism and Falconry through Five Millennia on a Global Scale*. Schleswig, Centre of Baltic and Scandinavian Archaeology Monographs.

Mägi, M., Allmäe, R. and Maldre, L. (1998) Viking Age Graveyard at Piila, Saaremaa. In Ü. Tamla (ed.) *Arheoloogilised välitööd Eestis 1997*, 99–116. Tallinn, Muinsuskaitseinspektsioon.

O'Shea, J. and Zvelebil, M. (1984) Oleneostrovski mogilnik: reconstructing the social and economic organization of prehistoric foragers in northern Russia. *Journal of Anthropological Archaeology* 3, 1–40.

Ots, M. (2010) Loomakujukesed Valma keskneoliitilises kaksikmatuses. In Ü. Tamla (ed.) *Ilusad asjad. Tähelepanuväärseid leide Eesti arheoloogiakogudest*. Muinasaja teadus 21, 11–22. Tallinn, Ajaloo Instituut.

Paaver, K. (1965) = Паавер, К. (1965) *Формирование териофауны и изменчивость млекопитающих Прибалтики в голоцене*. Тарту, Академия наук Эстонской ССР, Институт Зоологии и ботаники.

Primiano, L. N. (1995) Vernacular Religion and the Search for Method in Religious Folklife. *Western Folklore* 54, 37–56.

Rainio, R. and Mannermaa, K. (2014) Bird calls from a Middle Neolithic burial at Ajvide, Gotland? Interpreting tubular bird bone artefacts by means of use-wear, sound analysis and ethnographic analogy. In J. Ikäheimo, A-K. Salmi and T. Äikäs (eds.) *Sounds like Theory. XII Nordic Theoretical Archaeology Group Meeting in Oulu 25.–28.4.2012. Monographs of the Archaeological Society of Finland* 2, 85–100. Helsinki, Archaeological Society of Finland.

Rannamäe, E. and Lõugas, L. (in press) Animal exploitation in Karksi and Viljandi (Estonia) in the Late Iron Age and medieval period. In A. G. Pluskowski (ed.) *The Ecology of Crusading, Colonisation and Religious Conversion in the Medieval Eastern Baltic: Terra Sacra II.* Turnhout, Brepols.

Roio, M., Baburin, A. and Kriiska, A. (2016) Archaeological survey in the northern and north-western parts of Lake Peipsi. In E. Russow and A. Haak (eds.) *Archaeological fieldwork in Estonia 2015*, 225–34. Tallinn, Muinsuskaitseamet.

Stark, L. (2015) Magic and witchcraft in their everyday context: childhood memories from the nineteenth-century Finnish countryside. In A. Ohrvik and A. Guðmundsdóttir (eds.) *ARV. Nordic Yearbook of Folklore* 70, 125–46. Sweden, Uppsala.

Strassburg, J. (2000) *Shamanic Shadows. One Hundred Generations of Undead Subversion in Southern Scandinavia, 7,000–4,000 BC.* Stockholm, Stockholm Studies in Archaeology 20.

Sørensen, B. H. (2013) Dogs in the Danish Viking Age – the Ladby Ship and other finds. In O. Grimm and U. Schmölcke (eds.) *Hunting in Northern Europe until 1500 AD. Old Traditions and Regional Developments, Continental Sources and Continental Influences*, 163–74. Neumünster, Wachholtz Verlag.

Tamla, Ü. and Maldre, L. (2001) Artefacts of bone, antler and canine teeth among the archaeological finds from the hill-fort of Varbola. In A. Choyke and L. Bartosiewicz (eds.) *Crafting Bone: Skeletal Technologies through Time and Space: Proceedings of the 2nd Meeting of the (ICAZ) Worked Bone Research Group Budapest.* British Archaeological Reports International Series 937, 371–82. Oxford, Archaeopress.

Tolley, C. (2006) The bear in Germanic legend. In C. Tolley (ed.) *Karhun Kannoilla: In the Footsteps of the Bear*, 94–104. Pori, School of Cultural Production and landscape Studies/ University of Turku.

Tvauri, A. (2001) *Muinas-Tartu: uurimus Tartu muinaslinnuse ja asula asustusloost. Prehistoric Tartu: a study of the settlement history of the Tartu prehistoric hillfort and settlement.* Muinasaja teadus 10. Tallinn, Teaduste Akadeemia Kirjastus.

Valk, H. (2001) *Rural Cemeteries of Southern Estonia 1225–1800 AD.* CCC Papers 3. Visby-Tartu, University of Tartu.

Valk, H. and Laul, S. (2014) *Siksälä kalme I. Muistis ja ajalugu.* Tartu, University of Tartu.

Vassar, A. (1943) Nurmsi kivikalme Eestis ja tarandkalmete areng. PhD thesis, University of Tartu. Available at: http://www.arheo.ut.ee/docs/Vassar19431.pdf (accessed 10 October 2017).

Willerslev, R. (2007) *Soul Hunters: Hunting, Animism and Personhood among the Siberian Yukaghirs.* Berkeley CA, University of California Press.

Wood, R. (1957) Perforated elk teeth: a functional and historical analysis. *American Antiquity* 22(4), 381–7.

Zachrisson, I. and Iregren, E. (1974) *Lappish Bear Graves in Northern Sweden. An Archaeological and Osteological Study.* Early Norrland 5. Stockholm, Kungl. Vitterhets Historie och Antikvitets Akademien.

Zagorska, I. and Lõugas, L. (2000) The Tooth Pendant Head-Dresses of Zvejnieki cemetery. In V. Lang (ed.) *De Temporibus Antiquissimis ad Honorem Lembit Jaanits.* Muinasaja Teadus 8, 223–44. Tallinn, Teaduste Akadeemia Kirjastus.

Zariņa, A. (2006) *Salaspils laukskolas kapulauks: 10.–13. gadsimts.* Rīga, Latvijas Vēstures Institūta apgāds.

Zvelebil, M. (2003) People behind the lithics. Social life and social conditions of Mesolithic communities in temperate Europe. In L. Bevan and J. Moore (eds.) *Peopling the Mesolithic in a Northern Environment.* British Archaeological Reports International Series 1157, 1–26. Oxford, Archaeopress.

Zvelebil, M. and Jordan P. (1999) Hunter fisher gatherer ritual landscape – questions of time, space and representation. In J. Goldhahn (ed.) *Rock Art as Social Representation. Papers from a Session Held at the European Association of Archaeologists Fourth Annual Meeting in Göteborg 1998.* British Archaeological Reports International Series 794, 101–27. Oxford, Archaeopress.

Chapter 13

Birds in Death: Avian Archaeology and the Mortuary Record in the Scottish Islands

Julia Best and Jacqui Mulville

Introduction

The incorporation of avian material in archaeological mortuary contexts takes a variety of forms and occurs in a wide range of temporally, geographically, and culturally differentiated contexts. Through these death-related deposits researchers have the opportunity to examine avian–human relationships exhibited in life and beyond, particularly when examined alongside other contemporary site types. This paper considers the role of birds in ideas and actions surrounding death through traditional zooarchaeological study, applied in detail to avian remains; an area which has sometimes been overlooked. Focusing primarily (but not exclusively) on a species group – the eagles – and one region – the Scottish islands – data pertaining to birds in mortuary environments are explored through time and contextualised within a wider European setting. These data were collated as part of a large zooarchaeological and biomolecular investigation of new and pre-existing bird bone and eggshell data from Mesolithic to Norse sites in North Atlantic islands (Best 2014).

Avian remains from the mortuary record are examined here and considered within the wider social and cultural framework of avian-human interactions in their geographic and temporal context. Both wild and domestic bird species are included in order to explore more fully the numerous reasons for avian presence in mortuary settings, which range from the ritualistic to the accidental. The active involvement of birds in processing the dead through scavenging and defleshing is also discussed. The ultimate aim is to employ such bioarchaeological data to facilitate interpretation of the social context of avian-human interactions in life and death, and in the creation of worldviews.

The paper is novel in focusing on this theme, as generally only single site case studies or very limited datasets have been used to explore birds in mortuary contexts in the Scottish islands, despite the area having a rich avian record, both external to and within mortuary settings. By bridging this gap, this paper is an initial exploration of the dataset and aims to act as a springboard for future research, which may then facilitate deeper consideration of social implications.

Birds in mortuary contexts: examples from Continental Europe

The Scottish islands (which are the primary focus of this paper) did not exist in isolation, and thus, studies targeting other areas and periods can be used as a basis for starting to consider avian-human interactions in the North Atlantic islands. Research on birds in mortuary environments has largely focused on spectacular and evocative case studies that highlight avian-human interactions in death, particularly such as within Mesolithic and Neolithic burials of the Baltic and Scandinavian region (Mannermaa 2006; 2007; 2008a; 2008b). Comparative examples exist in other areas of mainland Europe, such as Mesolithic and Neolithic Dudka in Poland, however, they have often received less study and as such limited datasets are available for comparative research (Gumiński 2005). The reduced representation of birds in other areas may, therefore, be an accurate reflection of past burial rites, but may also partially result from a recovery and research, or an ecological bias. For example, many of the sites which do produce birds in mortuary contexts are from ecotonal or coastal environments that often house large numbers of bird species, and this may be a factor in the distribution of their use.

Examples of birds in mortuary contexts are found at multiple sites in the Baltic/Scandinavian region during several phases of prehistory. At Late Mesolithic Vedbæk

Bøgebakken in Denmark a young woman and newborn baby were buried together, with the infant situated on the wing of a whooper swan (*Cygnus cygnus*). Interpretations of this arrangement have considered that the avian body part may have been chosen to cradle and protect the child, to aid the child's movement from one realm (of living) to another (death or another life), or to symbolise a specific relationship with swans in that site/context (Mannermaa 2008a).

At Middle Neolithic Ajvide on Gotland, Sweden, there are multiple examples of birds in funerary contexts. One probable female (Grave 62) was buried with numerous worked bird bone tubes, interpreted as decorative items or potential instruments, such as wind chimes/whistles, which were possibly used to create music within ritual or social events. Her grave also included cut crane (*Grus grus*) tarsometatarsi, whilst in the same mortuary site another adult female (Grave 2) was accompanied by similarly cut cormorant (*Phalacrocorax carbo*) tarsometatarsi, and an infant (Grave 20) had similar artefacts made from an unidentified bird. These finds could represent talismans or items of personal adornment. A double child burial (Grave 52) contained paired bird bone beads (mainly Anatidae and Charadriiformes), suggesting that birds in this context were used in personal adornment during life and that these items could have continued this role in death. Such beads may have held symbolic associations with the birds that they were created from (Mannermaa 2008b; Mannermaa and Storå 2006).

Within Middle Neolithic Zvejnieki in Latvia at least three burials contained parts of wings from the jay (*Garrulus glandarius*). One adult male (Grave 164) was buried with over 40 carpometacarpi spread across his body. These have been interpreted as dress decoration, with the jay's blue wing feathers considered as attractive or holding a specific totemic significance for this group (Mannermaa 2006; 2008a; 2008b).

Avian presence in death may be found in varied forms and materials, for example an Early Neolithic child burial at Tamula in Estonia included bird bones and figurines (see also Jonuks and Rannamäe this volume). The child had parts of crane wings placed in their hands, and rich grave goods including a bird bone whistle-like tube and two bone bird figurines (Albrethsen and Brinch Petersen 1976; Mannermaa 2008b; Mannermaa and Storå 2006; Mannermaa *et al.* 2007). Other Neolithic burials at Tamula also produced bird wing bones, including a golden eagle (*Aquila chrysaetos*) radius in an adult male grave, and a capercaillie radius in a female burial. Wing elements may have been included to represent movement between states of being, or the properties of the birds themselves (including flight). Their inclusion may also have been related to the use of feathers in fletching arrows and hunting, or, as has recently been suggested for the Ajvide swan wing, the importance of the wing in avian communication (Mannermaa 2008a; Overton and Hamilakis 2013). Case studies such as these can be used to help consider the social and cultural interpretations of bird use in the Scottish islands.

Birds in mortuary contexts: the Scottish islands

The coast of Scotland is surrounded by hundreds of islands with rich archaeological records. They contain a temporally, geographically, culturally and contextually diverse body of sites, and today are home to some of Britain's largest and most important avian populations. Past human populations living in these ecotonal environments exploited a wide range of resources from the land, sky and sea. The zooarchaeological analysis of animal bone assemblages from these islands from the Mesolithic into the Norse period (and sometimes beyond) has revealed that birds played an important, repeated but also flexible and varied role in food and resource provisioning, providing meat, eggs, oil, and fat, but also skins, feathers and bones for tools (Best 2014; Best and Mulville 2010; 2013; 2014; 2016; in press; Serjeantson 1988). Interactions with birds, however, were not limited to subsistence activities, but comprised a much broader part of the human experience that in some instances extended beyond life-use and entered into the realm of death and the mortuary environment. Such occurrences have the potential to reveal social aspects of bird use that are not focused on their calorific contributions. Through combining pre-existing datasets with new in-depth analyses, a comprehensive and complex picture of bird usage and the roles they occupied in this location can be formed. Evidence is organised by phase in the sections that follow.

Mesolithic: pre-farming bird use

Evidence for pre-Mesolithic activity is sparse and contested, with only a handful of recognised Late Upper Palaeolithic sites (Saville and Wickham-Jones 2012, 9–21) but avian-human interactions are present from the earliest substantive periods of occupation in the Mesolithic. This period starts at the beginning of the Holocene warming *c.* 9700 BC and is ending in Scotland by *c.* 4300/4000 BC with the introduction of farming in the Neolithic (Finlayson and Edwards 2003, 112; Saville and Wickham-Jones 2012, 108). The main body of animal and human remains comes from the Inner Hebridean Oronsay midden sites (Caisteal nan Gillean, Cnoc Sligeach, Cnoc Coig and Priory Midden), with additions from An Corran on Skye, Sand in the Applecross area, and West Voe on Shetland. Human mortuary deposits for this period are very limited but remains have been recovered from within several of the middens in the Inner Hebrides, including Cnoc Coig and An Corran (Schulting and Richards 2002). Therefore, in the Mesolithic, human and avian bone remains are often commingled in midden contexts.

The Mesolithic Scottish island bird bone assemblage contains at least 52 species, but the majority of the bone comes from seabirds, such as razorbill/guillemot (*Alca torda/ Uria aalge*), great auk (*Pinguinus impennis*), puffin (*Fratercula arctica*), and gannet (*Morus bassanus*). These birds visit the islands to breed in spring/summer and would have been captured for meat, oil and fat, and probably for other products, such as feathers. Cnoc Coig produced many species of bird including two raptors: the sparrow hawk (*Accipiter nisus*) and the buzzard (*Buteo buteo*). An Corran contained bones from the white-tailed eagle (*Haliaeetus albicilla*) (also known as the white-tailed sea eagle). This eagle is the largest bird of prey in Britain; it became extinct in Scotland and the rest of Britain in the early 20th century and the present-day population derives from reintroduced birds (Bramwell 1983, 164; Serjeantson 2010, 152).

The presence of raptors indicates that Mesolithic fowlers were utilising birds for more than purely food products and were either capturing these birds or recovering naturally deceased specimens. Intentional killing of large raptors has been proposed for later farming societies since these birds can carry off young or small livestock (Lockie and Stephen 1959, 43–50), but within a hunter-gatherer-fisher society this would not have occurred. These raptors with which the populations shared their landscape may have held many different symbolic meanings and associations and could have been used in personal ornamentation and decoration.

The majority of the human remains within the Oronsay middens are partially or fully disarticulated (Meiklejohn *et al.* 2005, 88–9; Nolan 1987, 251–7). This is the case at Cnoc Coig, where human bone was found in distinct groupings comprising more than one individual, and as isolated fragments, suggesting that this profile resulted from a range of practices (Meiklejohn *et al.* 2005, 88–93). This included the intentional use of the midden as a mortuary environment which resulted in the incorporation of human bone within the midden either by deliberate deposition, or as a result of partial disarticulation, decomposition or defleshing on site, by avian or other agents (Meiklejohn *et al.* 2005, 98–103; Nolan 1987, 251–7). An exposed disposal tradition would have attracted scavenging birds to the area (Pollard 1996), and may have resulted in the subsequent movement and loss of defleshed body parts. No direct evidence of avian scavenging has been identified by raptor modifications on Scottish island human remains. However, since the identification of such modifications can be complex further work is needed in this area to elucidate practices (Berger 2006; Lawrence 2006). Overall the very limited corpus of human remains from the Scottish island Mesolithic means that interpretations of animal/ avian-human interactions in mortuary environments are at present restricted.

Neolithic and Bronze Age: raptors and tombs

The Neolithic (*c.* 4300/4000–2500 BC) avian dataset from all site types includes material from both the Northern Isles (Shetland and Orkney) and the Inner and Outer Hebrides. Although the dataset is relatively small in size it demonstrates a continued use of several key seabirds, such as auks and gannet, for food and other resources, but also an increased use of species available at multiple points of the years such as cormorant/shag (*Phalacrocorax aristotelis*), gulls (Laridae), ducks (*e.g. Anas*) and geese (here *Anser/ Branta*) (Best and Mulville 2016). The Bronze Age (*c.* 2500–700 BC) dataset reveals a similar pattern of bird use to the Neolithic, with some targeted exploitation of summer visiting seabirds, as part of a year-round fowling calendar.

In addition to these species, both the Neolithic and Bronze Age material also included a large number of raptors. This is mainly due to the recurrent presence of large raptors in prehistoric Scottish island mortuary settings, particularly the white-tailed eagle in the Neolithic and Bronze Age tomb culture, suggesting a symbolic avian association with the dead in some contexts (Bramwell 1983; Harman 1997; 2009; Platt 1934a; 1936; Serjeantson 2010). These tombs contain the largest quantity of archaeological white-tailed eagle bones in Britain through time. The Hebrides and the Northern Isles exhibit some differences, with the latter displaying a much larger raptor presence, which forms nearly a sixth of the total bird remains from all sites. However, the assemblages continue to be taxonomically diverse with an important seabird contribution (see Best and Mulville 2016 for data).

In the Neolithic and Bronze Age Northern Isles the white-tailed eagle is the most commonly occurring bird in terms of NISP (Number of Identified Specimens). A large proportion of the Neolithic bones (139 NISP, see Table 13.1 and Fig. 13.1) derives from the Point of Cott tomb, a chambered cairn on Westray, Orkney, where they represent at least eight individual birds (Harman 1997). A single bone is recorded in the small assemblage from the Neolithic tomb site Knowe of Ramsay on Rousay, Orkney, and it is possible that more bones had been present here but were lost (Platt 1936). A specimen was also present at the Holm of Papa Westray Neolithic tomb, and although the bird dataset is not yet available for this site, it will add another line of evidence in the future (Harman 2009). The significance of the white-tailed eagle in particular is highlighted at Isbister (also known as the Tomb of the Eagles), a Neolithic tomb site on South Ronaldsay, Orkney. Here, between 10 and 20 white-tailed eagles, represented by over 600 bones, were recovered (Bramwell 1983). However, these birds date to the Bronze Age some 1000 years after the site's Neolithic construction and primary period of human internment (Pitts 2006; Sheridan 2005), although some episodes of human deposition occurred over an extended period (Lawrence 2012, 103–11; Renfrew *et al.* 1983).

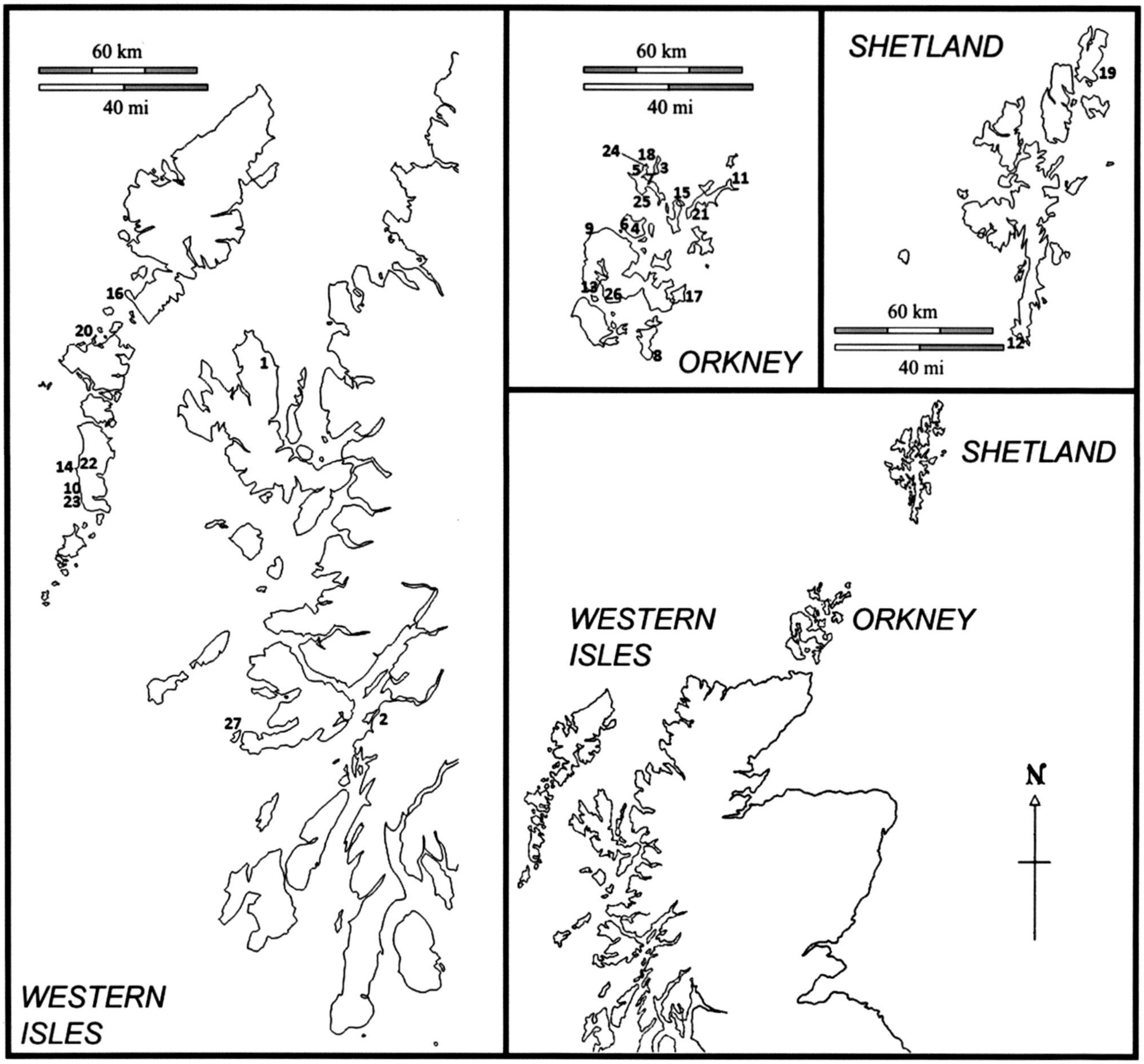

Figure 13.1 Map of Scottish islands showing sites which produced eagle bones. Site details are provided in Table 13.1 by corresponding numbers. Base maps by Ian Dennis and WWW1.

Further eagle specimen/s were present in another Neolithic tomb assemblage from Midhowe Cairn, also on Rousay, Orkney, but the analyst did not differentiate between white tailed and golden eagle or provide the NISP (Platt 1934a, 348–50). An immature eagle was recovered from later deposits at this site, which may indicate that these birds were choosing to nest on the structure. Interestingly, at this site (which has no full quantification data) the birds are purported to be 'both varied and more numerous than either sheep or pig' and included seabirds and raptors, indicating that a variety of different birds may have played an important role in this setting (Platt 1934a, 348–50).

In the Scottish island Neolithic, white-tailed eagle remains are not only found in tombs; another 98 fragments were recovered from the settlement site Links of Noltland, Westray, Orkney (Armour-Chelu 1985). It is interesting that in this instance, unlike the tombs, the bones are likely to have come from a small number of birds, probably a single articulated individual (Armour-Chelu 1985). In the Bronze Age, single bones were found at settlements, such as Tofts Ness, Sanday, Orkney, and there is also a probable example from Cladh Hallan, South Uist, Outer Hebrides (Best and Mulville 2013; 2014; Serjeantson 2007b).

The quantity and range of bird species present in the overall corpus of Neolithic and Bronze Age sites suggest that

Table 13.1 Details of sites producing eagle bones by NISP (Number of Identified Specimens).

Map ID	Site Name	Island Group	Island	Period	White-tailed eagle	Golden eagle	Eagle cf white-tailed	White-tailed/ Golden eagle	Avian analyst reference
1	An Corran	IH	Skye	Mesolithic	2 (*1)				Bartosiewicz 2012
2	Carding Mill Bay I	ML	N/A	Early Neolithic	1				Hamilton-Dyer & McCormick 1993
3	Holm of Papa Westray	OR	Holm of Papa	Neolithic	1				Harman 2009
4	Knowe of Ramsay	OR	Rousay	Neolithic	1				Platt 1936
5	Links of Noltland	OR	Westray	Neolithic	98 (*1)				Armour-Chelu 1985
6	Midhowe Cairn	OR	Rousay	Neolithic				1	Platt 1934a
7	Point of Cott	OR	Westray	Neolithic/ Mixed	139 (*8+)				Harman 1997
8	Isbister	OR	South Ronaldsay	Early Bronze Age	641 (*10-20)				Bramwell 1983; Jones 1998
9	Point of Buckquoy (5&6)	OR	Mainland	Middle Bronze Age		1			Rackham & Nicholson 1989
10	Cladh Hallan	OH	South Uist	Late Bronze Age			1		Best & Mulville 2013; 2014
11	Tofts Ness Phase 4	OR	Sanday	Later Bronze Age	1				Serjeantson 2007a; 2014
12	Jarlshof	SH	Mainland	Late Bronze/ Early Iron Age				1	Platt 1933; 1934b; 1956
13	Howe	OR	Mainland	Middle Iron Age	1				Bramwell 1994
14	Dun Vulan	OH	South Uist	Middle–Late Iron Age	1				Cartledge and Grimbly 1999
15	Calf of Eday	OR	Calf of Eday	Iron Age	1				Platt 1937
16	Northton	OH	Harris	Iron Age	1				Best Unpublished; Finlay 1984
17	Skaill Deerness	OR	Mainland	Iron Age	2 (*1)				Allison 1997
18	St Boniface's Church	OR	Papa Westray	Iron Age			3 (*1+)		Hamilton-Dyer 1998
11	Tofts Ness Phases 5 & 6	OR	Sanday	Iron Age	1				Serjeantson 2007a; 2014
19	Milla Skerra Sandwick	SH	Unst	Iron Age	1				Smith 2008

(Continued)

Table 13.1 Details of sites producing eagle bones by NISP (Number of Identified Specimens). (Continued)

Map ID	*Site Name*	*Island Group*	*Island*	*Period*	*White-tailed eagle*	*Golden eagle*	*Eagle cf white-tailed*	*White-tailed/ Golden eagle*	*Avian analyst reference*
20	Udal XI-XIII	OH	North Uist	Late Iron Age	1				Serjeantson 2013
13	Howe	OR	Mainland	Late Iron Age	7 (*1+)			1	Bramwell 1994
21	Pool	OR	Sanday	Late Iron Age/Viking	1				Serjeantson 2007b; 2014
22	Bornais M2	OH	South Uist	Pictish	1				Best In Prep
17	Skaill Deerness	OR	Mainland	Viking	1				Allison 1997
20	Udal Ixc-X	OH	North Uist	Viking	2				Serjeantson 2013
22	Bornais M2A	OH	South Uist	Early Norse	1				Best In Prep
22	Bornais M2	OH	South Uist	Middle Norse	7 (*1+)		1		Best In Prep
23	Cille Pheadair	OH	South Uist	Middle–Late Norse	1				Best and Cartledge In Press
24	Quoygrew Farm Midden	OR	Westray	Middle–Late Norse			2 (*1)		Harland 2006; Harland *et al.* 2012
25	Tuquoy	OR	Westray	Norse	1				Hamilton-Dyer 1991
26	Earl's Bu	OR	Mainland	Late Norse				1	Mainland 1995
27	Iona Abby / Monastery	IH	Iona	Medieval/ recent		1?		1	Bramwell 1981
13	Howe	OR	Mainland	Recent	4 (*1+)				Bramwell 1994

Figures with (*) show MNI (Minimum Number of Individuals) where data allow. IH = Inner Hebrides; OH = Outer Hebrides; OR = Orkney; SH = Shetland; ML = Mainland.

bird use was varied and flexible, with some contexts, such as tombs, favouring increased or species-specific avian input due to their ritual, symbolic or dietary significance. Within the tomb contexts, however, the lack of recorded butchery marks on the eagles, the broad range of skeletal elements present, as well as the presence of several apparently articulated bone groups suggest that consumption was highly unlikely. Indeed, it is not practical, in terms of optimality, to target raptors for food. They are both fewer in number than the plentiful colony-breeding seabirds and flocking waders and waterfowl that inhabit these landscapes, and are also hard to catch. However, if raptors such as the eagles were consumed then the difficulties of capture may have rendered them a special food, potentially imbued with the attributes of the bird. White-tailed eagle bones associated with Late Mesolithic burials at Yuzhniy Oleniy Ostrov in Russia were found to have cut marks indicating filleting, implying that in some instances their symbolic mortuary value could have included ritual consumption or food-based offerings (Mannermaa 2008a; 2016).

The recurrent presence of these large raptors (Plate 21) in the mortuary context is intriguing and suggests that these birds may have occupied a symbolic role in the avian-human relationships of these prehistoric island populations. This is discussed further below.

Iron Age and beyond: domestic arrivals and wild continuations

During the Bronze Age, a transition in mortuary practice is observed including a shift towards cremation (Downes 2012). Despite this, multiple methods of disposing of the dead were employed in the Bronze Age, including some reuse of older tomb sites and mortuary locations; as seen at the above mentioned Isbister. However, after this period, during the Iron Age, Pictish/Pre-Norse, and Norse periods, occurrences of birds in Scottish island mortuary contexts are generally rarer and less visible in the archaeological and zooarchaeological record. This may indicate changes in human-avian interactions and related social roles, or different modes of expressing these relationships. Such variations may have also resulted from changes in the bird populations in the Scottish islands.

Iron Age bird use in the Hebrides was very seabird focused, involving mainly auks (Alcidae), cormorant/shag and gannet, whilst the Northern Isles exhibited a strong focus on gannet and cormorant/shag but also a greater use of ducks, geese and red grouse (*Lagopus lagopus*). In the Norse period bird exploitation in the Hebrides was less summer focused with exploitation of gulls, and a potentially opportunistic diversification of fowling that included more waterfowl, such as ducks and geese, and wading birds. In the Northern Isles, Norse fowling was focused on colony breeding seabirds, suggesting more targeted fowling trips (Best 2014; Best and Mulville in press). After the Bronze Age, white-tailed eagle bones only occur in very small numbers in all Scottish islands assemblages. They have been identified on 18 post-Bronze Age sites, with the NISP never exceeding seven fragments (Table 13.1) and may also be present in 'Eagle sp.' identifications. The data demonstrate that although the population might have been small there was a constant presence of these birds in the Scottish islands in the Iron Age and Norse periods and beyond. However, these largely come from settlement contexts, indicating that eagles no longer held the mortuary role that appears to have been associated with them in the Neolithic and Bronze Age. Thus, human choice was an active agent in their changing societal role and use. In England, a decline is visible from the Middle Ages (Serjeantson 2010, 153) but within the Scottish islands the data suggest that a potential decline may be visible much earlier in time, post-Bronze Age. The presence of butchered white-tailed eagle talons at Norse Bornais (South Uist, Outer Hebrides), and a bone tube possibly made from an eagle bone at Iron Age Midhowe broch (Rousay, Orkney) indicate that eagle bone material may have been transformed into items that moved across the landscape via trade or gift giving. Such objects may have had special associations with the animals they came from and the landscape in which they were captured, although these were not necessarily mortuary focused.

The Scottish island Iron Age zooarchaeological data contrast with the wider British pattern of avian material in mortuary settings, particularly in regards to the introduction of the chicken (*Gallus gallus domesticus*). The Iron Age in Scotland covers a far longer period of time – running from *c.* 700 BC–AD 800 – than in England and Wales, due to the limited impact of Roman expansion (Downes and Ritchie 2003; Harding 2004, 3; Hunter and Carruthers 2012, 9; Parker Pearson and Sharples 1999, 15, 359; Parker Pearson *et al.* 2004, 84). However, even with this extended Iron Age period, the Scottish islands' avian mortuary profile varies from that seen in England during the same timeframe (English Iron Age, Roman and Early Medieval Period).

Chickens entered Britain, primarily England, during the Iron Age, and are found in Iron Age and Roman period graves in England and the Continent. Chickens are represented in the mortuary context via material iconography, such as the bronze cockerel figurine from a mid-2nd century AD child burial in Cirencester, by eggshell, and as physical bone remains, including whole chickens, disarticulated remains, and cremated bone (Foster 2012; WWW4). In these areas, the chicken was perceived as an appropriate animal for deposition with the dead for multiple reasons. The cockerel was an animal companion to several deities including the gods Mithras and Mercury (Lentacker *et al.* 2004; Serjeantson 2009, 351). Its repeated use in graves may have been related to Mercury's status as a psychopomp, and his role in heralding the new-dawn and rebirth at sunrise (Parker 1988; Serjeantson 2009, 344). However, the chicken remains could alternatively represent a meal for the deceased, which depending on the context may have either been considered a luxury or a regular component of the diet (Foster 2012; Lauwerier 1993; Lentacker *et al.* 2004). The deposited chickens may also have still held associations of exoticism, with connotations of introduction from the Continent and long-distance trade; attributes which would also have been relevant in the Scottish islands (Poole 2010; Sykes 2010). The evidence therefore indicates that chickens could hold numerous roles in society and in death.

However, although the chicken may be present in the Scottish islands in small numbers from the later Scottish Iron Age, there are none known from mortuary contexts in this or in later periods (Best 2014; Best and Mulville 2014; 2016; in press). This indicates that the chicken did not occupy a comparable role in the Scottish island Iron Age burial tradition. This may have resulted from the delayed and limited scale of chicken introduction during the Scottish Iron Age, and restricted Roman influence. Their absence in mortuary contexts after their introduction to the Scottish islands indicates that the cultural and social human-chicken interactions associated with them in other areas of Britain were not at play in Orkney, Shetland and the Hebrides. However, interpretation of this difference in domestic fowl use in the mortuary setting during their period of primary introduction is limited by issues of burial survival and recovery in these island locations. The variable preservation conditions in the Scottish islands may have also biased bone recovery in all periods, again limiting full understanding (Parker Pearson *et al.* 2004).

The Scottish island mortuary environment in these later periods is not entirely devoid of avian representations. These are not restricted to zooarchaeological evidence, but include visual representations, such as a Pictish symbol stone with eagle motif which was found covering one of the cists containing burnt bone and ashes at Oxtro on Mainland, Orkney. Unfortunately, excavation occurred in 1847 and the eagle stone was removed, built into a farm building at Boardhouse, and has subsequently gone missing (Curle 1932;

Petrie 1859). Bird bones do still occur in the Scottish island grave environment, but often as worked bone items. Seven Viking/Norse burials were excavated at Cnip, Isle of Lewis, and one of these, an adult female, was buried with a bird bone needle case alongside a wide range of other grave goods (Welander 1987, 157). Humans and birds were still interacting within the context of death, just as they continued to be very intimate within the living contexts of food and resource provision. However, this relationship may have been taking different forms of expression and need not be represented by bone or material remains in the archaeological record.

Interpreting the prehistoric Scottish island raptors

Due to the prominence of raptors, particularly the white-tailed eagle, in the Scottish island mortuary record, and their apparent significance, their presence will now be explored further.

It is thought that people made no distinction between the white-tailed eagle and the similar golden eagle until the 17th century (Serjeantson 2009, 152); however, the archaeological evidence may suggest otherwise. At Isbister and Point of Cott, multiple individual birds are present within the tomb environment and these are exclusively white-tailed rather than golden eagle. Whilst today the golden eagle is rare in Orkney and Shetland, which would explain its absence in these assemblages, one of the only two reported archaeological examples in the Scottish islands is from Orkney: a single bone from the Bronze Age settlement Point of Buckquoy, identified by Don Bramwell. This suggests that the two species did overlap in this area in the past, although the white-tailed may have been more numerous. To capture many individuals of the same species indicates that differences between the two eagle species were recognised as early as the Neolithic and Bronze Age, prompting selective targeting. Alternatively, the white-tailed eagle, a scavenger, may have been easier to catch (Yalden 2007, 471–3).

As previously noted, after the Bronze Age white-tailed eagles were less numerous in the bone assemblages and only occur in small numbers (Table 13.1). This suggests that these birds were familiar but relatively rare, with a possible population decline from earlier prehistory. Therefore, in the Neolithic and Bronze Age the eagle population available for exploitation may have been larger. Alternatively, these earlier tomb cultures of the Scottish islands may have made a substantial effort to acquire the birds, perhaps even curating them for a period of months or years for inclusion in tomb deposits. The occurrence of a seemingly whole white-tailed eagle in the aforementioned settlement site of Links of Noltland could feasibly provide support for the curation of eagles, but could also indicate that these birds occupied multiple roles within society and were used in the contexts of the living and the dead. Further taphonomic study of the bird bone is needed to investigate these possibilities, and to exclude the possibility of intrusive specimens.

This inclusion of multiple birds of one species in mortuary settings suggests a social investment and involvement over an extended period of time. Aside from Armour-Chelu's (1985) detailed work on the birds from the Neolithic settlement Links of Noltland, little taphonomic data have been recorded for birds from Scottish island sites, and in this instance her work was not focused on the eagle bone. New taphonomic work is currently being undertaken for the Cnoc Coig remains (Boyle pers. comm.), but further profiling is needed for avian bone in these settings, particularly in tombs. Eagle remains may have also been traded. However, the range of skeletal elements represented, combined with the lack of any definite examples of worked eagle bones in this context, suggests that generally these were locally captured fleshed birds.

The contexts in which these birds occur, the comparative zooarchaeological material and the later historical sources indicate that they may have been incorporated within Neolithic and Bronze Age contexts of the living and the dead for multiple reasons. The following sections examine the possible roles of raptors as pests, in falconry, in symbolism and as scavengers.

Pests

As noted above, eagles, and particularly the white-tailed eagle, can be regarded as competitors for food and a hazard to livestock and even human life. One account from Skye (recorded by Martin 1716, 299) indicates that these birds were capable of carrying off human infants; in this case the child's wails were heard by sheep-herders who rescued him. On St Kilda, although the white-tailed eagle was a rare visitor, it was dissuaded from settling or successfully breeding on the island by the burning of its nests (Connell 1887, 125). This was achieved by descending down the cliff face whilst under attack from these ferocious birds, and the unfortunate individual who had to perform this dangerous task was selected by lot (Connell 1887, 125). It was also possible to kill the white-tailed eagle by hand (*i.e.* with a club) after the animal had gorged heavily on prey and had difficulty getting off the ground (Cowles 1978, 63). Whilst these birds may have been killed as pests their inclusion within mortuary contexts is likely to relate more to themes associated with mastery over nature rather than the mere disposal of pest carcasses. Consequently, difficulties associated with their capture may have resulted in bones (or the whole carcass) being kept as trophies.

Falconry

The birds may have been kept for falconry but this is unlikely for a number of reasons. Firstly, the Neolithic and Bronze Age raptor concentrations pre-date the known introduction of bird-assisted hunting in Britain, which although unclear in origin only appears to enter Europe in the 3rd and 4th centuries AD (Prummel 1997, 335). Research in Sweden has identified falconry birds in over 30 graves dated between AD 500–1000 (Tyrberg 2002). Secondly, the majority of exploited raptors in the Hebrides, Orkney and Shetland are ill-suited for this purpose. Whilst eagles can be used for hunting, the white-tailed eagle, and some other raptors present in the Scottish island assemblages, such as the red kite (*Milvus milvus*), are prolific scavengers and less suitable for this task. This behaviour would, however, have attracted them to human settlements, thus presenting opportunities for them to be killed (Baxter 1993, 78–80; Hull 2001, 131–6; Yalden 2007, 471–3). Finally, no pathologies associated with falconry or captivity have been identified on the Scottish island birds, and although rare, these have been found on individuals from later English sites (Cherryson 2002, 307–12). The high numbers of individuals in Scottish island tombs are thus unlikely to represent trained birds.

Symbolism in life and death

These large and spectacular birds would have been prominent within the island skyscape, and their feathers, bones and talons may have been imbued with associations, such as strength, beauty and prowess in the sky. Avian items are likely to have been incorporated in both personal adornment and/or ritual activities. Unfortunately, direct evidence of feather usage only survives in exceptional preservation conditions, but feathers may have been present in mortuary contexts and other settings as decoration, for insulation, and as arrow fletchings. There are many ethnographic and historic examples of eagle feather usage. Within Native American societies the Plains Indians used eagle feather headdresses to demonstrate prowess as a warrior, and among the Cherokee only warriors dared to wear and carry its feathers (Mooney 1902, 281–3), whilst the Sioux chose eagle feathers for fletching to imbue the arrow with the eagle's flight and hunting qualities (Serjeantson 2009, 184–7). Historical texts, such as Hesiod's 'Shield of Heracles', also refer to fletching special arrows with eagle feathers (Hesiod, Sh. 130–5). Archaeological examples, including the jay wing bones at Zvejnieki, mentioned above, suggest that feathers could have been used for decoration both on and off the bone. Items made from large birds' bones, interpreted as pins, awls and a bead, were found at Bronze Age Cladh Hallan in the Outer Hebrides, indicating that bird bone was used for tools, and potentially decoration and personal adornment in the Scottish islands.

The movement of these spectacular, large birds around the sky may have also held symbolic significance in transcending boundaries and the movement between life and death. Such concepts are familiar in societies that viewed (and view) birds as carriers of messages or souls, or that perform shamanism. Archaeologically familiar examples include the Roman belief that the eagle associated with Jupiter transported dead emperors' souls into the realm of the gods, whilst in Norse mythology Odin's ravens (*Corvus corax*) Huginn (meaning thought) and Muninn (meaning mind) carried information of the world to the god (Moreman 2014, 6–7). Recent and current examples include the belief shared by several Native American groups that vultures can carry the dead to the spirit world, whilst in Tibetan folklore birds are generally regarded as divine messengers and include 'untouchable' birds, such as ravens and vultures, which are emissaries for death (Moreman 2014, 1–7). To the Cherokee (Southeastern United States) the chickadee (*Poecile*) is traditionally a bird that tells the truth, whilst the tufted titmouse (*Baeolophus bicolor*) brings false messages (Mooney 1902, 285–6). In the context of shamanism WaiWai shamans (Hawaii) use feathers to be transformed into birds and become winged messengers (Serjeantson 2009, 338–9), and the Mongol Buryats (Siberia) believe that deceased ancestors can return in bird form (Moreman 2014, 9; Tate 2007, 32–4).

In the Scottish islands, the distribution and frequency profiles of species in mortuary settings compared with contemporary settlement sites shows a raptor concentration within tombs and suggests that eagles in particular occupied a symbolic role in avian-human mortuary relationships. This could have included transcending boundaries between life and death, symbolically imbued associations, decoration, totemism and potentially involvement in excarnation.

Scavengers: engaging with the dead and their landscape

Although the white-tailed eagle mainly feeds on fish and marine foods it is a scavenger that will strip human flesh and scavenge the dead; in later periods, it was associated with the dead of the battlefield, such as in Anglo-Saxon and Norse poetry (Baxter 1993, 78–80; Yalden 2007, 471–3). As such, it is reasonable to propose, as has previously been suggested by others, that in some instances these large birds may have been involved in defleshing excarnated or exposed burials (Chesterman 1983; Hedges 1983; 1984; Henshall 1985). It should not be assumed, however, that all prehistoric tombs with multiple human occupants acted as depositories for disarticulated or partial remains following exposure/excarnation (Barber 1988; 1997; Henshall 2004; Lawrence 2012, 489–90). Indeed, Isbister has a relatively proportional abundance of skeletal elements, and tombs such as Midhowe have evidence for articulation and direct internment. The original interpretation that the Isbister human bones were weathered and sun-bleached through long exposure has recently been challenged by thorough reanalysis, which did not find evidence for exposure (Lawrence 2006). Instead, after extensive taphomomic observation Lawrence (2006; 2012, 571–97) concluded that at Isbister the bodies were most likely directly interred within the tomb. The minimum estimated number of human individuals at Isbister has been dramatically revised from 341 to 85 and the role of recovery practices has been recognised as a factor influencing the skeletal profile at this and at other sites (Chesterman 1983; Davidson and Henshall 1989, 54; Hedges 1983; Lawrence 2006; 2012, 489–90; Reilly 2003, 144–6).

However, it would be incorrect to assume that excarnation automatically equates to the full destruction and disarticulation of skeletal remains or that the process necessarily needs a long exposure time. In avian excarnation rites practiced today, such as Tibetan sky burials, the skeleton often remains largely articulated, except when intentionally dismembered to aid consumption, even though huge numbers of birds (in this instance vultures) process the body (Plate 22) (Logan 1997; MarcoM 2014). After extensive defleshing, special hammers are used to smash the skeleton and turn the bones into a fine powder (Goss and Klass 1997, 385; Logan 1997). As such, it is plausible that cadavers may have been exposed to these avian predators in a symbolic part of a Scottish island mortuary ritual without the body being fully processed in the environment outside the tomb.

The recovered human bone assemblage demonstrates that skeletal remains have not been located or survived for a large proportion of the prehistoric population and are not present in the tombs (Lawrence 2012, 571–2). This implies that other means of disposal would have also been employed for the wider population, with options including aqueous disposal and exposure/excarnation (Fowler 2010, 7; Lawrence 2006; 2012, 489–90, 579). Disarticulated human remains, including cranial fragments, have been discovered at several Orcadian Neolithic settlement sites, including Tofts Ness, Knap of Howar and Ness of Brodgar. These finds could suggest that exposure sites existed within the island landscape, from which such bone fragments could have been intentionally or accidentally integrated into settlement sites (Barber 1988; 1997; Henshall 2004; Lawrence 2007; 2012, 571). However, such bones may also result from a range of other disposal practices, such as exhumation and transportation of skeletal remains.

The dating of the eagle bones from Isbister makes it unlikely that these birds were directly involved with defleshing the humans deposited in the tomb. However, it is likely that the eagles (and other scavengers) would have been closely associated with the general landscape of death and the transitioning of the bodies from one form to another. Their scavenging behaviour would have attracted them to both human and animal dead including any excarnation/exposure sites handling human cadavers, as well as dead livestock and middens with processed animal components. These activities would have reinforced the eagle's relationship with the dead and potentially merged distinctions between human and animal in the state of death.

Whist so far in this paper the inclusion of these raptors in the tomb environment has been discussed as an active anthropogenic deposition, this is not the only interpretation. For example, at Isbister it is possible that these eagles are an intrusive deposit, since they are later in date than all bar perhaps the last human burials in the tomb (Lawrence 2012, 103–11; Sheridan 2005). It is unlikely that these eagles would have entered into a roofed tomb structure of their own volition, indicating that these birds are anthropogenically introduced materials. However, their presence may indicate that by the Bronze Age an eyrie had been established upon the now partially collapsing tomb structure. The radiocarbon dates of the selected eagle bones are similar to the date given by a male femur that comes from a later burial in the tomb, suggesting that the eagle bones may have accidentally entered the tomb at its reopening, or that they were intentionally deposited within it at the time of this later burial. However, even if the eagle bones were not intentionally incorporated into the burial environment, their presence may still be significant. Their occurrence in the landscape would have been notable and may have attracted Bronze Age residents to the area. Thus, if an eyrie had been built on the tomb then the avian relationship with this mortuary setting in the landscape may have been one of the reasons for choosing this location for a later internment. The birds' choice to occupy this location for their eyrie may have become entangled with human ideologies of death. This status is not an unfamiliar one for the eagles (and other carrion birds such as ravens); as already mentioned, even in much later periods they often remain associated with death and the battlefield, for example in the Anglo-Saxon poem Battle of Brunanburh, and the Norse Icelandic Höfuðlausn from Egil's Saga (Baxter 1993, 78–80; Yalden 2007, 471–3).

As previously noted, raptors were not the only birds to occur in the tomb environment. A variety of other species were present, including seabirds, landbirds, waders and small passerines. Whilst some of these, and in particular the latter, may be intrusive, several of the larger seabirds are unlikely to have entered such locations of their own will. As with the eagles, the possible reasons for their inclusion are manifold and range from talismanic status to being food for the dead. The combined Neolithic dataset from all site types also contained other raptor species, including buzzard, goshawk (*Accipiter gentilis*), short-eared owl (*Asio flammeus*), barn owl (*Tyto alba*), kestrel (*Falco tinnunculus*), and peregrine falcon (*Falco peregrinus*), some of which occasionally also occurred in tomb environments in low numbers. Whilst some of these raptors may have entered the archaeological record non-anthropogenically (*i.e.* through inhabiting/scavenging from a site at a time contemporary with its human occupants, or using an abandoned structure as a perch), a butchered falcon tarsometatarsus from the Neolithic phase of Outer Hebridean settlement Northton clearly demonstrates that people were interacting with and processing these predatory birds. As such the incorporation of raptors in the mortuary setting should be viewed as part of the wider interaction with these birds. Whether natural or anthropogenic, the remains reveal the range of predatory birds encountered within the Neolithic environment, which through their hunting and scavenging behaviours would have been intimately associated with areas of animal (and also human) death. These birds may have been captured for their feathers, symbolic associations, or in the case of larger eagles, to protect livestock. It would be interesting, should comparable assemblages from mortuary settings become available from the Outer Hebrides, to see if similar patterns of raptor use are present.

Conclusion

This paper has shown that in the Mesolithic human and avian deposits are found comingled within middens. The Mesolithic inhabitants were interacting with a variety of birds in life, and the evidence supporting an exposed burial tradition suggests that scavenging birds would have been attracted in death. In the Neolithic and Bronze Age, birds, and particularly the white-tailed eagle, may have played a socially and culturally significant role in Scottish island mortuary practices. Their presence in the tomb environment could have been directly or indirectly associated with human death rites. If directly incorporated into funerary activity they may have been used for personal adornment, acting as a messenger, moving between states of being, or in harnessing the power of the animal. If indirectly associated, the eagles may have used the tombs as an eyrie and become connected with the landscapes of the dead through their presence. After the Bronze Age the occurrence of eagles in the archaeological record declines, and the use of birds in the mortuary record is less evident. Changes in the bird populations exploited and in the role of birds in human engagement with death are apparent. Differences in bird use are also visible between the Scottish islands and other areas of Britain, such as England, particularly in the uptake of the chicken.

This paper has also demonstrated the need for further work to continue being conducted in this area. The picture of bird use in the mortuary environment needs to be contextualised within the wider archaeologically and historically documented tradition of fowling and avian-human interaction within the Scottish islands, which persisted into the recent past. Furthermore, when considering birds and death, or birds and the mortuary record, there is a need to look beyond the physical record. The role of birds in death can manifest itself in a multitude of ways. Forty-one miles to the west of the Outer Hebrides of Scotland lies the St Kilda archipelago. Here birds remained an integral part of life up until the last century, providing the inhabitants with food, feathers, oils, clothes and a myriad of other products. Such is their prominence in life that their role in death should be expected. A person does not need to be buried with birds for them to be part of the death rite and a feature of the mourning process. The following words of a lamenting mother in the form of a St Kildan elegy for a lost son reveals how birds were bound to death through the importance and significance that they held in life for the inhabitants of this location, and for other areas of the Scottish islands, through time and space:

> My share of eggs I shall never receive; the strong and alive will have them, for thou my son art gone. My share of the fowls now fly in the air, up to the clouds they ascend; there they sport and flutter; but I am sad and forlorn. (Campbell 1799)

Such examples serve to remind us that past avian-human relations in these locations were complex and often extended beyond the provision of food. Birds, in a myriad of forms, were a part of not just life, but also death, within the Scottish and other North Atlantic islands from their earliest occupation into the modern world.

Acknowledgements

Special thanks go to those who generously provided data and insights for this research project including Dale Serjeantson, Sheila Hamilton-Dyer, Miranda Armour-Chelu and Ingrid Mainland. Caroline Grigson and Katherine Boyle are also thanked for their communication and contribution. We are very grateful to Kristiina Mannermaa for her comments and information. Mark Maltby is also thanked for reading the draft paper. We wish to thank Ian Dennis who drew part of the base map for Figure 13.1. This research formed a part of Julia's PhD, supervised by Jacqui Mulville. As such Julia would like to thank the AHRC who funded her study. Development of the domestic fowl research was facilitated by the AHRC (Grant No AH/L006979/1).

References

Albrethsen, S. E. and Petersen, E. B. (1976). Excavation of a Mesolithic cemetery at Vedbæk, Denmark. *Acta Archaeologica* 47, 1–28.

Allison, E. (1997) Bird bones. In S. Buteux (ed.) *Settlements at Skaill, Deerness, Orkney. Excavations by Peter Gelling of the Prehistoric, Pictish, Viking and Later Periods* 1963–81, 247–8. BAR British Series No. 260. Oxford, Archaeopress.

Armour-Chelu, M. J. (1985) The Taphonomic and Cultural Significance of an Assemblage of Neolithic Bird Bones from the Links of Noltland, Orkney. Masters dissertation, University of London.

Barber, J. (1988) Isbister, Quanterness and the Point of Cott: the formulation and testing of some middle range theory.

In J. C. Barrett, and I. A. Kinnes (eds.) *The Archaeology of Context in the Neolithic and Bronze Age: Recent Trends*, 57–62. Sheffield, Department of Archaeology and Prehistory, University of Sheffield.

Barber, J. (1997) *The Excavation of a Stalled Cairn at the Point of Cott, Westray, Orkney*. Edinburgh, Scottish Trust for Archaeological Research.

Bartosiewicz, L. (2012) Vertebrate remains. In A. Saville, K. Hardy, R. Miket, T. Ballin (eds.) *An Corran, Staffin, Skye: A Rockshelter With Mesolithic and Later Occupation*, 47–59. Scottish Archaeology Internet Reports 51. Edinburgh, Society of Antiquaries of Scotland/Historic Scotland/Council for British Archaeology.

Baxter, I. (1993) Eagles in Anglo-Saxon and Norse poems. *Circaea* 10(2), 78–80.

Berger, L. (2006) Predatory bird damage to the Taung type-skull of *Australopithecus africanus* Dart 1925. *American Journal of Physical Anthropology* 131, 166–8.

Best, J. (2014) Living in Liminality: *an Osteoarchaeological Investigation into the Use of Avian Resources in North Atlantic Island Environments*. PhD thesis, Cardiff University.

Best, J. (in prep.) The bird bone. In N. Sharples (ed.) *A Norse Farmstead in the Outer Hebrides. Excavations at Mound 2/2A, Bornais, South Uist*. Oxford, Oxbow Books.

Best, J. and Cartledge, J. (in press) The bird bone. In M. P. Pearson, H. Smith, J. Mulville and M. Brennand (eds.) *Cille Pheadair: A Norse Period Farmstead on South Uist*. Oxford, Oxbow.

Best, J. and Mulville, J. (2010) The fowling economies of the Shiant Isles, Outer Hebrides: resource exploitation in a marginal environment. In W. Prummel, J. T. Zeiler and D. C. Brinkhuizen (eds.) *Birds in Archaeology, Proceedings of the 6th Meeting of the ICAZ Bird Working Group in Groningen 2008*, 87–96. Groningen, Barkhuis.

Best, J. and Mulville, J. (2013) Between the sea and sky: the archaeology of avian resource exploitation in Scottish island environments. In M. Y. Daire, C. Dupont, A. Baudry, C. Billard, J. M. Large, L. Lespez, E. Normand and C. Scarre (eds.) *Ancient Maritime Communities and the Relationship Between People and Environment Along the European Atlantic Coasts*, 395–405. BAR International Series No. 2570. Oxford, Archaeopress.

Best, J. and Mulville, J. (2014) A Bird in the hand: data collation and novel analysis of avian remains from South Uist, Outer Hebrides. *International Journal of Osteoarchaeology Special Issue: Birds and Archaeology: New Research* 24(3), 384–396.

Best, J. and Mulville, J. (2016) Birds from the water: reconstructing avian resource use and contribution to diet in prehistoric Scottish Island environments. *Journal of Archaeological Science Reports* 6, 654–64.

Best, J. and Mulville, J. (in press) Flock together: large-scale analysis of avian resource use in the North Atlantic Islands. *Open Quaternary. Special collection: Meta-analyses in zooarchaeology.*

Bramwell, D. (1981) Report on bones of birds. In R. Reece (ed.) *Excavations in Iona 1964 to 1974*, 45–6. Institute of Archaeology Occasional Publication No. 5. London, University of London.

Bramwell, D. (1983) The bird remains. In J. W. Hedges (ed.) *Isbister: A Chambered Tomb in Orkney*, 159–70. BAR British series No. 115. Oxford, Archaeopress.

Bramwell, D. (1994) The bird remains. In B. Ballin-Smith (ed.) *Howe: Four Millennia of Orkney Prehistory Excavations 1978–1982*, 153–7. Edinburgh, Society of Antiquaries of Scotland.

Campbell, R. (1799) *An Account of the Island of St Kilda and Neighbouring Islands, Visited August 1799*. National Library of Scotland, Manuscript 3051.

Cartledge, J. and Grimbly, C. (1999) The bird bone. In M. Parker Pearson and N. Sharples (eds.) *Between Land and Sea: Excavations at Dun Vulan, South Uist, SEARCH Volume 3*, 282–8. Sheffield, Sheffield Academic Press.

Cherryson, A. K. (2002) The identification of archaeological evidence for hawking in Medieval England. *Acta zoologica cracoviensia* 45, 307–14.

Chesterman, J. T. (1983) The human remains. In J. W. Hedges (ed.) *Isbister: A Chambered Tomb in Orkney*, 73–132. BAR British Series No. 115. Oxford, Archaeopress.

Connell, R. 1887. *St Kilda and the St Kildans*. London and Glasgow, Thomas D. Morison.

Cowles, G. S. (1978) Bird bones. In J. Mercer (ed.) The Investigation of the King's Cave, Isle of Jura, Argyll. *Glasgow Archaeological Journal* 5, 44–70.

Curle, J. (1932) An inventory of objects of Roman and provincial Roman origin found on sites in Scotland not definitely associated with Roman constructions. *Proceedings of the Society of Antiquaries of Scotland* 66, 227–397.

Davidson, J. L. and Henshall, A. S. (1989) *The Chambered Cairns of Orkney*. Edinburgh, Edinburgh University Press.

Downes, J. (ed.) (2012) *Chalcolithic and Bronze Age Panel Report*. Scottish Archaeological Research Framework, Society of Antiquaries of Scotland. Available at: http://tinyurl.com/clxgf5s (accessed accessed 30 October 2017).

Downes, J. M. and Ritchie, A. (eds.) (2003) *Sea Change: Orkney and Northern Europe in the Later Iron Age AD 300–800*. Forfar, The Pinkfoot Press.

Finlay, J. (1984) Faunal Evidence for Prehistoric Economy and Settlement in the Outer Hebrides to *c*. 400 AD. PhD Thesis, University of Edinburgh.

Finlayson, B. and Edwards, K. J. (2003) The Mesolithic. In K. J. Edwards, and I. B. M Ralston (eds.) *Scotland after the Ice Age: Environment, Archaeology and History 8000BC– AD 1000*, 109–26. Edinburgh, Edinburgh University Press.

Foster, A. (2012) *3 Driffield Terrace, York: Vertebrate Remains Analysis*. York, York Archaeological Trust.

Fowler, C. (2010) Pattern and diversity in the Early Neolithic mortuary practices of Britain and Ireland: contextualising the treatment of the dead. *Documenta Praehistorica* 37, 1–22.

Goss, R. E. and Klass, D. (1997) Tibetan Buddhism and the resolution of grief: the Bardo-Thodol for the dying and the grieving. *Death Studies* 21(4), 377–95.

Gumiński, W, (2005) Birds for dinner: Stone Age hunters of Dudka and Szczepanki, Masurian lakeland, NE-Poland. *Acta Archaeologica* 76, 111–48.

Hamilton-Dyer, S. (1991) *The Bird Bones from Tuquoy, Westray, Orkney*. Unpublished report, Ancient Monuments Scotland.

Hamilton-Dyer, S. and McCormick, F. (1993) The animal bones. In K. D. Connock, B. Finlayson and C. M. Mills (eds.) Excavation of a shell midden site at Carding Mill Bay near

Oban, Scotland. *Glasgow Archaeological Journal* 17, 34 (and Fiche M90–97).

Hamilton-Dyer, S. (1998) Bird bone. In C. Lowe (ed.) *St Boniface Church, Orkney: Coastal Erosion and Archaeological Assessment*, 155. Stroud, Sutton Publishing and Historic Scotland.

Harding, D. W. (2004) *The Iron Age in Northern Britain: Celts and Romans, Natives and Invaders*. Abingdon, Routledge.

Harland, J. (2006) Zooarchaeology in the Viking Age to Medieval Northern Isles, Scotland: An Investigation of Spatial and Temporal Patterning. PhD thesis, University of York.

Harland, J. F., Bennett, R. A., Andrews, J. I., O'Connor, T. and Barrett, J. H. (2012) Fowling: The Bird Bones. In J. H. Barrett (ed.) *Being an Islander: Production and Identity at Quoygrew, Orkney, AD 900–1600*, 155–60. Cambridge, McDonald Institute for Archaeological Research.

Harman, M. (1997) Bird bone. In J. Barber (ed.) *The Excavation of a Stalled Cairn at the Point of Cott, Westray, Orkney*, 49–51. Edinburgh, Scottish Trust for Archaeological Research.

Harman, M. (2009) Animal bone. In A. Ritchie (ed.) *On the Fringe of Neolithic Europe: Excavation of a Chambered Cairn on the Holm of Papa Westray, Orkney*, 48–59. Edinburgh, Society of Antiquaries of Scotland.

Henshall, A. S. (1985) The chambered cairns. In A. C. Renfrew (ed.) *The Prehistory of Orkney*, 83–117. Edinburgh, Edinburgh University Press.

Henshall, A. S. (2004) Scottish passage-graves: some confusions and conclusions. In A. Gibson and A. Sheridan (eds.) *From Sickles to Circles: Britain and Ireland at the Time of Stonehenge*, 78–91. Stroud, Tempus.

Hesiod (nd) *Shield of Heracles*. Transl. H. G. Evelyn-White (1914) *The Homeric Hymns, Epic Cycle, Homerica*. Loeb Classical Library Vol. 57. London, William Heinemann Ltd.

Hull, R. (2001) *Scottish Birds: Culture and Tradition*. Edinburgh, Mercat Press.

Hunter, F. and Carruthers (eds.) (2012) *Iron Age Panel Report*. Scottish Archaeological Research Framework, Society of Antiquaries of Scotland. Available at: http://tinyurl.com/cx4nlt8 (accessed 30 October 2017).

Jones, A. (1998) Where eagles dare: landscape, animals and the Neolithic of Orkney. *Journal of Material Culture* 3(3), 301–24.

Lawrence, D. (2006) Neolithic mortuary practice in Orkney. *Proceedings of the Society of Antiquaries of Scotland* 136, 47– 60.

Lawrence, D. (2007) The Human Remains from The Ness of Brodgar Excavations 2006. Unpublished report, Orkney Archaeological Trust.

Lawrence, D. (2012) Orkney's First Farmers. Reconstructing Biographies from Osteological Analysis to Gain Insights into Life and Society in a Neolithic Community on the Edge of Atlantic Europe. PhD thesis, University of Bradford.

Lauwerier, R. (1993) Bird remains in Roman graves. *Archaeofauna* 2, 75–82.

Logan, P. (1997) *Witness to a Tibetan Sky-Burial. A Field Report for the China Exploration and Research Society*. http://alumnus.caltech.edu/~pamlogan/skybury.htm.

Lentacker, A., Ervynck, A. and Van Neer, W. (2004) The symbolic meaning of the cock: the animal remains from the Mithraeum at Tienen (Belgium). In M. Martens, and G. De Boe (eds.) *Roman Mithraism: The Evidence of the Small Finds*, 57–80. Archeologie Vlaanderen Monografie 4. Brussels, Instituut voor het Archaeologisch Patrimonium.

Mainland, I. (1995) *A Preliminary Discussion of the Animal Bone Assemblage from the 1979–1993 Excavations at the Earl's Bu, Orphir, Orkney*. Sheffield, Sheffield Environmental Facility Report.

Mannermaa, K. (2006) Bird remains in the human burials at Zvejnieki, Latvia. Introduction to bird finds and a proposal for interpretation. In L. Larsson, and I. Zagorska (eds.) *Back to the Origin. New Research in the Mesolithic-Neolithic Zvejnieki Cemetery and Environment, Northern Latvia*, 289–300. Acta Archaeologica Lundensia Series 8, 52. Stockholm, Almqvist and Wiksell.

Mannermaa, K. (2008a) The Archaeology of Wings. Birds and people in the Baltic Sea region during the Stone Age. PhD thesis, University of Helsinki.

Mannermaa, K. (2008b) Birds and burials at Ajvide (Gotland, Sweeden) and Zvejnieki (Latvia) about 8000–3900 BP. *Journal of Anthropological Archaeology* 27, 201–25.

Mannermaa, K. (2016) Good to eat and good to think? Evidence of the consumption of ospreys (*Pandion haliaetus*) and white-tailed eagles (*Haliaeetus albicilla*) in the Late Mesolithic at Yuzhniy Oleniy Ostrov, NW Russia. In J. Grünberg, B. Gramsch, L. Larsson, J. Orschiedt and H. Meller (eds.) *Mesolithic Burials – Rites, Symbols and Social Organisation of Early Postglacial Communities*, 781–92. Halle-Saale, Tagungen des Landesmuseums für Vorgeschichte.

Mannermaa, K. and Storå, J. (2006) Stone Age exploitation of birds on the Island of Gotland, Baltic Sea: a taphonomic study of the avifauna on the Neolithic site of Ajvide. *International Journal of Osteoarchaeology* 16, 429–52.

Mannermaa, K., Zagorska, I., Jungner, H. and Zarina, G. (2007) New radiocarbon dates of human and bird bones from the Zvejnieki Stone Age burial ground in northern Latvia. *Before Farming* 2007(1), 1–12.

Marco, M. (2014) *Enterrement Tibétain Larung Gar Gompa*. Available at: https://vimeo.com/101074263 (accessed 13 May 2016).

Martin, M. (1716) A Description of the Western Islands of Scotland. London, privately published, available at http://bit.ly/2xGC6iF (accessed 30 October 2017)

Meiklejohn, C., Merrett, D. C., Nolan, R. W., Richards, M. P. and Mellars, P. (2005) Spatial relationships, dating and taphonomy of the human bones from the Mesolithic site of Cnoc Coig, Oronsay, Argyll, Scotland. *Proceedings of the Prehistoric Society* 71, 85–105.

Mooney, J. (1902) *Myths of the Cherokee*. Washington DC, Government Printing Office.

Moreman, C. (2014) On the relationship between birds and spirits of the dead. *Society and Animals* 22(5), 1–22.

Overton, N. J. and Hamilakis, Y. (2013) A manifesto for a social zooarchaeology. Swans and other beings in the Mesolithic. *Archaeological Dialogues*, 20(2), 111–36.

Parker, A. J., (1988) The birds of Roman Britain. *Oxford Journal of Archaeology* 7(2), 197–226

Parker Pearson, M., Sharples, N. and Symonds, J. (2004) *South Uist: Archaeology and History of a Hebridean island.* Gloucester, Tempus.

Parker Pearson, M. and Sharples, N. (1999) *Between Land and Sea: Excavations at Dun Vulan, South Uist*. Sheffield, Sheffield Academic Press.

Petrie, G. (1859) Description of antiquities in Orkney recently examined, with illustrative drawings. *Proceedings of the Society of Antiquaries of Scotland* 2, 59–60.

Pitts, M. (2006) Flight of the eagles. *British Archaeology* 86, 6.

Platt, M. (1933) Report on the animal bones from Jarlshof, Sumburgh, Shetland. In A. O. Curle (ed.) An account of further excavation in 1932 of the prehistoric township at Jarlshof, Shetland, on Behalf of H.M. Office of Works. *Proceedings of the Society of Antiquaries of Scotland* 67, 127–36.

Platt, M. (1934a) Report on the animal bones. In J. G. Callander and W. G. Grant (eds.) A long stalled chambered cairn or mausoleum (Rousay Type) near Midhowe, Rousay, Orkney. *Proceedings of the Society of Antiquaries of Scotland* 68, 348–50.

Platt, M. (1934b) Report on the animal bones from Jarlshof, Sumburgh, Shetland. In A. O. Curle, (ed.) An account of further excavation at Jarlshof, Sumburgh, Shetland in 1932 and 1933, on Behalf of H.M. Office of Works. *Proceedings of the Society of Antiquaries of Scotland* 68, 313–9.

Platt, M. (1936) Report on the animal bones found in the chambered cairn, Knowe of Ramsay, Rousay, Orkney. In J. G. Callander and W. G. Grant (eds.) A stalled chambered cairn, the Knowe of Ramsay, at Hullion, Rousay, Orkney. *Proceedings of the Society of Antiquaries of Scotland* 70, 415–9.

Platt, M. (1937) Report on the animal bones. In S. T. Calder (ed.) A Neolithic double-chambered cairn of the stalled type and later structures on the Calf of Eday, Orkney. *Proceedings of the Society of Antiquaries of Scotland* 71, 152–4.

Platt, M. (1956) Report on the animal bones. In J. R. C. Hamilton (ed.) *Excavations at Jarlshof, Shetland: Ministry of Works Archaeological Reports No. 1*, 212–5. Edinburgh, Her Majesty's Stationary Office.

Pollard, T. (1996) Time and tide: coastal environments, cosmology and ritual practice in early Prehistoric Scotland. In T. Pollard and A. Morrison (eds.) *The Early Prehistory of Scotland*, 198–210. Edinburgh, Edinburgh University Press.

Poole, K. (2010). Bird introductions. In T. O'Connor and N. Sykes (eds.) *Extinctions and Invasions: A Social History of British Fauna*, 156–65. Oxford, Windgather Press.

Prummel, W. (1997) Evidence of hawking (falconry) from bird and mammal bones. *International Journal of Osteoarchaeology* 7(4), 333–8.

Rackham, J. D. and Nicholson, R. (1989) Excavations 1960: the biological assemblage. In C. Morris (ed.) *The Birsay Bay Project Volume 1: Coastal Sites Beside the Brough Road, Birsay, Orkney, Excavations Between 1976–1982*, 77–8. Department of Archaeology Monograph Series 1. Durham, University of Durham.

Reilly, S. (2003) Processing the dead in Neolithic Orkney. *Oxford Journal of Archaeology* 22(2), 133–54.

Renfrew, A. C., Stenhouse, M. J. and Switsur R. (1983) The radiocarbon determinations. In J. W. Hedges (ed.) *Isbister: A Chambered Tomb in Orkney*, 61–71. BAR British Series No. 115. Oxford, Archaeopress.

Saville, A. and Wickham-Jones, C. (eds.) (2012) *Palaeolithic and Mesolithic Panel Report*. Scottish Archaeological Research Framework, Society of Antiquaries of Scotland. Available at: http://tinyurl.com/d86dgfq (accessed 30 October 2017).

Serjeantson, D. (1988) Archaeological and ethnographical evidence for seabird exploitation in Scotland. *Archaeozoologia* 2(1/2), 209–24.

Serjeantson, D. (2007a) The bird bones. In J. Hunter (ed.) *Excavations at Pool, Sanday. (Investigations in Sanday, Orkney 1)*, 279–85. Kirkwall, The Orcadian.

Serjeantson, D. (2007b) Bird bones. In S. J. Dockrill (ed.) *Tofts Ness, Sanday. (Investigations in Sanday, Orkney 2)*, 216–27. Kirkwall, The Orcadian.

Serjeantson, D. (2009) *Cambridge Manuals in Archaeology: Birds*. Cambridge, Cambridge University Press.

Serjeantson, D. (2010) Extinct birds. In T. O'Connor and N. Sykes (eds.) *Extinctions and Invasions: A Social History of British Fauna*, 146–55. Oxford, Windgather Press.

Serjeantson, D. (2013) *Farming and Fishing in the Outer Hebrides AD 600 to 1700: The Udal, North Uist*. Southampton, Highfield Press.

Serjeantson, D. (2014) Diverse assemblages, diverse origins: bird bones from coastal sites and the implications for the distribution of birds in the past. *International Journal of Osteoarchaeology Special Issue: Birds and Archaeology: New Research* 24(3), 279–88.

Sheridan, A. (2005) The National Museums Scotland radiocarbon dating programmes: Results obtained during 2004/5. *Discovery and Excavation in Scotland*, New Series 6, 182–3.

Schulting, R. J. and Richards, M. P. (2002a) The wet, the wild and the domesticated: the Mesolithic–Neolithic transition on the west coast of Scotland. *European Journal of Archaeology* 5, 147–89.

Smith, C. (2008) The Mammal and Bird Remains from Milla Skerra, Sandwick, Unst. Unpublished Report. GUARD Project 1969/2325.

Sykes, N. (2010). Worldviews in transition: the impact of exotic plants and animals on the Iron Age/Romano-British landscape. *Landscapes* 10(2), 19–36.

Tate, P. (2007) *Flights of Fancy*. New York, Delacorte.

Tyrberg, T. (2002) The archaeological record of domesticated and tamed birds in Sweden. *Acta Zoologica Cracoviensia* 45, 215–31.

Yalden, D. (2007) The older history of the white-tailed eagle in Britain. *British Birds* 100, 471–80.

Welander, R. D. E., Batey, C. and Cowie, T. G. (1987) A Viking burial from Kneep, Uig, Isle of Lewis. *Proceedings of the Society of Antiquaries of Scotland* 117, 149–74.

WWW1: http://d-maps.com/carte.php?num_car=18245&lang=en (accessed 30 October 2017).

WWW2: https://commons.wikimedia.org/wiki/File:White_tailed_Eagle_DG.JPG (accessed 30 October 2017).

WWW3: https://commons.wikimedia.org/wiki/File:Vulture_-_Sky_burial.jpg (accessed 30 October 2017).

WWW4: http://www.cotswoldarchaeology.co.uk/cirencester-roman-cockerel-back-from-its-clean-up/ (accessed 30 October 2017).

Chapter 14

Beyond Bones: Ritual and Social Secrets in Archaeological Remains

Brian Hayden

Introduction

The editors and contributors of this volume have strived to take bioanthropology beyond tabulations of bones and plant parts in order to address social issues and processes, especially in regards to rituals. In addition to the chapters that have gone before, I would like to raise another topic that deals with ritual, social processes, and biological materials; that is, secret societies. Secret societies have been almost entirely overlooked or neglected in European and Near Eastern archaeology although I have tried to emphasise their importance (Hayden 2003). In fact, they were completely overlooked in the recent massive *Oxford Compendium on Ritual and Religion in Archaeology* (Insoll 2011). Yet secret societies were very prevalent ethnographically in tribal and chiefdom level societies on most continents as well as among complex hunter-gatherers (Hayden, in press). One might well expect them to be represented in many prehistoric cultures, including some of the cases dealt with by authors in this volume. Therefore, it is worth keeping secret societies in mind when undertaking our interpretations of past ritual behaviour.

What are secret societies?

Since many readers may be unfamiliar with secret societies, it will be helpful, first, to explain what they are. Secret societies can be defined as associations with internal ranks in which membership, especially in the upper ranks, is exclusive, voluntary, and associated with secret knowledge (Hayden, in press). Typically, there are high costs for admittance and advancement in these societies, and the secret knowledge is usually of a ritual nature, such as the knowledge required to keep order in the cosmos, to control dangerous spirits, to heal people, and to procure abundance. Initiations generally involve severe physical trials that are meant to induce altered and ecstatic states of consciousness in the initiates, thereby confirming the contentions that the society holds keys to contacting supernatural forces and creating numinous experiences. Claims of members being transformed into animal or other patrons are common (*e.g.* Boas 1897; Drucker 1940; Harley 1941a; 1941b). Contrary to what many people imagine, membership is not secret, and in fact, there are usually major public demonstrations of the putative supernatural powers of secret society members. The secret is the knowledge that members – especially high-ranking members – possess and which non-members do not.

I argue that when societies began to produce surplus foods on a regular basis, ambitious individuals started to develop a range of strategies based on surplus production to concentrate political and economic power in their own hands. These strategies included: holding feasts that incurred debts, creating prestige objects that were then promoted as necessary for burials and other social transactions, developing high prices for marriages, and creating secret societies requiring wealth for admission. In general, only the most wealthy and powerful members of communities were able to attain high ranks within secret societies, and these high ranks conferred great power.

It is currently fashionable in many parts of the world to view the first indications of ritual in the prehistoric record as functioning to enhance community solidarity and group identity, whether in the Near East, in Europe, or in the American Southwest (*e.g.* Goring-Morris and Belfer-Cohen 2008; Kuijt 1996; 2000, 145, 148; Saitta 2013; but see critiques for the Southwest by Ware 2014, 93). As may be appreciated from the introduction, a number of ethnographers have pointedly emphasised that secret societies do *not* enhance social identities or community integration beyond their own membership. Rather secret societies exist to promote the self-interests and power of those in the highest positions in secret societies, and hence they are a major means of creating socioeconomic inequalities and social divisions. More than one ethnographer has referred to the use of terror to enforce the will of secret society leaders, or to the societies as 'terrorist organizations' (Drucker 1940). Anyone with ambitions or who wanted to protect their interests, eagerly tried to become members and advance as far as they could.

Implications for bioanthropology

Secret society practices affect bioanthropological remains in a number of ways. However, as I hope is evident, if secret societies can be identified as having existed in the past, the implications for prehistoric social organisations and dynamics are profound and create a very different kind of framework from the functionalist models for reconstructing past cultures that are currently in fashion (*e.g.* Goring-Morris and Belfer-Cohen 2008; Kuijt 2000; Saitta 2013; Sterelny and Watkins 2015).

Secret society materiality can express itself in many ways: in special structures, in special burials for high ranking members, in special ritual paraphernalia, in special practices or foods (notably soups, alcoholic beverages, finely ground flours or meals, and anthropophagy), and in special iconographies. There are some basic commonalities in the manifestations of these materialities, but the specific contents, of course, vary with the materials regionally available and with specific culture histories. For the purposes of this paper, I will focus on some examples of ritual paraphernalia, iconographies, and burials from secret societies with a brief mention of some specialty foods and the use of caves.

Ritual paraphernalia

In order to make claims of ritual supernatural power more believable, exotic, usually costly, materials generally were more effective, coming as they did from unknown realms with unknown origins, combined with claims for great powers. Specially made, complex, items were also part of ritual paraphernalia. Thus it was that members of the Midewiwin society of the Ojibway Indians around the Great Lakes of North America 'shot' cowrie shells into the bodies of initiates and killed them, only to have them revived by removing the cowrie shells from their bodies (Hoffman 1891) – the source of cowrie shells being from oceans thousands of miles away. Shells from distant sources occur frequently in other parts of the world as secret society paraphernalia. On the Northwest Coast, members of secret societies used quartz crystals instead of cowrie shells (McIlwraith 1948b). More generally in North America, it was a common practice to use bone and wood whistles and bullroarers in secret society ceremonies and to claim that the sounds they produced were the voices of spirits – secrets that were never to be revealed upon pain of death. Bone drinking tubes were also commonly used in the Americas during the seclusion of new initiates into secret societies (Loeb 1929). Specially raised pigs with circular tusks were required for initiations and advancements in the Banks Islands (Vanuatu, Melanesia) as well as being required for ceremonies (Deacon and Wedgewood 1934). Such pigs had to be fed by hand using tubes and were therefore of great value (*ibid*). Special bird feathers or down formed part of secret society costumes in many parts of the world as did wooden or fibre masks and animal skins with horns, antlers, and/or claws or hoofs attached (*e.g.* Boas 1897; Deacon and Wedgewood 1934; Drucker 1940; Harley 1941a; 1941b).

Some of the papers in this volume provide potential examples of items that could have been secret society paraphernalia. Specifically, it can be wondered if the bird bone tubes and flutes reported by Best and Mulville could represent secret society activities as well as the many uses of bird wings cited, especially the Neolithic child buried in Estonia with a bone flute and swan wing (Jonuks and Rannamaë this volume). A child would not be expected to be a shaman, but elite children were often initiated into secret societies. Shamans could have been using the carved bone and tooth pendants reported by Jonuks and Rannamaë for Estonia (this volume), however, they could also have been part of secret society paraphernalia especially since shamans were often members of secret societies and these organisations frequently adopted shamanic costume and other elements. The use of frankincense in Roman Britain as reported by Brettell *et al.* (this volume) might simply reflect the wealth of a buried individual, but such use of an exotic ritual material would also be consistent with membership in a mystery cult and resulting special burial treatment. I will suggest that the Classic mystery cults may have developed from secret societies or may be special cases of them.

Iconography

One of the most insightful areas of materiality of secret societies is iconography. Because the primary goal and lure of secret society membership is obtaining power, secret societies generally demonstrated and publicised their powers at public feasts near their ritual sites, such as could have been the case at High Pasture Cave (McKenzie this volume). Elaborate costuming reflecting secret society ideology and iconography of the secret society were generally part of these performances. The animal patrons of the societies and their members typically took the form of what can be called 'power animals' – types of animals or birds that exhibit power, domination, fierceness, or danger (*e.g.* Boas

1897; Talbot 1923). These were also the animals into which powerful members of the societies said that they could transform themselves and which featured prominently in the iconography of the societies or in the households of their members (*ibid*). Such animals most commonly included: bears, felids, canids, bovids, boars, crocodiles, elephants, raptors, snakes, scorpions, and similar species. These could be carved in wood, as with Northwest Coast totem poles, or in bone or antler; they could be painted; they could be modelled in clay; or parts of these animals could be used in society costumes or paraphernalia (*ibid*). In particular, secret society members on the American West Coast often donned bear skins for dances or predatory exploits (*ibid*). Secret society members in Africa often used lion or panther skins for similar purposes. While these skins generally did not preserve very well archaeologically, the claws, and sometimes cranial bones, were usually left attached with the third phalanges, thus providing good archaeological indicators of secret society practices. Raptor talons or feathers, or those of other symbolic birds, were frequently part of costumes (*ibid*).

In addition to this, various bones of these power animals could be kept as parts of medicine bundles or for display. This not only explains the rather strange occurrence of isolated bones from power animals at sites, like the single panther bone at Çatalhöyük (Hodder 2011, 359), but also explains the predominance of 'aggressive', dangerous, or fierce types of animals in the walls of the ritual areas of houses in locations like Çatalhöyük (a source of puzzlement for the excavators), and the dominance of power animal iconography in sites like Göbekli Tepe (bulls, scorpions, boars, snakes). It also explains the dominant importance of aurochs in the iconography and ritual instalment of bucrania at sites like Çatalhöyük, Hallan Çemi, and Qermez Dere.

It should be noted that on the American Northwest Coast, secret society meetings, ceremonies, and initiations were held in the homes of leaders, which may have also been the case in Çatalhöyük. In fact, McIlwraith (1948b, 22) reported that important individuals who belonged to the *Kusiut* secret society often killed and buried slaves in their houses to give more power to the secret society paraphernalia kept there. Archaeologists generally interpret burials in households as foundation deposits or family burials, but such occurrences might also represent 'power' burials by secret society members as described by McIlwraith, especially where neonates or children were involved, or perhaps even involving young domestic animals as substitutes for human lives (as mentioned by Iborra reporting on the various theories on the subject for her study area, this volume). The predominance of 'aggressive' power animals in early ritual contexts has puzzled archaeologists for over a hundred years, and clearly did not fit the predominant views of hunting-related rituals focused on game animals or social integration, or of other peaceful, nurturing, matriarchal models of early Neolithic religion and society. However, the emphasis on fierce aggressive animals is entirely understandable and expectable in the context of secret societies.

Special food

While it is entirely expectable that feasts hosted by secret societies would feature unusual, and perhaps symbolic foods, such occurrences might be difficult to distinguish from other kinds of feasts, such as those for marriages, funerals, or alliances, especially those hosted by elites. Animal sacrifices, as well, could have been demonstrations of elite wealth or power as well as secret society performances. Nevertheless, extreme cases, such as the many perinatal piglets reported by Silvestri *et al.* (this volume) and the Greek Thesmophoria, seem more suggestive of ritual symbolism characteristic of secret societies.

Burials

One other striking feature of many secret societies that is generally ignored by archaeologists is the special burial treatment often accorded to high ranking members of secret societies. Sometimes skulls are modelled in clay or other plastic materials and kept in sacred structures (Speiser 1996). Sometimes members are buried lavishly in their houses or house compounds. However, most often, high ranking members are buried in secret locations or under large stones or stone heaps so that those who seek power cannot get access to the bones of the deceased and use them as sources of supernatural power (*ibid*). Thus, a number of the burials discussed in this volume might be candidates for secret society members, including the Bronze Age cave burials in central Italy (Silvestri *et al.* this volume), the burials in the Balearic Isles caves or hypogeums (Picornell *et al.* and Riera *et al.* this volume), and the Roman burial with frankincense (Brettell *et al.* this volume). One must also be wary about attributing disarticulated human remains to intentional burials. They might also be the remains of human sacrifices or even cannibalism, both of which are relatively common in the more powerful secret societies (*e.g.* Deacon and Wedgewood 1934; Harley 1941a; 1941b; Talbot 1923).

Caves

There is considerable ethnographic documentation for the use of caves by secret societies either for holding rituals or as locations for secluding new initiates (Hayden, in press). Good examples come from California and the Northwest Coast. In the Southwestern Pueblos, secret societies used both kivas and remote caves for some of their special rituals. Archaeologists have often recovered important caches of ritual paraphernalia from such caves including feathers, scalps, prayer sticks, and weapons (Ellis and Hammack

1968). However, remarkably, with the exception of Ruth Whitehouse (1992) and a few others, archaeologists in the Old World and the New World, by and large, have ignored this in their interpretations of the ritual uses of caves. Given the suitability of caves for conducting ritual activities in secret, for use by exclusive groups, for inducing numinous emotions, and for creating altered states of consciousness, caves are eminently adapted for secret society activities. It seems to me that the first issue that any archaeologist dealing with cave rituals ought to deal with is whether secret societies could have been involved in the use of the caves being studied. I think that a prime candidate would be High Pasture Cave in Scotland (McKenzie this volume) given the restricted, probably hierarchical, access to the cave, the elaborate modifications at the entrance (which secret societies would be very capable of undertaking given their typical wealth resources), evidence of wealth offerings and sacrifices (including a lyre), and the evidence of large-scale feasting outside the cave as might be expected at secret society display feasts.

Similarly, I would expect many of the Bronze Age caves in Italy reported by Silvestri *et al.* (this volume) and others discussed by Villa (1986; 1992) to have been used by secret societies. Whether the human bones were from disturbed burials, primary burials, or perhaps from human sacrifices or even cannibalism is not clear in the examples discussed by Silvestri *et al.* (this volume), but Villa makes a good case for these practices. Human sacrifices and cannibalism tend to be common occurrences in many of the most powerful secret societies. Whether the hypogeum cave in Mallorca described by Picornell *et al.* (this volume) can be considered in such a context is difficult to say. The many urns with infant remains (natural deaths or sacrifices?) is a strange feature for any interpretive scenario. As other authors have observed, infants are not usually highly valued and often not even buried. However, the Mallorcan caves display an unusual investment in the creation of hypogeums, terraforming, and offerings that do not seem commensurate with unvalued infant deaths. What then are the alternatives?

Other archaeological examples

Because secret societies have not featured in most discussions of early ritual life, I would like to provide a few examples of what I consider to be good candidates for prehistoric secret societies. The earliest possible indication of secret society practices (or perhaps proto-secret society practices) comes from the Middle Palaeolithic in Europe. The deep cave at Bruniquel has two circles of broken stalagmites 200m into the dark zone dating to 200,000 years ago (Jaubert *et al.* 2016), indicating that small groups of people entered the cave for ritual purposes, plausibly including the inducing of altered states of consciousness. In a similar context, a mammoth tusk (an important part of a power animal) was brought into a dark cave chamber unfit for habitation, but left on the Middle Palaeolithic floor of the Gallerie Schoffelin at Arcy sur Cure (Farizy 1990, 307; Girard 1976, 53), while a complete bear was buried adjacent to a Neanderthal body at Regourdou (Bonifay 1965; 2002). In addition, the Grotte de Hortus (an obscure cave 200m above the valley bottom) yielded sets of panthers' third phalanges, probably part of a costume, as well as an articulated bear foot that may indicate a similar use of a bear skin (Pillard 1972, 170, 174). The excavated occupation floor is at the bottom of a narrow rock fissure totally hidden from outside views and so small that only a few people could have participated in any activities, although the shelter may have been used as a storage location for ritual paraphernalia. The occupation also included human remains from at least 20 individuals with an emphasis on skulls (associated with skulls of Alpine ibexes) as well as evidence of cannibalism (Lumley *et al.* 1972, 616–9). Given the fauna present, the analysts felt certain that the cannibalism was not engaged in for subsistence purposes, but for ritual purposes. All these features fit well with secret society characteristics. Eagle talons and feathers appear to also have been used by Neanderthals as part of costumes or ornaments (Peresani *et al.* 2011; Radovcic *et al.* 2015). These finds hint at some kind of ritual practices similar to secret society organisations, but by themselves are not conclusive.

Much more convincing arguments can be made for the use by secret societies of the elaborately painted Upper Palaeolithic caves in Southwest France, such as Lascaux and Font-de-Gaume (Hayden, in press). The deep cave locations lend themselves to secret meetings and the induction of altered states of consciousness. Many of the decorated areas are not suitable for large groups of people, and certainly not entire communities. The cave iconography is predominantly of power animals (mammoths, bison, aurochs, felids, ursids), the iconography includes carved examples of costumed or transformed humans (therianthropes), and bone whistles and bullroarers are found predominantly in caves (Morley 2013). Bear skulls were even carefully positioned in some caves like Chauvet (Clottes 2003). In other caves or rock shelters, the use or storage of raptor wings occurred (Bouchud 1953) and undoubtedly constituted parts of ritual costumes, such as those characteristically used in secret society performances or by shamans or elites.

Another good candidate for secret society activity is the burial of a Late Natufian ritual specialist in Hilazon Tachtit Cave in the Levant, including the copious remains of a funerary feast comprising 86 tortoises together with two marten skulls, a wild boar forearm, an aurochs tail, a leopard pelvis, and the articulated remains of a human foot (Grosman and Munro 2016). These remains indicate the burial of someone with exceptional status (*ibid*). The cave is small and 150m up a steep incline above the bed of the wadi. Secret societies often buried their most important members in remote and secret locations, and Hilazon Tachtit seems like an ideal location (Hayden, 2017). In my opinion, it is difficult to explain the highly unusual treatment and location of this burial in other terms.

The PPNA site of Jerf el Ahmar strongly resembles the kiva structures found in Pueblo communities in the American Southwest, and kivas were constructed for secret society uses. Jerf el Ahmar also has good evidence of human sacrifice and the cooking of human heads associated with the ritual structures (Stordeur and Abbès 2002). Ethnographically, human sacrifice and anthropophagy are common means of displaying power by secret societies and perhaps inducing altered emotional states of consciousness.

The other major PPNA candidate for a special secret society meeting place and specialised secret society structure is Göbekli Tepe in Eastern Anatolia. There is no indication of any domestic structure or settlement in the vicinity of the site, which is located on the top of a regionally prominent mountain with no water (Peters and Schmidt 2004). The remote location together with the massive scale of the megalithic buildings and the power animal iconography, all indicate a wealthy, powerful exclusive group, largely concerned with the exercise of power. The semi-subterranean structures, similar to kivas, would have kept proceedings private and secret. It is a remarkable site, and one that is difficult to explain without appealing to a secret society type of organisation. Unfortunately, the bioanthropological analyses have not been completed, but do include human remains mixed with the animal remains (Dietrich pers. comm.).

Secret societies undoubtedly continued to exist into the Neolithic and Bronze Ages in the Near East and in Europe, as exemplified by the PPNB site of Nahal Hemar Cave far from any known settlement (probably a storage location for secret society paraphernalia, replete with stone masks adorned with human hair – Bar-Yosef and Alon 1988), and Nakovana Cave in Croatia with its copious evidence of clandestine drinking surrounding a phallic shaped stalagmite (Kaiser 2005; Kaiser and Forenbaher 2012).

In North America, archaeological examples of kivas constitute some of the best candidates for secret society structures (Ware 2014), as well as caves used for ritual purposes. On the American Northwest Coast, no ritual structures have been identified, although this is probably due to the ethnographic pattern of hosting secret society events in large residential longhouses (Boas 1897; Drucker 1940). In the Northwest Interior, however, I have been excavating what I believe to be secret society structures at the Keatley Creek site, dating back 1,000–2,000 years. These are modest size structures, only 7–8m in diameter, located 100–200m from the main winter village. One interesting aspect in one of these structures are the copious amount of articulated salmon fins, heads, and tails left on the floor. This was very different from normal residential food waste, and I think these bone segments probably represented the remains of fish soups, which would have been specialty dishes highly relished during the cold winters.

Evolved possibilities

I would like to raise the possibility that some secret societies succeeded in centralising power to the extent that they were able to establish regional cult centres that eventually were transformed into some of the Classical cults, such as those at Delphi, and later into most of the 'mystery' cults, such as the Eleusinian mysteries and Isis cults that claimed to hold the keys to rebirth after death as mentioned by Koch *et al.* (this volume). The Apollo sanctuary described by Veropoulidou and Nikolaidou (this volume) might even qualify, depending on how its administration and internal hierarchy was organised. In the New World, Chavín de Huántar and the Chacoan Great Houses may represent evolved forms of secret societies that developed into regional cults. It is far from unusual for the major early centres of 'civilisations' to be ritual centres. This is difficult to explain unless one appeals to the 'power of belief' or to some way of linking ritual life with political power. While I do not think that belief was powerful enough to create Stonehenge or any major centre, secret societies provide just such a link between ritual and power.

Secret societies also share many characteristics of elite 'state' religions and cults including: voluntary membership; hierarchical organisations; high costs of membership and advancement; self-serving practices; claims to provide important benefits for the communities through their supernatural contacts (controlled by high-ranking members); displays of power in the form of elaborate temples, rich furnishings, and lavish public performances or feasts with impressive costumes representing supernatural beings and special foods. Thus, I think that secret societies need to be carefully considered wherever elaborate remains of ritual life are encountered. They provide a far different perspective on the nature of early communities and polities than the dominant functionalist paradigms in which rituals acted to sanctify and unify society as a whole, providing members with a social identity. Ethnographic reality is far different. Should we expect prehistoric reality to be otherwise?

Conclusions

Ethnographically, secret societies were prominent in many transegalitarian and chiefdom societies. We should expect them to have been similarly prominent in the archaeological record of complex hunter/gatherers, horticulturalists, and chiefdoms. Yet, with a few exceptions, archaeologists have not interpreted any of their evidence for rituals in these terms. Cave and kiva-like occurrences (outside the Southwest) are prime candidates for potential activities of secret societies, and it is remarkable that this possibility is not addressed in the archaeological literature. Thus, I would encourage archaeologists to re-examine their data with the possibility of secret society interpretations in mind. However, I would also urge due caution so as to avoid blind interpretations of virtually any and every indication of ritual activity as evidence for secret societies. Certainly, there were alternative forms of ritual behaviour including ancestor worship, tribal initiations, age grade rituals, individual shamanism and sorcery, and many social dance or festive rituals that lack the power or other components of secret societies. The task for the future will be to try to distinguish between the many forms of rituals and their different implications for social dynamics of the past.

Acknowledgements

I would like to thank the editors for the opportunity to present some of my thoughts on secret societies. I am very grateful to the Sir John Templeton Foundation for supporting my research on secret societies although the opinions expressed do not necessarily reflect the views of the Foundation.

References

Bar-Yosef, O. and Alon., D. (1988) *Nahal Hemar Cave*. Jerusalem, Atiqot 18.

Boas, F. (1897) The social organization and the secret societies of the Kwakiutl Indians. *Annual Report, U.S. National Museum*, 313–738.

Bonifay, E. (1965) Un Ensemble Rituel Moustérien à la Grotte du Régourdou. Paper presented at the *VI Congrès International de l'Union Internationale des Sciences Préhistoriques et Protohistoriques*, Rome.

Bonifay, E. (2002) L'Homme de Néandertal et l'Ours (Ursus Arctos) dans la Grotte du Régourdou. In T. Tillet and L. Binford (eds.) *L'Ours et l'Homme*, 247–54. Liège, ERAUL.

Bouchud, J. (1953) Les Paléolithiques utilisaient-ils les plumes? *Bulletin de la Société Préhistorique Française* 50, 556–60.

Clottes, J. (2003) *Chauvet Cave*. Salt Lake City UT, University of Utah Press.

Deacon, A. B. and Wedgewood, C. (1934) *Malekula*. London, Routledge.

Drucker, P. (1940) Kwakiutl Dancing Societies. *University of California Anthropological Records* 2(6), 201–30.

Ellis, F. and Hammack, L. (1968) The inner sanctum of Feather Cave, a Mogollon Sun and Earth shrine linking Mexico and the Southwest. *American Antiquity* 33, 25–44.

Farizy, C. (1990) Du Mousterien au Châtelperronien à Arcy-sur-Cure. *Mémoire, Musée de Préhistoire, Ile-de-France* 3, 281–9.

Girard, C. (1976) L'Habitat et la Mode de Vie au Paléolithique Moyen. Paper presented at the *Les Structures d'Habitats Paléolithique Moyen*, Nice.

Goring-Morris, N. and Belfer-Cohen, A. (2008) A roof over one's head: Developments in Near Eastern residential architecture across the Epipaleolihtic-Neolithic transition. In J. Bocquet-Appel and O. Bar-Yosef (eds.) *The Neolithic Demographic Transition and its Consequences*, 239–85. New York, Springer.

Grosman, L. and Munro, N. (2016) A Natufian ritual event. *Current Anthropology* 57, 311–31.

Harley, G. (1941a) *Notes on the Poro in Liberia*. Vol. XIX(2), Peabody Museum Papers, Cambridge, Harvard University.

Harley, G. (1941b) *Native African Medicine*. Cambridge MA, Harvard University.

Hayden, B. (2003) *Shamans, Sorcerers, and Saints: A Prehistory of Religion*. Washington DC, Smithsonian Books.

Hayden, B. (2017) Are burials socially integrative? The Natufian case. *Current Anthropology* 58, 410–11.

Hayden, B. (in press) *The Power of Ritual in Prehistory: Secret Societies and the Origins of Social Complexity*. Cambridge–New York, Cambridge University Press.

Hodder, I. (2006) *The Leopard's Tale*. London, Thames and Hudson.

Hoffman, W. J. (1891) The Midewiwin or 'Grand Medicine Society' of the Ojibwa. *Smithsonian Institution, Bureau of American Ethnology, Annual Report* 7, 149–299.

Insoll, T. (ed.) (2011) *The Archaeology of Religion*. Oxford, Oxford University Press.

Jaubert, J. Verheyden. S., Genty, D., Soulier, M., Cheng, H., Blamart, D., Burlet, C., Camus, H., Delaby, S., Deldicque, D., Edwards, R. L., Ferrier, C., Lacrampe-Cuyaubère, F., Lévêque, F., Maksud, F., Mora, P., Muth, X., Régnier, E., Rouzaud, J.-N. and Santos, F. (2016) Early Neandertal construcions deep in Bruniquel Cave in southwestern France. *Nature* 534, 111–4.

Kaiser, T. (2005) Archaeological caving in Croatia. *Expedition* 47(3), 25–9.

Kaiser, T. and Staso, F. (2012) Recognizing ritual in the dark: Nakovana Cave anbd the end of the Adriatic Iron Age. In H. Moyes (ed.) *Sacred Darkness*, 263–74. Boulder CO, University Press of Colorado.

Kuijt, I. (1996) Negotiating equality through ritual: a consideration of Late Natufian and Prepottery Neolithic A period mortuary practices. *Journal of Anthropological Archaeology* 15, 313–36.

Kuijt, I. (2000) Keeping the peace: ritual, skull caching, and community integration in the Levantine Neolithic. In I. Kuijt (ed.) *Life in Neolithic Farming Communities*, 137–63. New York, Kluwer Academic/Plenum.

Loeb, E. (1929) Tribal initiations and secret societies. *University of California Publications in American Archaeology and Ethnology* 25(3), 249–88.

Lumley, H. d., Lumley, M.-A. d., Brandi, R., Guerrier, E., Pillard, F. and Pillard, B. (1972) Haltes et Campements de Chasseurs Néandertaliens dans la Grotte de l'Hortus. In H. d. Lumley (ed.) *La Grotte de l'Hortus*, 527–624. Marseille, Laboratoire de Paléontologie Humaine et de Préhistoire.

McIlwraith, T. (1948) *The Bella Coola Indians, Vol. II*. Toronto, University of Toronto Press.

Morley, I. (2013) *The Prehistory of Music*. Oxford, Oxford University Press.

Peresani, M., Flore, I., Gala, M., Romandini, M. and Tagliacozzo, A. (2011) Late Neandertals and the intentional removal of feathers as evidenced from bird bone taphonomy at Fumane

Cave 44 ky BP, Italy. *Proceedings, National Academy of Sciences* 108, 3888–93.

Peters, J. and Schmidt, K. (2004) Animals in the symbolic world of Pre-Pottery Neolithic Göbekli Tepe, south-eastern Turkey. *Anthropozoologica* 39, 179–221.

Pillard, B. (1972) La faune des grands mammifères du Würmien II de la grotte de l'Hortus. In H. d. Lumley (ed.) *La Grotte Moustérienne de l'Hortus*, 163–205. Marseille, CNRS.

Radovcic, D., Srsen, A., Radovcic, J. and Frayer, D. (2015) Evidence for Neandertal jewelry: modified white-tailed eagle claws at Krapina. *PLoS One* 10(3), e0119802.

Saitta, D. (2013) Agency and collective action. In D. Carballo (ed.) *Cooperation and Collective Action*, 129–49. Boulder CO, University Press of Colorado.

Speiser, F. (1996) *Ethnology of Vanuatu: An Early Twentieth Century Study*. Honolulu HI, University of Hawaii Press.

Sterelny, K. and Watkins T. (2015) Neolithization in Southwest Asia in a context of niche construction theory. *Cambridge Archaeological Journal* 25, 673–705.

Stordeur, D. and Abbès, F. (2002) Du PPNA au PPNB: Mise en Lumière d'une Phase de Transition à Jerf el Ahmar (Syrie). *Bulletin, Société Préhistorique Française* 99, 563–95.

Talbot, A. (1923) *Life in Southern Nigeria: Magic, Beliefs and Customs of the Ibibio Tribe*. New York, Barnes and Noble.

Villa, P. (1986) Cannibalism in the Neolithic. *Science* 233, 431–4.

Villa, P. (1992) Cannibalism in prehistoric Europe. *Evolutionary Archaeology* 1(3), 93–104.

Ware, J. (2014) *A Pueblo Social History*. Santa Fe NM, School for Advanced Research.

Whitehouse, R. (1992) *Underground Religion: Cult and Culture in Prehistoric Italy*. London, Accordia Research Centre, University of London.

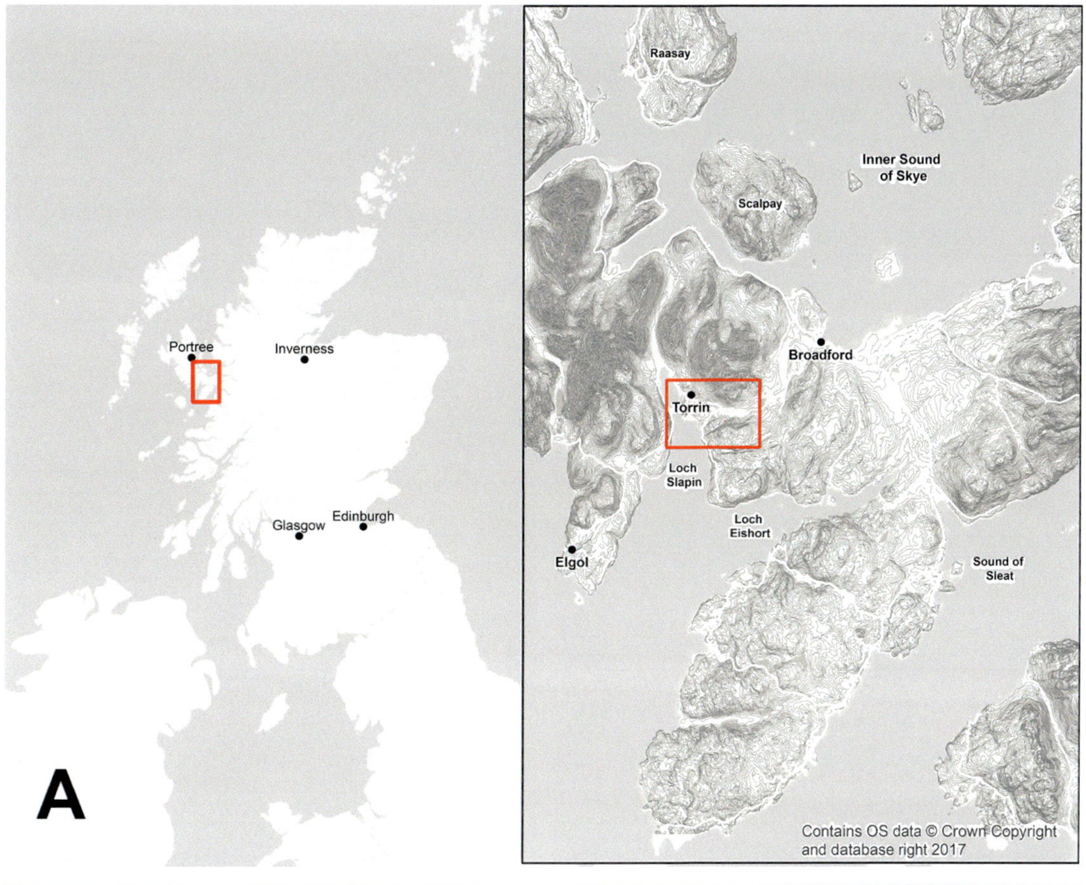

Plate 1 A) the location of High Pasture Cave, south-east Isle of Skye, Western Isles of Scotland; B) the site under excavation.

Plate 2 A) the cave entrance with precinct fully excavated and primary walkway visible, looking NE; B) view N along the walkway towards the cave. Unexcavated precinct deposits and the in-situ stairwell obscuring the cave entrance; C) SW precinct view from the cave, showing boulder infill into sections of the walkway; D) excavation within the cave passage.

(a)

(b)

(c)

(d)

Plate 3 a) view NNE through the deep and complex precinct deposits, showing the partially revealed primary walkway and the natural cave entrance at the base of the limestone face; b) fuel residue and midden accumulation near the precinct centre; c) N view of the Phase 3 stairwell, showing retaining wall and paved landing; d) NE view of the Phase 4 stairwell inserted within the Phase 3 structure.

a

b

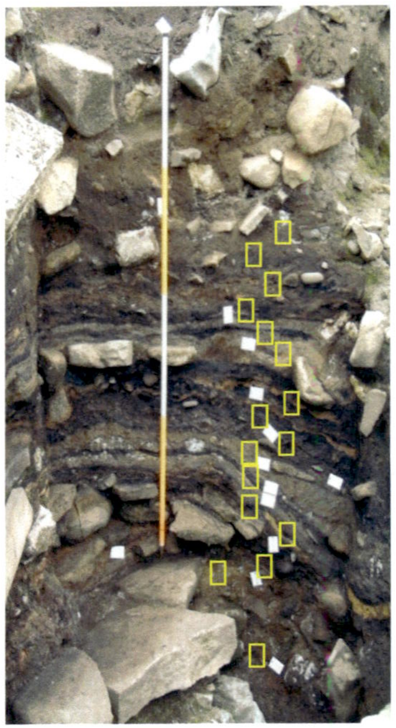

c d e

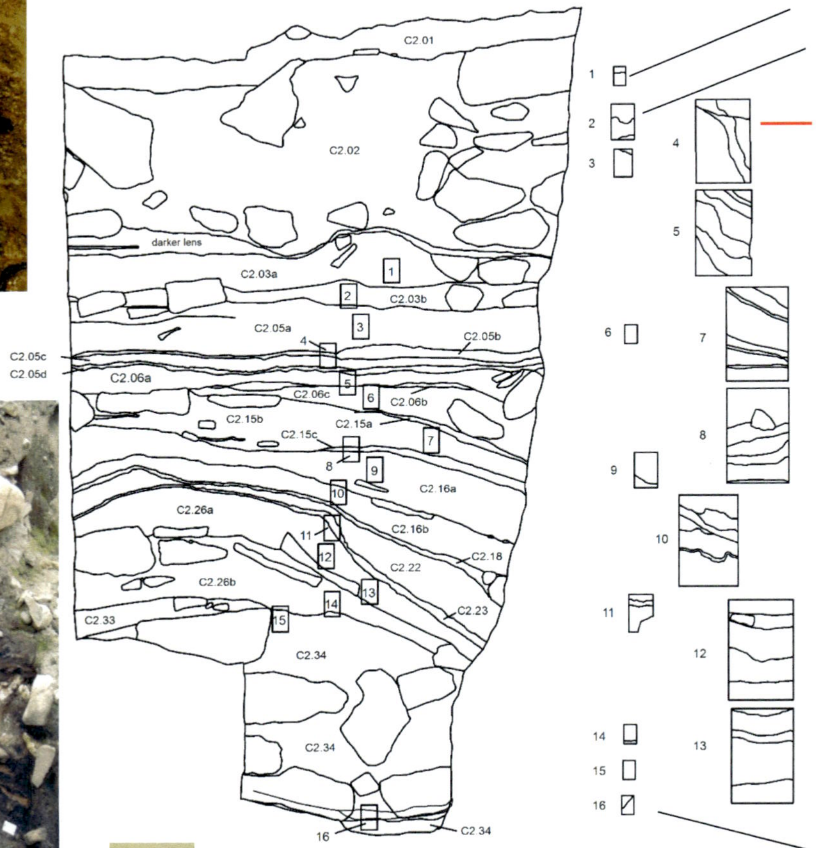

Context	Strat block	Associated contexts/material
C2.03a	11	Phase 7 (end P6 c. 200-50 cal. BC*)
C2.03b	10	Phase 6 (start c. 400-150 cal. BC*)
C2.05a	9c	**F2.07** hearth: grain rich
C2.05b C2.05c	9b	**F2.13** hearth: grain rich **C15.25** associated fuel residues/midden: grain rich
C2.05d C2.06a-c	9a	**F2.05** hearth: grain rich
C2.15a C2.15b C2.15c	8c	**F15.20** hearth: bone rich **C15.38** associated fuel residues/midden: grain rich
C2.16a	8a	**F2.27** hearth: bone rich
C2.16b	7e	**F2.15** and **F15.28** hearths **C15.56** associated fuel residues/midden: v. bone rich
C2.18 C2.22 C2.23	7c	**F2.10** hearth: grain rich **C2.27** associated fuel residues/midden: bone rich
C2.26a	7b	**F15.30** hearth: grain rich **C15.64** associated fuel residues/midden: v. bone rich
C2.26b	7a	**F15.35** hearth: grain rich **C2.32** associated fuel residues/midden: bone rich
C2.34	6	Phase 2 (start c. 800-650 cal BC*)
C2.35	2	Phase 1a: Bronze Age deposits

Phase 5 (start c. 500-400 cal. BC*)

Phase 4 (start c. 650-500 cal. BC*)

Phase 3 (start c. 750-450 cal. BC*)

Truncation point

** Dates equate to c. 95% probability, but are derived from preliminary modelling and therefore not presented in italics or with an associated probability range*

Plate 4 Micromorphology in context: a) fuel residues in Slide 7 (C2.15b); b) section (W), showing Kubiëna positions; c) section at (b), alongside d) schematic illustration of thin section micro-stratigraphies (complex sections enlarged), and; e) chart summarising bioarchaeological features of related stratigraphic blocks.

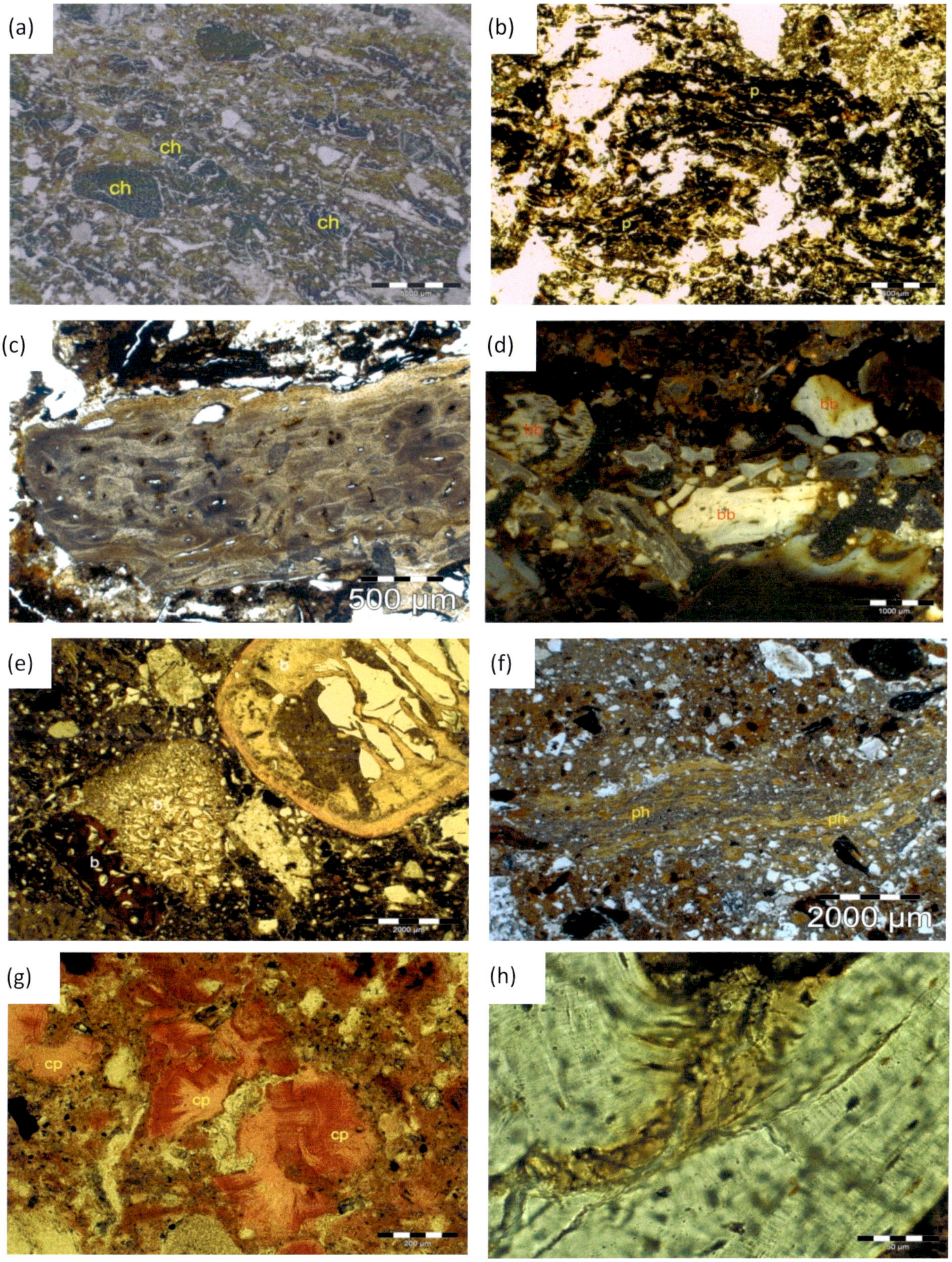

Plate 5 Photomicrographs: a) micro-lenses showing horizontal lamination of charcoal fragments (ch) in C2.15b; b) burnt peat (p) in C2.26a; c) detail of bone fragment in C2.15b; d) burnt bone concentration (bb) in C2.15b; e) bone fragments (b) in C2.06a; f) amorphous phosphatic accumulations (ph) in C2.26a; g) calcium-iron-phosphate pedofeatures showing radial crystallisation (cp) in C2.16b; h) detail showing dissolution of bone in C2.03b. All viewed in plane polarised light; image d in crossed polarised light.

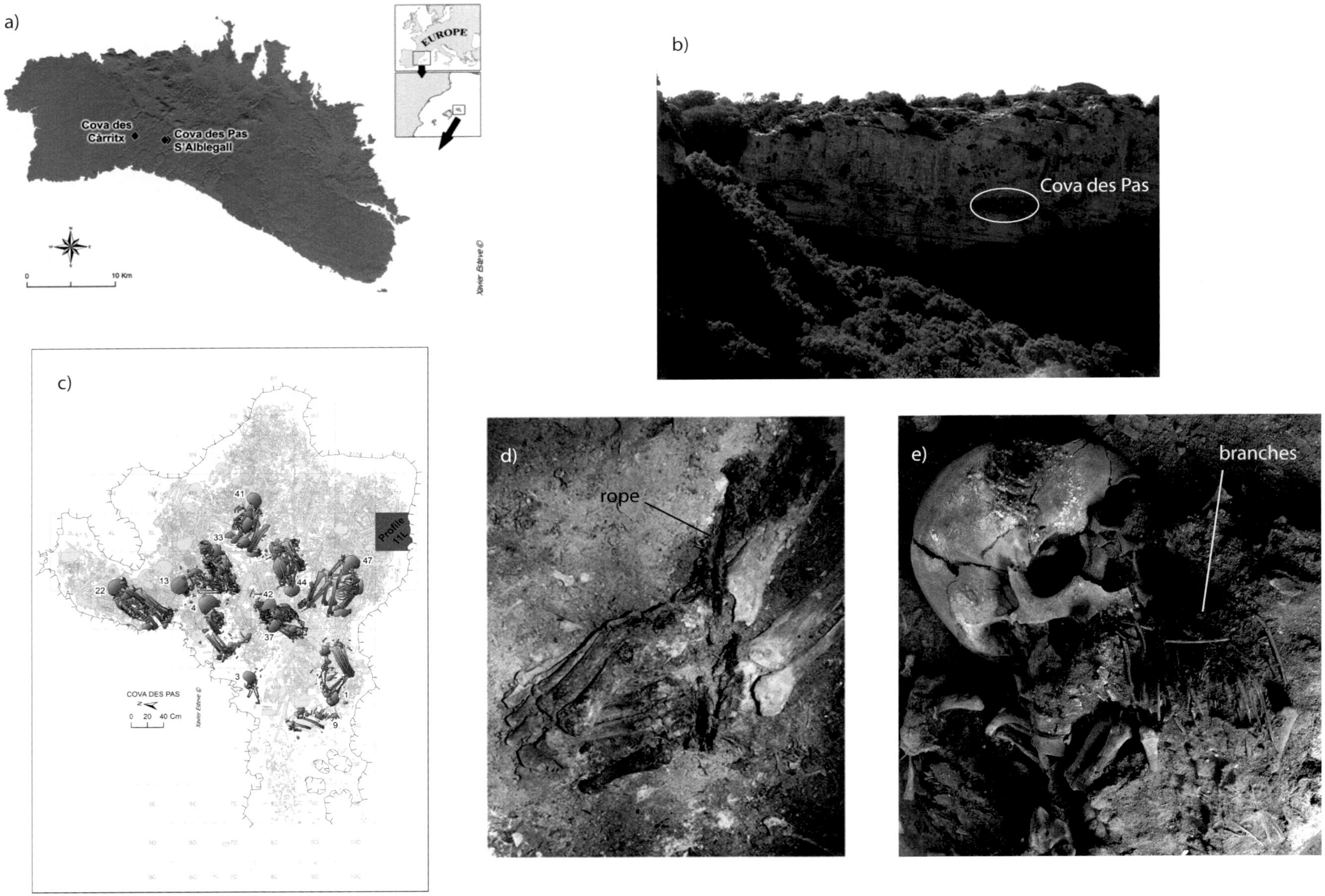

Plate 6 a) the island of Minorca and the location of Cova des Pas and other funerary caves and hypogea mentioned in the text; b) view of the Trebalúger cliff-wall and the entrance of the cave; c) plan of the cave showing the location of the studied corpses and the sedimentological profile; d) view of a plant fibre rope tying the feet of individual 33; and e) view of individual 33 with plant branches covering the hands and the skull. Photos taken by Víctor Guerrero.

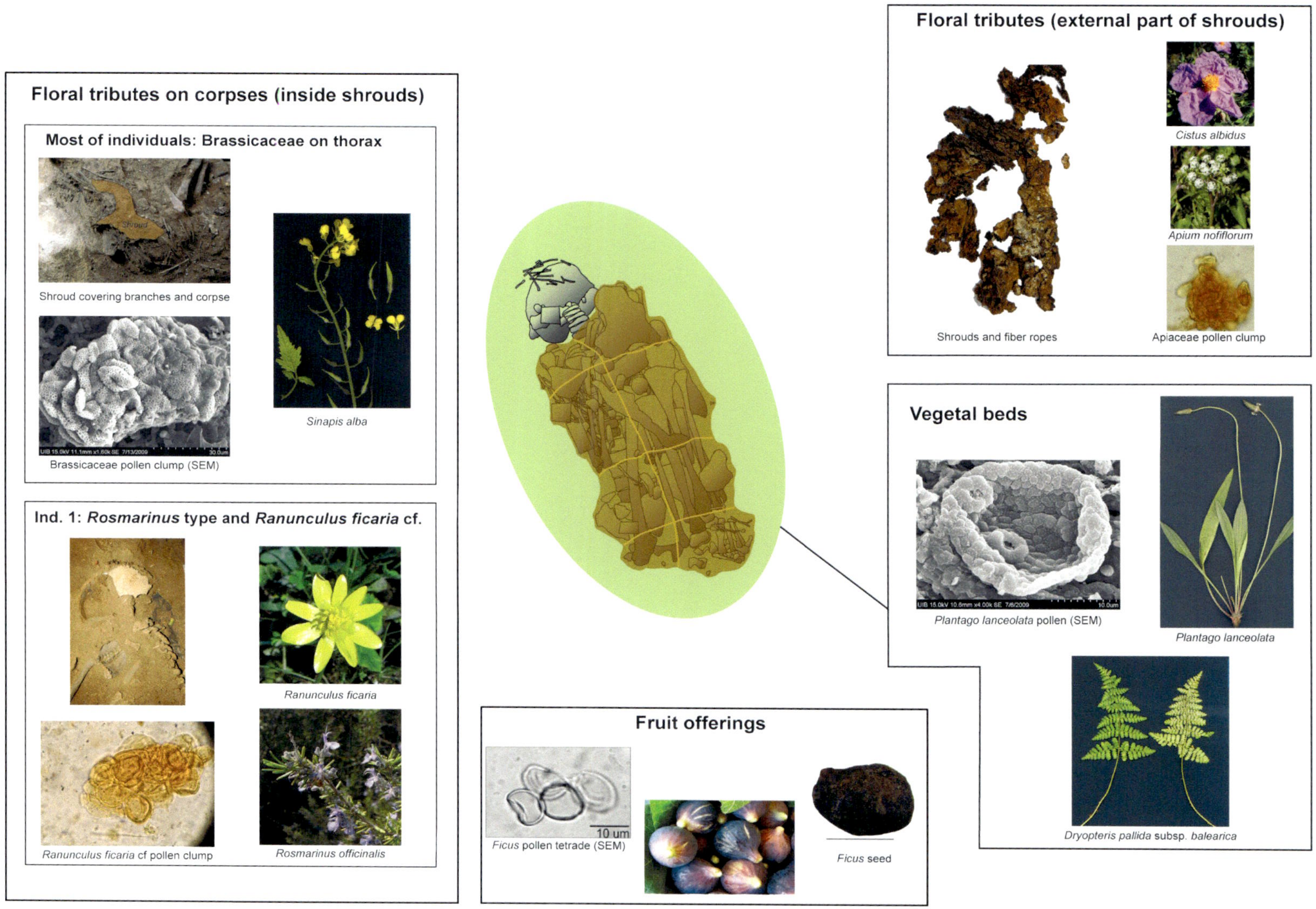

Plate 7 Reconstruction of the successive actions that configure the funerary ritual at Cova des Pas and the main plants used in each step. Selected examples of common Minorcan plants whose pollen grains could correspond to the identified pollen morphotypes are also shown. The plant images were obtained from the Herbario Virtual del Mediterráneo Occidental (UIB, UB, UV) with their kind permission. Drawing of the skeleton by Xavier Esteve.

Plate 8 Late Roman inhumation in stone sarcophagus (Grab 107), St. Maximin, Trier, Germany. a) skeletal remains encased in gypsum; b) fabric (linen?) solidified by embalming substances including Pistacia *spp. resin; c) resin fragments embedded in the enshrouding textiles. Images courtesy of Nicole Reifarth. Figure created with the assistance of Hector Orengo.*

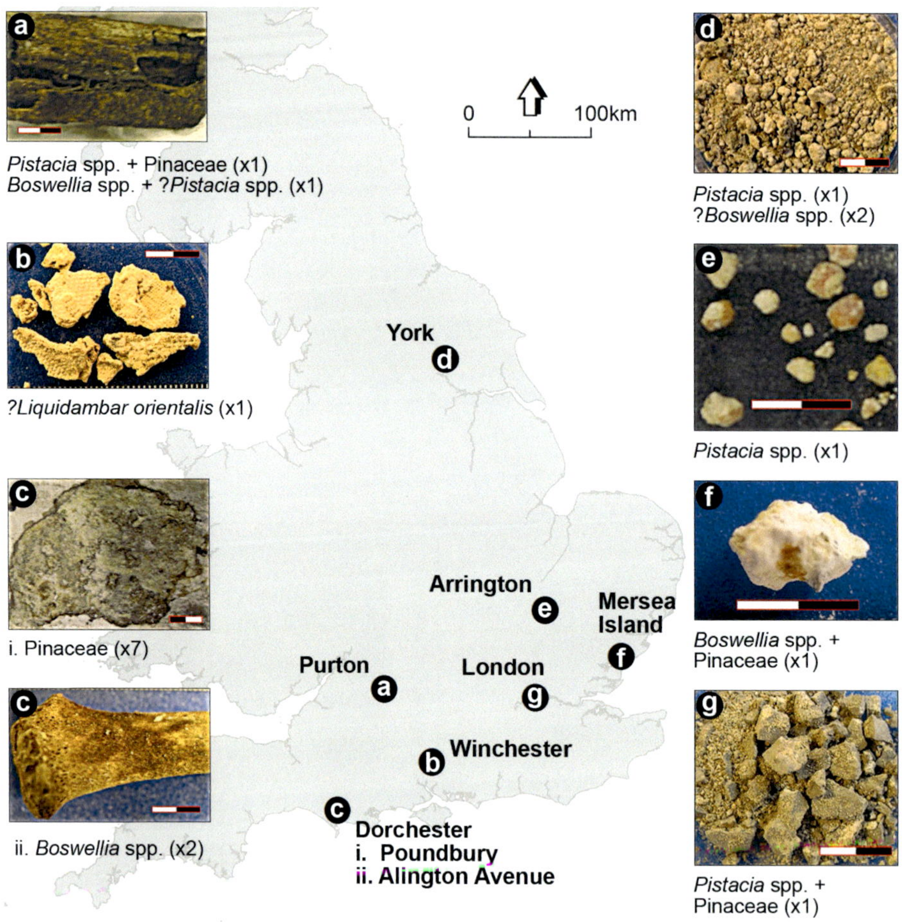

Plate 9 Examples of the range of samples and the distribution of Roman burials from Britain analysed as part of this project. a) residue-coated femur, lead-lined sarcophagus, Grave 1, Purton, Wiltshire; b) mineral-replaced textiles, lead-lined coffin burial, G336, Eagle Hotel, Winchester; ci) dark residue adhering to inner surface of plaster, lead-lined coffin burial, B530, Poundbury, Dorchester; cii) organic residue, proximal humerus, Grave 4378, Alington Avenue, Dorcester; d) grave deposits, base of sarcophagus, plaster burial YORYM:2010.1219, York; e) resinous fragments, lead-lined coffin, infant burial, Arrington, Cambridgeshire; f) amorphous white mass with crystalline orange inclusions, glass cremation urn, Mersea Island barrow, Essex; g) grave deposits, basal layer, intact lead-lined sarcophagus, SK15903, Spitalfields lady, London. Scale bars in cm. Figure created with the assistance of Hector Orengo.

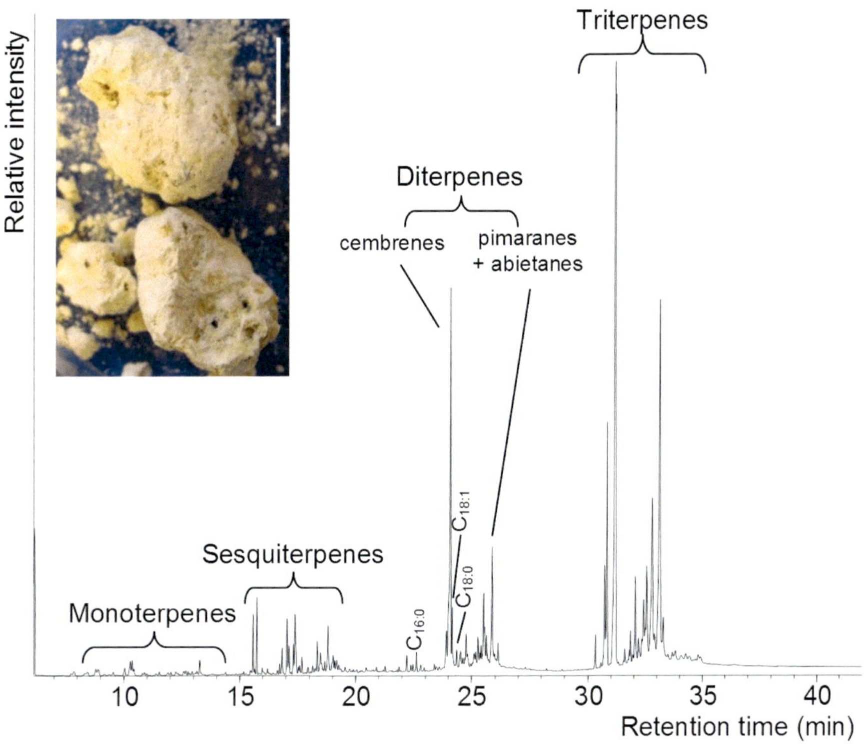

*Plate 10 Total ion current chromatogram showing the range of terpenic compounds characteristic of frankincense (*Boswellia *spp. gum-resin) combined with a lesser amount of a Pinaceae resin, in samples from the Roman period cremation burial, Mersea Island Barrow. Essex, UK. For full details of the compounds present, see Brettell* et al. *2015b. Inset. A portion of the yellow-white amorphous material analysed. Scale bar: 1 cm.*

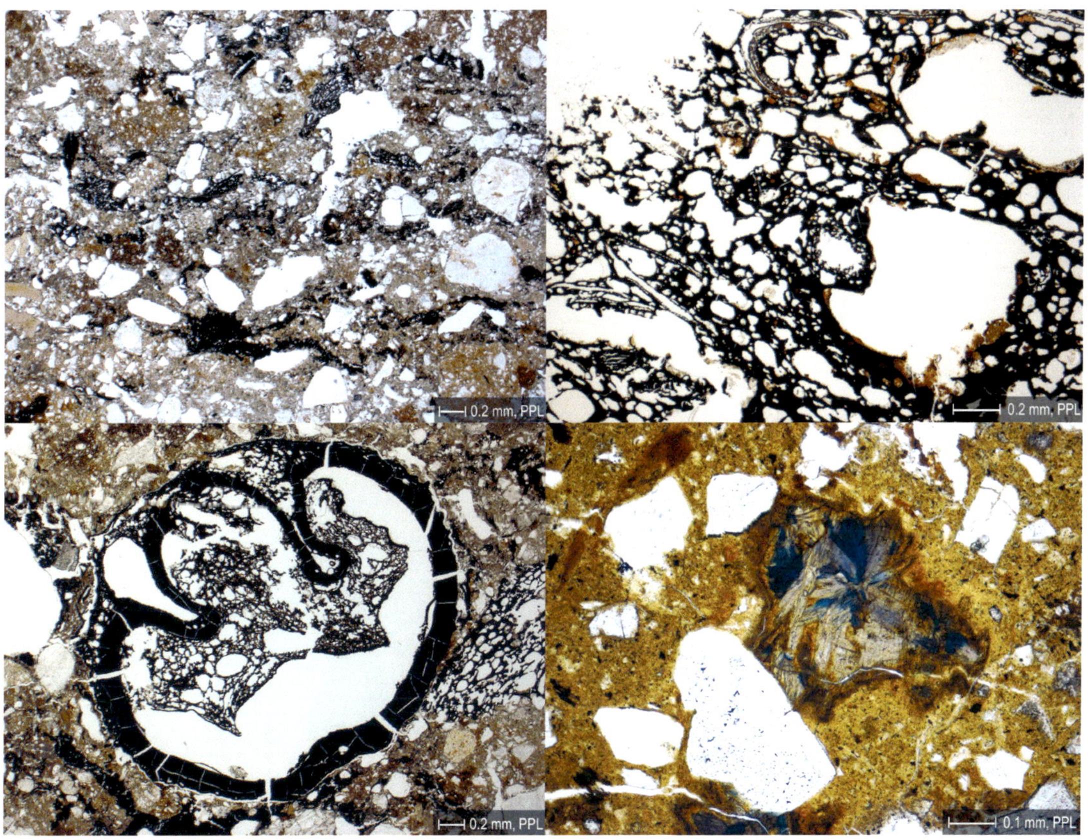

Plate 11 Rapperswil-Jona, Kempraten Seewiese. Details of thin sections. Top left: Section of the compacted occupation layer with horizontally oriented components and fragmented charcoal due to trampling. Top right: Charred remains of possible bread or porridge. Bottom right: Possible charred grape kernel. Bottom left: High concentration of vivianite near the altar (Pos. 1).

Plate 12 Aerial photograph of the ritual complex of Apollo (reproduced with the kind permission of Mrs. Tsatsopoulou).

Plate 13 Shells: a) the most common molluscan species (numbers refer to Table 7.2); b) examples of scorched scallop shells.

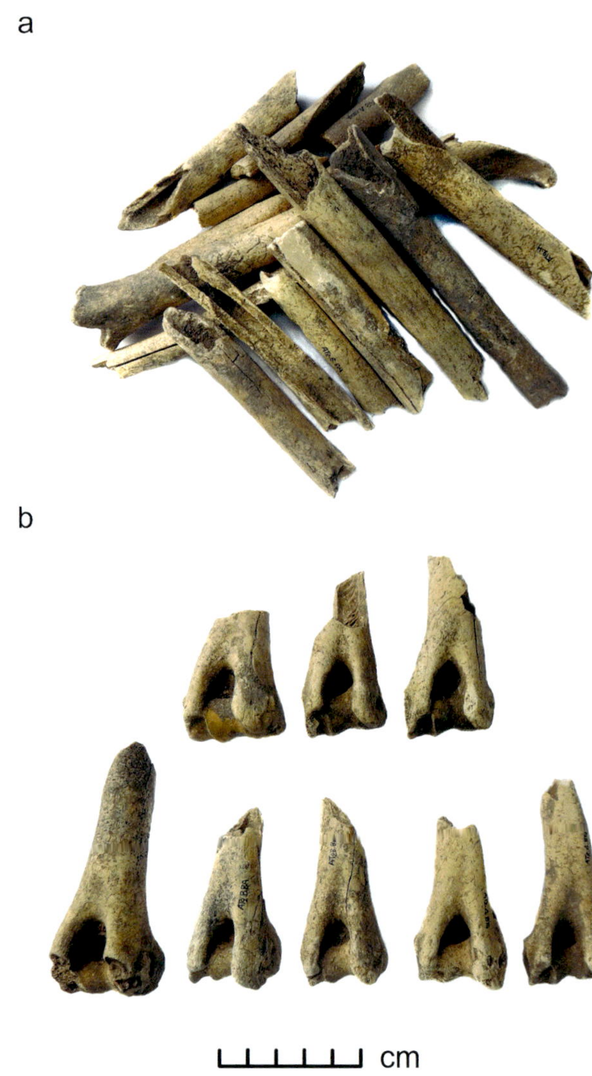

Plate 14 Animal bones: a) cylinder fragments, indicating gnawing by dogs; b) shaft and epiphysis fragments of left humeri from domesticated species.

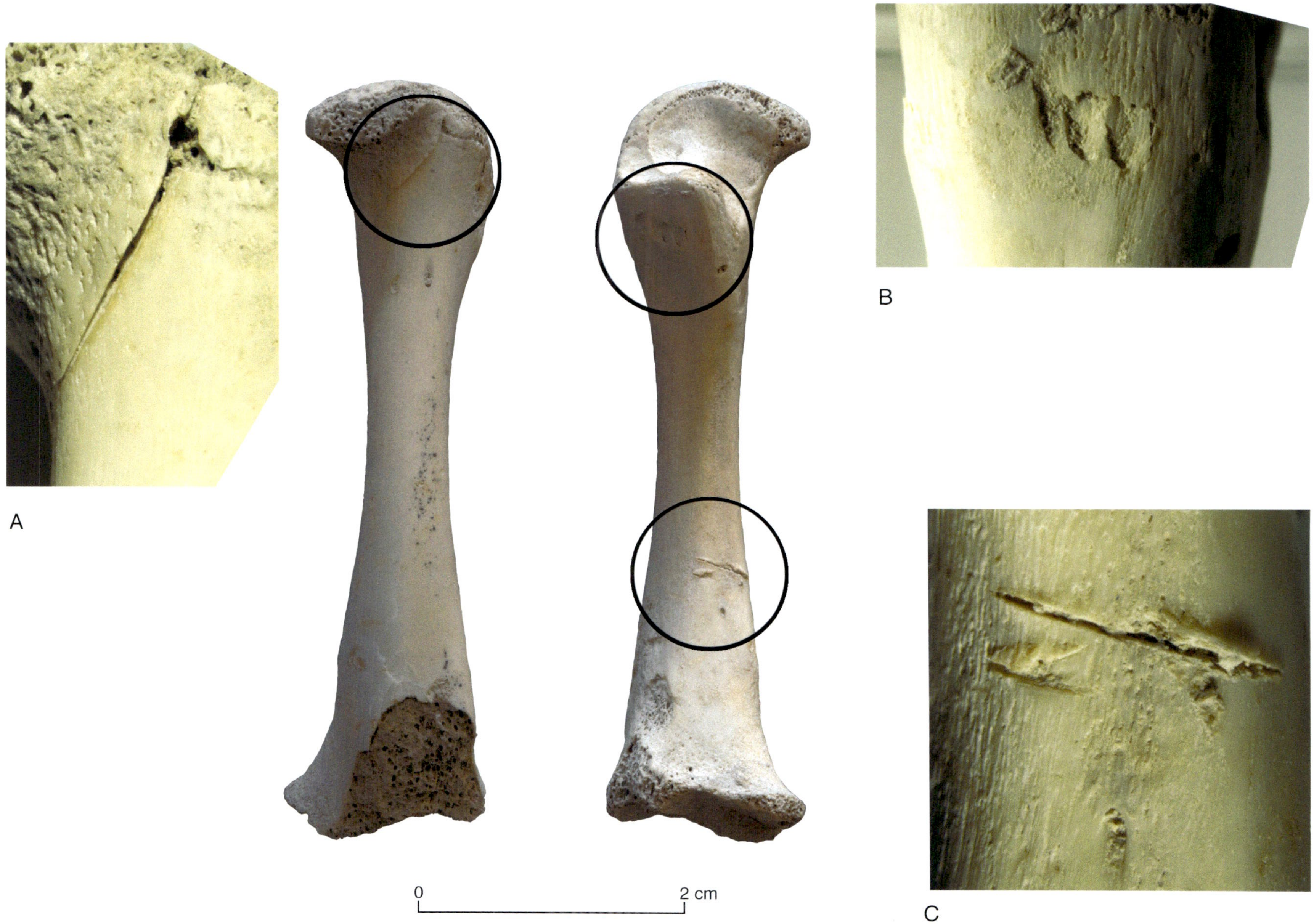

Plate 15 EL Molón. Cut marks and possible human bite marks on pig's left femur. A) dismembering cut on proximal end; B) possible human teeth imprints; and C) cuts on the shaft for meat extraction.

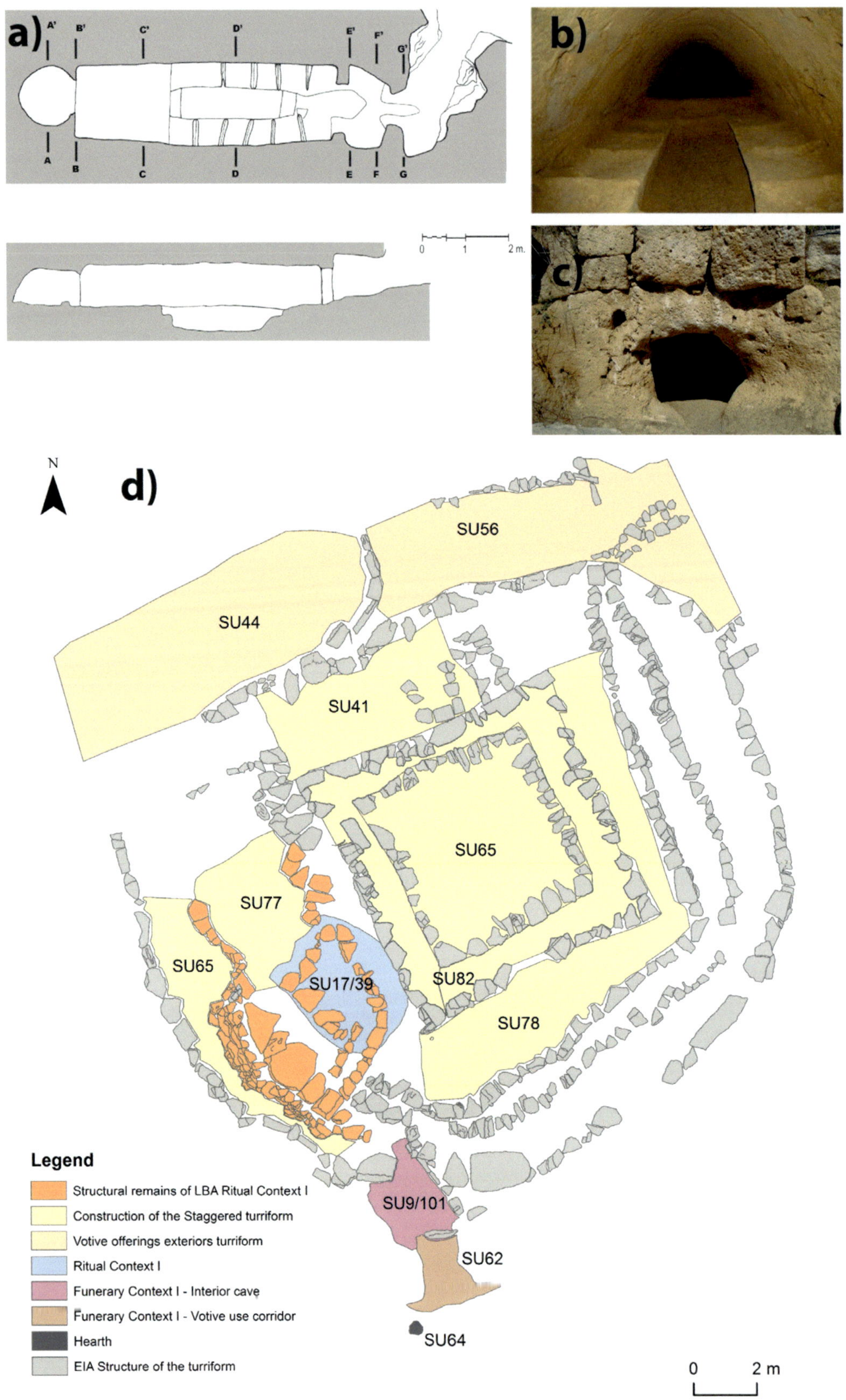

Plate 16 a) plan and section of the hypogeum, which remains under the staggered turrifom and, thus, it is not visible in the main plan (d); b) image of the interior of the hypogeum after excavation; c) image of the access of the hypogeum from the entrance corridor; d) plan of the staggered turriform of Son Ferrer, drawn by Alejandra Galmés. Highlighted areas: the remaining walls of the LBA Ritual Context I and the area occupied by the SU17 and 39 corresponding to the LBA use of this previous monument; walls conforming the staggered turriform erected during the EIA; structural fillings of this EIA construction that produced bioarchaeological remains (SU 27, 41, 65, 77, 78, 82); the area of votive offerings in the structural fillings of the turriform during LIA (SU 44 and 56); the entrance of the hypogeum reused as collective necropolis during the LIA (SU9/101); and the access corridor to the hypogeum (SU62 and hearth SU64).

Plate 17 Current view of the staggered turrifom of Son Ferrer (Calvià, Mallorca) after excavation and restoration.

Plate 18 Pisum sativum *(pea) seed identified in the Funerary Context I (SU 9/101).*

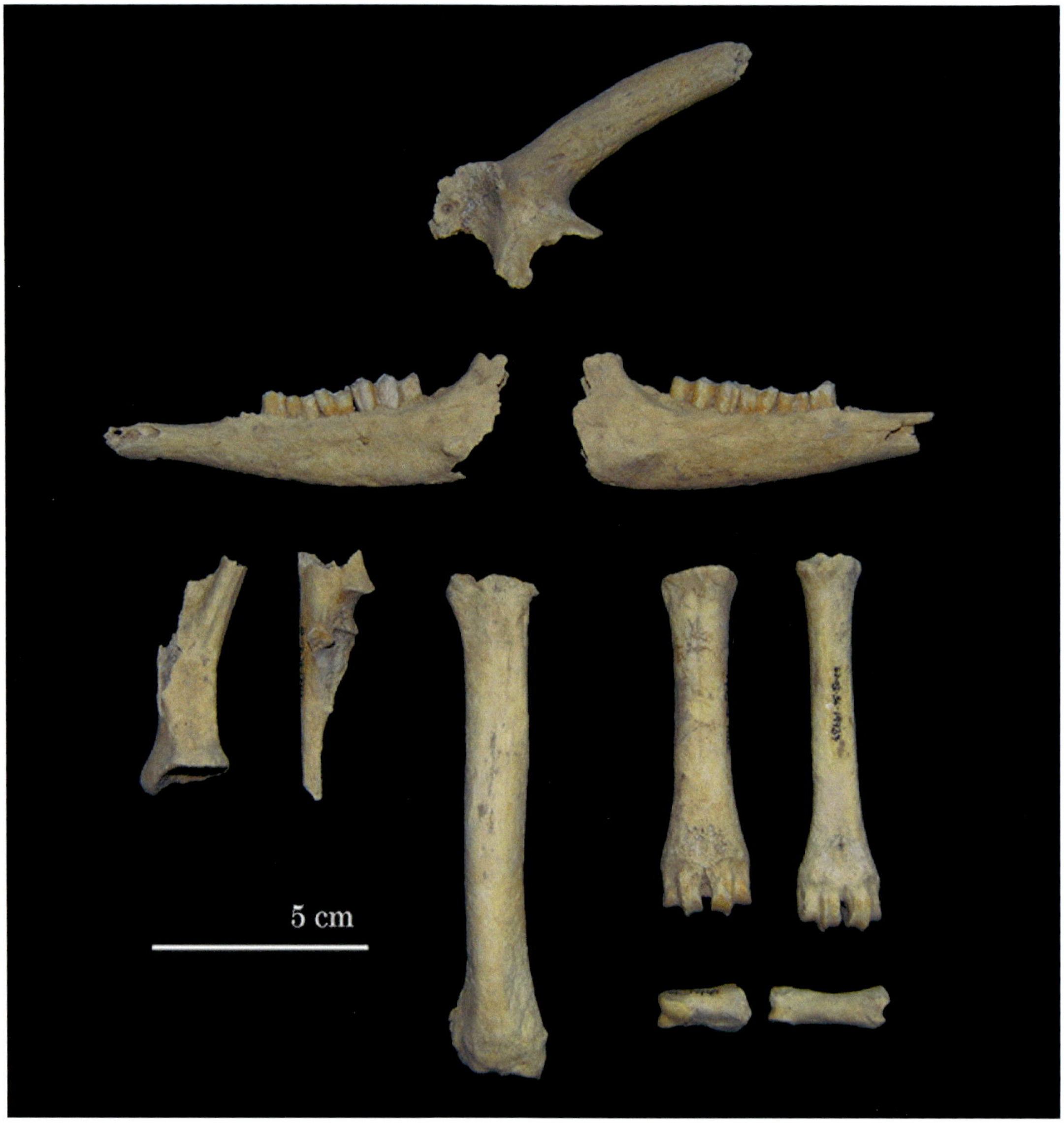

Plate 19 Faunal remains of the SU 56, Late Iron Age votive offerings at the exterior of the turriform: almost complete adult goat of small size, represented by skull parts, one complete radium and both complete metacarpals.

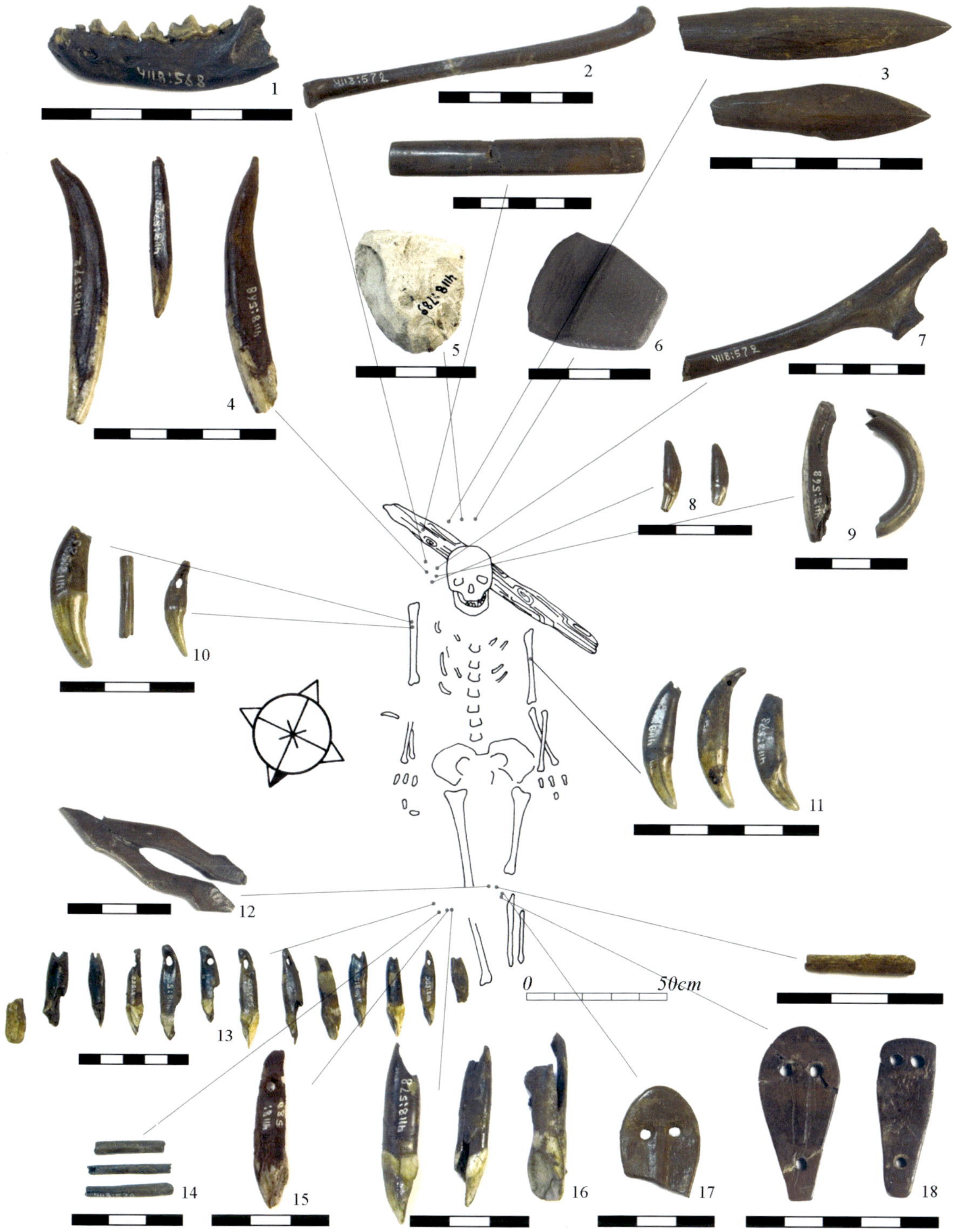

Plate 20 Female inhumation from the settlement-cemetery of Tamula I. Note the tooth pendants at the knees, fastened with human head shaped pendants, and probably once attached to clothing or an apron. A selection of specific animal bones has also been deposited at the head of the burial. 1) marten jaw; 2) grouse radius and a polished bone; 3) bone arrowheads; 4) wild boar incisors; 5) quartz; 6) adze; 7) elk hyoid bone; 8) marten canines; 9) wild boar and beaver incisors; 10) pendants made of dog and marten canines, and a bead made of a bird's tubular bone; 11) pendants made of dog canines; 12) figurine of a double-headed snake; 13) pendants made of elk incisors; 14) beads made of bird tubular bones; 15,16) pendants made of elk incisors; 17,18) figurines of human heads. Drawing by Marko Usler and Kristel Roog.

Plate 21 White-tailed eagle in flight (WWW2).

Plate 22 Image shows vultures defleshing a human cadaver in Tibet (WWW3).